U.S. Hydrographic Office

The coasts of Chile, Bolivia and Peru

U.S. Hydrographic Office

The coasts of Chile, Bolivia and Peru

ISBN/EAN: 9783742866196

Manufactured in Europe, USA, Canada, Australia, Japa

Cover: Foto ©berggeist007 / pixelio.de

Manufactured and distributed by brebook publishing software
(www.brebook.com)

U.S. Hydrographic Office

The coasts of Chile, Bolivia and Peru

U. S. HYDROGRAPHIC OFFICE.

No. 59.

THE COASTS

OF

CHILE, BOLIVIA, AND PERÚ,

COMPILED AT THE

U. S. HYDROGRAPHIC OFFICE,

WASHINGTON, D. C.

WASHINGTON:
GOVERNMENT PRINTING OFFICE.
1876.

The basis of these directions and the description of the coasts of Chile, Bolivia, and Perú have been the works of Capt. F. Chardonneau, of the French navy, and of Capt. Aurelio Garcia y Garcia, of the Peruvian navy, *Les Instructions nautiques sur les Côtes du Chile et de la Bolivie*, and *El Derrotero de la Costa del Perú.* In the compilation, extracts have been made from the Surveys on the Coast of Chile, by Captains Simpson and Gormaz, of the Chilean navy; from Notices of the Examination of the Coast of Chile, now in progress, under the direction of Lieutenant Uribe, Chilean navy; from *El Annuario Hydrografico de Chile;* the British Admiralty South America Pilot; the American Cyclopædia, and the remark-books of officers of the U. S. Navy, together with the most recent British Admiralty, French, and Chilean charts, notices, &c.

R. H. W.

U. S. HYDROGRAPHIC OFFICE,
      Washington, D. C.

NOTE.

All courses and bearings are true, unless otherwise stated.
The distances are expressed in nautical miles.
The longitudes are from the meridian of Greenwich.

TABLE OF CONTENTS.

CHAPTER VIII.

CHAPTER IX.

BOLIVIA.

CHAPTER X.

PERÚ.

CHAPTER XI.

CHAPTER XII.

CHAPTER XIII.

CHAPTER XIV.

VIEWS.

ERRATA.

Page 6, for " coffee " read " copper."
Page 29, for " Tatayo " read " Taytao."
Page 30, for " Melcher " read " San Melchor."
Page 37, for " Tatayo " read " Taytao."
Page 37, for " Yeuche Mo " read " Ynche Mo."
Page 86, for " Abato " read " Abtao."
Page 88, for " Nahuelhuahi " read " Nahuelhuapi."
Page 93, for " Quniched " read " Quinched."
Page 118, for " Pachuapi " read " Paehuapi."
Page 170, for " Fort Viel " read " Port Viel."
Page 196, for " Tapolcama " read " Topolcama."
Page 199, for " Fort San Antonio " read " Port San Antonio."
Page 343, for " Salzar " read " Salazar."
Page 357, for " Carmotal " read " Camotal."
Plate X, for " Victor " read " Vitor."
Plate XVI, for " Legarto " read " Lagarto."

THE COAST OF CHILE,

FROM THE

GULF OF PEÑAS TO THE BOUNDARY OF BOLIVIA,

WITH THE

OFF-LYING ISLANDS.

THE COAST OF CHILE, FROM THE GULF OF PEÑAS TO THE BOUNDARY OF BOLIVIA, WITH THE OFF-LYING ISLANDS.

CHAPTER I.

GENERAL DESCRIPTION.

The republic of Chile* is bounded on the north by Bolivia; Boundaries. the 24th parallel of south latitude, which is seven miles to the southward of cape Jara, is the dividing line between these two states from the Pacific coast to the Andes. The tract of land comprised between the 23d and 24th parallels is in fact considered neutral territory. † The boundary line then follows the crest of the Andes to the Rio Negro, and thence the course of this river to the Atlantic ocean; these two natural boundaries separate this country from the Argentine Republic.

Chile claims the whole of Patagonia, though the Argentine Republic contests a portion of it, and by establishing a settlement at Punta Arenas in the strait of Magellan, Chile pretended to secure the entire possession, and a protest filed by the consul-general of Chile at London in 1872 clearly established this claim.

These directions will treat only of that portion of the Chilian coast comprised between cape Tres-Montes and the river Loa.

In 1872 the population of Chile was about two millions. Population.

The capital is Santiago. Capital.

The provinces on the coast, commencing at the north, are Maritime provinces. as follows: Atacama, Coquimbo, Aconcagua, Valparaiso, Maule and Colchagua, Concepcion, Valdivia, Llanquihue, Chiloé, Magallanes.

---

* According to Malina, the word Chile or Tchili is derived from the cry of a thrush which is very common in this country.

† This tract, which comprises the desert of Atacama, was considered without value. Since the discoveries of mines of precious metals and of copper, numerous difficulties have arisen between these two countries, brought about by the explorers of this territory.

1 C

The maritime governors of these provinces reside as follows: of Atacama, at Caldera; there are subdelegates at Sarco, Peña Blanca, Huasco, Carrizel-Bajo, Chañaral, Pan de Azucar, Taltal, and Paposo; of Coquimbo, at Coquimbo; subdelegates at Guyacan, Tongoy, and Totoralillo; of Aconcagua, at Papudo; subdelegates at los Vilos and Pichidanqui; of Valparaiso, at Valparaiso; subdelegate at San Antonio; of Maule and Colchagua, at Constitucion ; subdelegates at Llico, and Buchupureo, and at Curanipe ; of Conception, at Talcahuano; subdelegates at Tomé, Coronel, Lota, and Lebu ; of Valdivia, at Corral; a subdelegate at Queule; of Llanquihue, the maritime governor resides at Puerto Montt, a subdelegate at Calbuco; of Chiloé, at Ancud, a subdelegate at Melinka, (Guaiteca Grande ;) of Magallanes, the governor resides àt Punta Arenas.

History.

Chile to the Rio Maule was part of the empire of the Incas. Diego Almagro, the lieutenant of Pizarro, first led the Spaniards to this country in 1535. It was conquered by Pedro de Valdivia during the years 1541 to 1554, he having laid the foundation of Santiago in December, 1540 ; of Séréna in 1543, of Valparaiso and Penco (old Concepcion) in 1550, and of Imperial and Valdivia in 1552. He commenced the struggle against the Araucanians, who, after many conflicts, captured him and put him to death in 1559.

For two centuries and a half the history of Chile is full of the events of this struggle. During this period the Araucanians defended their independence, and often with success. All the cities of the southern part of Chile were frequently destroyed by them, notably during the period from 1599 to 1604, and all excepting Imperial were obstinately rebuilt by the Spaniards. They were often forced to treat with the Araucanians and to observe the treaties.

The Spaniards were also several times disturbed in their possession by the English buccaneers. Drake plundered Valparaiso in 1578, Hawkins ravaged Chile in 1594, Narborough in 1668, Sharp in 1680. The Hollanders also plundered the coasts of Concepcion, Valdivia, and Chiloé in 1600, 1615, and 1643. The selfish policy of the Spanish government produced the same effect in Chile as in its other colonies. Profiting by the revolution of 1808, the Chilians made their first efforts toward separation in 1810, but with very little success. General San Martin, of Buenos Ayres, aided by Argentine troops and the remainder of the independent party, defeated the Spaniards at Chacabuco; on the 15th of February, 1817, took possession of Santiago, and by the victory of Maipo, April 5, 1818, assured the independence of Chile, although the war was continued until 1822. Valdivia was held by the Spaniards until 1820, when it was captured by the English Admiral Cochrane, who had formed a Chilian navy, placing himself at its head. Chiloé was finally incorporated in the republic in 1826. The republic then gave its assistance to all the other Spanish colonies, especially to Perú, in their struggles against the mother country.

After the declaration of independence, Chile, like its sister South American republics, was rent by civil discord until 1830. The radicals, who had Pinto and Freire for their chiefs and generals, battled against the conservatives, who were led by O'Higgins and Prieto. The latter finally triumphed in the battle of Lircay, the 17th of August, 1830, and instituted the conservative progressive form of government which gave to Chile forty years of almost uninterrupted prosperity. The country became settled little by little, and owing to the wisdom of the inhabitants the revolutions were superseded by a peaceful constitution.

President Prieto, who was elected in 1832, and supported by a true statesman, Portales, was re-elected in 1836. General Bulnes, also twice president, 1841 to 1851, had a tranquil administration, but after his term of office was called to suppress a serious insurrection caused by the radicals. He then became the commandant of the troops raised by President Moutt, who had succeeded him in 1851, and was re-elected in 1856. During his second presidency he had to struggle against a combination formed by the radicals and ultra-conservatives. He was successful, and transmitted his office peaceably to Perez in 1861, who continued in office until 1871. Under his administration the most important event was the war against Spain, in which this republic had Perú, Bolivia, and Equador for its allies. The principal incidents of this war were the capture of the Spanish steamer la Covadonga by the Chilian corvette Esmeralda, the blockade and bombardment of Valparaiso in 1865, and the undecided battle of Abtao between two Spanish frigates and the Chilian squadron. Perez transmitted his charge to Errazuriz in 1871.

The province of Araucania is still independent, nominally under the government of Chile. It formerly extended from the Rio Biobio to the gulf of Ancud, but at present it comprises only the territory between the Rio Imperial to the north, and the low land of Valdivia to the south. Although the Araucanians have been driven to the interior, they do not allow the Chilians to penetrate into their country. The latter have occupied the coast during the last few years, and have erected forts at Tolten and at Queule, which have become the centers of this territory. In 1869 it was only with great difficulty that the Chilian commission could penetrate to the ruins of Imperial, fifteen miles from the coast.

The Araucanians only tolerate a few missionaries. During the last three centuries these people have decreased about 75 per cent.; at present they number about 70,000, alcohol having proved a more formidable enemy to them than the European race. In 1862 the Araucanians again gave trouble under the lead of a Frenchman, De Tonneins, who claimed to be king of Araucania and Patagonia, under the title of Orélie Anotoine I. He was captured, but subsequently released.

General aspect of the country and coasts.   Chile, which is the western slope of the Andes, varies in breadth from eighty to two hundred and ten miles, and is traversed by the branches of this great cordillera and by several smaller chains of mountains, running either parallel or transversely to the Andes. The island of Chiloé, with its undulating plateaux, is the only exception to the general aspect of the coast, which is formed of one continuous line of steep cliffs. Certain portions of the coast of Araucania and a few points to the north of Valparaiso are formed by dunes and low sandy beaches. But here also it is easy to see that the granite formation of the Andes is not far off. In the south these steep shores are generally wooded, but advancing to the northward they become more barren, sterile, and dry. At Concepcion the eye is charmed with the richness of the foliage; at Valparaiso, which is two hundred and forty miles farther north, the hills are covered with poor brushwood and a thin matting of grass.

mbo, one hundred and eighty miles farther north, all bushes take the place of the brushwood; and at no trace of vegetation is visible, excepting near the springs running from the snows of the Andes; plains and hills are all covered with sand.

Between Coquimbo and Copiapó there are some few farms and plantations, but at the latter point the desert of Atacama commences.

The cordillera of the Andes trends nearly north and south, and its summits often rise higher than 22,960 feet. The principal of these are, Tupungató, to the east of Santiago; Descabezado; the volcanoes of Antuco, Villarica, Osorno, and Aconcagua. Although many of these peaks are visible from the sea, it is seldom that they can be used as a guide to the navigator; they are often hidden by fogs, and when they are visible they can hardly be distinguished on account of their number and their almost equal heights. Their geographical positions are also inexact. The inferior ridges of the seacoast are, therefore, more useful.

Such a chain of mountains renders the communication between Chile and the Argentine Republic difficult. Of the low mountain-passes, called *puertos* in this country as in the Pyrénées, the principal ones are those of Los Patos; of Portillo, which is the shortest and most dangerous; of Upsallata or Cumbre, which is the one most generally used; and of Planchon, in the province of Curicó. These passes are all more or less dangerous.

Toward the south the Andes are lower, and the passes more numerous; but they are of no interest, as they only lead to the desert of Patagonia.

Many rivers have their sources in the Andes, and are fed by their snows; but on account of their rapid descent and their limited extent, they are of little importance. The principal ones are the Rio Maullin; the Rio Valdivia, with its numerous tributaries, which admits vessels drawing 13 feet of water; the Tolten; the Rio Imperial, which has 16 feet of water on its bar; the Biobio, which has the greatest volume of water, but its mouth is silted up; the Maule; and the Rio Salado. Toward the north they are fewer and less important. Some of these rivers are the outlets of considerable lakes, such as the Llanquihue and Nahuelguapi.

The coast of Chile, which is washed by the Pacific, in clines gradually to the eastward from Valdivia to the Loa, which is the eastern point of the west coast of South America. It has numerous indentations, which are protected from the southward by promontories, but open to northward. After leaving the archipelago of Chiloé and Chonos, where many sheltered bights are found, there are only two or three bays on the entire coast to Bolivia which offer the double advantage of protection both from the north and south.

Land fall.

The coast, as may be inferred from its nature, is nearly everywhere steep and clean; a few miles from the shore the water is deep, and the coast can be approached, in favorable weather, without fear. The shore, like that of Patagonia, is covered with large masses of sea-weed, which can be detected by their odor, especially at night, at a long distance from the land.

Produce.

The great difference of latitude between the extreme points of Chile insures a great variety of produce. The narrow strip of land which constitutes the territory of Chile is divided into two almost equal parts by the beautiful valley of Aconcagua. In the part to northward of this line the spurs run out from the Andes perpendicular to the coast. Between them are extensive and thickly-populated valleys. One of these, that of Copiapó, furnishes silver in large quantities; the next is that of Coquimbo, which exports half of the coffee used by the entire world. Next come those of Huasco, Ligua, and Petorca, celebrated, during the time of the Spaniards, for their production of gold. All this tract forms the mineral region of Chile, situated between 24° and 32° south latitude, comprising the provinces of Atacama, Coquimbo, and North Aconcagua.

The part to the south of the line of Aconcagua is again divided into two parts by the Biobio. To the northward of this river lies the agricultural region, between 32° and 36° south latitude, which consists of the following provinces: South Aconcagua, Valparaiso, Santiago, Colchagua, Curicó, Talca, Maule, Nuble, Concepcion, and North Arauco. Finally, to the south of the Biobio lies the forest region, which comprises the remaining territory, between 38° and 44°—the

provinces of South Arauco, Valdivia, Chiloé, and Llanquihue.

The mineral region covers 46 per cent. of the superficial
area, the agricultural region 28 per cent., and the wooded
region 26 per cent. The proportion of the population in the
first is 12 per cent., in the second 77 per cent., and in the
third 11 per cent.

Fishing is not carried on to any great extent. In the Animal king-
archipelagoes of Chiloé and Chonos the seal-fishery is a
source of profit, and many excellent shell-fish are in abundance. Near the coast, whales, cod-fish, and sardines are
taken; the great depth of the water, however, is an obstacle
to the development of fisheries.

The domestic animals, without being of a superior quality, prosper. Cattle thrive, but the sheep and hogs are
only passable; the horses, donkeys, and, above all, the
mules, are excellent; the llama, vicuña, and guanaco are
also raised for their wool. There are two species of hares,
and also two animals of the feline species, of moderate
size, the jaguar and puma. The chinchilla, whose fur is
so valuable, abounds in the uninhabited regions of the
northeast.

A large portion of the area of Chile is unproductive. The Vegetable pro-
extensive surface covered by mountains is too cold for vegetation; the deserts of the north, the large tracts covered by
the primitive forest, and the districts inhabited by warlike
Indians being deducted, leave but a comparatively small
area for cultivation. The soil, however, where capable of
cultivation, is fertile, and the yield so abundant that Chile
exports considerable amounts of cereals and meats. The
principal grains are wheat, barley, maize, and oats; but
little rye is raised, there being no demand for it. Superior
hemp is produced in the country north of Maipo; beans
and pease are a large and important crop. In the southern
provinces the potato is of excellent quality. Fruits are abundant; including apples, pears, peaches, oranges, limes,
nectarines, plums, figs, apricots, grapes, strawberries, and
cherries. Melons, squashes, and gourds grow in perfection.
The vine and olive tree are somewhat cultivated, but the
wine and oil were for a long time bad, on account of the little care given them and the want of experience in their

preparation. During the last few years some intelligent men have made it their business to improve these two products, and they have succeeded. In one word, agriculture is not far advanced in this country, and the soil of Chile in general is not better than that of other countries in which granite predominates. All these products are European. Among those which belong to Chile is the *Fucus antarcticus*, which serves as nourishment for the inhabitants of the South, as well as for the Chinese; the *Aristotelia maqui*, from which the Chilians extract the beverages called *cici* and *theca*, the valerian which serves for combustible in the Andes. The myrtle, the soap-tree, and tobacco grow in abundance. The tropical fruits are moderately good. The boldo must also be mentioned, which is celebrated for its property of healing all skin diseases. Finally, Chiloé, Chonos, and the part of the continent near them, is rich in magnificent woods for construction, of which the alerce, a species of larch, is the principal. From it the forests in which it is found are called *alerzales*. During the last few years commerce has used it extensively.

Mineral king-dom.

All the country abounds in mineral products, the working of which has largely increased in the last ten years. Mines of silver, gold, and, above all, copper, are worked, principally in the provinces of Coquimbo and Atacama. According to a popular saying, the copper mines always enrich the miner, whereas the result is doubtful in silver, and always ruinous in gold mines. This was due to the small capital which was at first put into these enterprises, and the high price of labor and transportation, for the veins are generally very rich. Thanks to the railroads recently constructed, which connect the mines with the ports of the country, such as Caldera, Huasco, Chañaral, mining industry is progressing. Enterprises with large capital have been formed, and, as usual, those which were wisely created and conducted have succeeded. The number of mines worked in Chile is, gold, 95; silver, 424; and copper, 1,638. Since 1852 coal also has been mined. It is most abundant around the bay of Arauco, at Coronel, Lota, and Lebu. Copper founderies have been established around the mines, where all the residue is utilized. Although the use of this coal presented serious inconveniences at first, its discovery and mining has added undeniable resources to Chile and to the strangers who visit its coast. The manufacture of bricks and pottery-ware is also conducted on a large scale.

Earthquakes, *tierramotos*, are very frequent in Chile. Those Earthquakes. of 1822, 1824, 1829, 1834, and 1835 are marked among the most disastrous. Santiago was destroyed four times in fourteen years. Penco, the ancient Concepcion, was never rebuilt. The earthquake of 1835 ruined the new Concepcion completely, and caused a rising of the ground of from four to five feet. The northern part of the island Santa Maria was submerged more than nine feet, and the southern portion eight feet; the water, however, subsided after a few months. The great earthquake of 1868 caused but little damage in Chile; in the ports, however, effects were experienced analogous to those felt at Perú, but less severe. The phenomenon was felt at Talcahuano, Constitucion, Coquimbo, Caldera, and at Cobija. At Talcahuano the sea destroyed a part of the houses built along the beach, and inundated six miles of the coast toward Concepcion; an Italian vessel was thrown ashore. At Caldera the sea receded, then came back suddenly, and rose nearly thirteen feet above its ordinary level. The vessels swung very rapidly, and much damage was done by fouling; an English three-masted vessel was thrown against the pier.

These phenomena do not generally assume so disastrous proportions; simple shocks, *temblores*, are felt, and these are so frequent that the inhabitants pay little attention to them. These shocks occur nearly every month, sometimes several days in succession, and sometimes even several times in each day. They are generally accompanied by subterranean noises. The constant expectation of these shocks has determined the mode of building the Chilean houses and cities. The houses are generally built of wood or adobe, (brick baked in the sun;) they have only one floor. The streets are wide, and cut each other at right angles and terminate in large squares, which the inhabitants use as places of refuge.

In addition to these occasional commotions of the ground, Rising of the ground. the coast of Chile is constantly rising. On the promontories there are old beaches, where shell-fish of modern times are found. These beaches are at different elevations, and resemble gigantic steps.

From this it may be concluded that the coast is constantly rising, but that it takes place at intervals, with periods of repose intervening. The Indian word *huapi*, which means island, is now applied to many places, which are at present part of the main-land. Darwin found shells on the hills of the island Chiloé, at elevations of more than 328 feet; to the northward of Concepcion he found marks of the sea-level at 722 feet above the present level. Near Valparaiso these marks are found at an elevation of 1,300 feet, but they lower again farther to the northward, and in Bolivia they are only 200 feet above the present level. It would seem that in Chile the coasts were elevated at the same time as the peaks of the Andes in the same latitude, but in a less degree. According to historical observations the ground rose six feet between the years 1614 and 1817, whereas from 1817 to 1834 it rose 8.5 feet. This slow action has the effect of changing the configuration of the coast slowly; all harbors which were used in former times are now inaccessible and others have been formed.

Climate. The geographical position of Chile causes the greatest variety of climate; rainy and stormy in the south; in the north it is subject to calms, the sky is clear, the country dry, with a great want of water. The mean of these two extremes lies between Concepcion and Coquimbo. In describing this climate it may be well to divide the coast into two parts, the first being the part comprised between the Chronos archipelago and Concepcion.

PART I.—FROM CHROMOS TO CONCEPTION.

Barometer. The changes in the barometer can be relied upon as indicating changes of the wind when it is to the north or south of west. The barometer falls with the first and rises with the latter.

The winds from the north to west bring air, saturated with the vapors of the ocean, from the hot latitudes near the equator; causing the fall of the mercurial column and rain. The winds from the southward of west originating in the antarctic zone, where the column of air is colder and more dense, cause the barometer to rise.

Observations made at Ancud at elevations of 16 and 90 feet above the ocean, during a space of nineteen months, show the greatest height of the barometer to be 30.25 inches and the minimum 28 inches. The greatest absolute range, reduced to the freezing-point and level of the sea, was 1.94 inches. The greatest annual range observed during the years 1857, 1863, 1866, and 1867 was 1.34 inches.

The mean annual range deduced from the greatest monthly range observed during 1863, 1866, and 1867 was 0.73 inch, and the daily range found from the same observations was 0.04 inch. These last observations were only taken during the day, and cannot, therefore, represent the extreme changes in the barometric column, nor its daily curve, and are therefore incomplete; but as the daily variation in the barometer at Chiloé can only be observed during the short intervals of fine weather, we can assume that these oscillations are never greater than 0.039.

The mean monthly height deduced from the observations of the above-mentioned years are as follows:

	Inches.		Inches.
January	29.876	July	29.808
February	29.864	August	29.790
March	29.869	September	29.879
April	29.756	October	29.915
May	29.706	November	29.803
June	29.739	December	29.816

The following are the mean heights for the different seasons:

	Inches.		Inches.
Summer	29.850	Winter	29.779
Fall	29.738	Spring	29.866

From which 29.870 is the annual mean when reduced to the level of the sea.

At Valdivia, Puerto Montt, and Ancud the greatest pressure takes place during the spring; at Ancud the least pressure was found in the fall, at Puerto Montt in the winter, and at Valdivia in the summer. If at Ancud 29.87 inches is taken as indicating "variable weather," and this figure increased or decreased by half the mean monthly range, then 30.24 will represent "fine weather" and 29.51 "bad weather." This agrees perfectly with the observations made. The

great rain and wind storms always take place between the
last-mentioned number and 28.30, and they are absolutely
certain when the pressure is less than 29.02 inches.

Temperature.   Although the statistics of the maximum and minimum tem-
peratures are not complete, owing to the hours of observation and the want of special thermometers, the extremes
observed may be considered as approximately correct:

	Maximum.	Minimum.
Valdivia........................	82°.40	30°.20
Puerto Montt	82°.85	32°.00
Ancud................................	68°.00	32°.00

These localities give great differences. Ancud, which is
near the sea, has a lower temperature and slighter oscillations than Valdivia and Puerto Montt, which are protected
from the cold winds by the spurs of the Andes and the
cordillera of the coast.

Daily observations taken during the year 1867 give a mean
daily range of 2°.073, which proves the regularity of temperature in this part of Chile. The warmest months are
January and February; June is the coldest.

The mean temperature at Ancud for 1867 was 51°.60, all
corrections having been applied. The difference of temperature between the winter and summer was 5°.96.

It is ascertained from observation that there are a greater
number of rainy days at Ancud than the combined number
of clear and cloudy days, and that they are much more
numerous than at Valdivia or at Puerto Montt.

Observations show that the number of rainy days increase with the latitude, which holds good even to the southward of Ancud. The greatest rain-fall at Ancud takes
place during the months of May and June.

The great rains of Valdivia and Chiloé are due to northwest currents of the atmosphere bringing hot vapors from the
tropics, which are arrested in their progress by the Andes,
and the cold current which the mountains engender condenses them into rain. This is principally the case to the
southward of the 38th parallel of latitude. This fact holds
true for all the region to the westward of the Andes; to the
eastward of them there is but little rain.

The uncultivated state of this country is the cause of the
small range of the thermometer, rendered still smaller by the
large quantities of rain. The country is but little cleared

of forest; on the contrary, it is observed that in the territory of the Araucanians, where the population is decreasing, the forests are constantly enlarging. The effect of clearing can be observed on the vegetation near the mouth of the Rio Imperial, where it is quicker, better, and more active than at Tolten and Queule, which are in the same plain, but where the ground is densely covered with worthless bushes and the mountains are barren and naked. The same may be observed in the vicinity of Valdivia, where places but short distances apart have very different daily ranges of the thermometer; for instance, Valdivia and Cutipai are favored with better weather, caused by the clearing for the benefit of agriculture. In all cases, however, the clearing of the land has but a feeble influence on the rainfall, as this is principally caused by the Andes.

Snow-storms are unknown on the sea-coast; hail-storms are only of short duration, and the hail-stones small. They occur mostly in the spring, and, contrary to the ordinary rule, during the night. Hail-storms never occur with north winds, or at least not until the wind is on the point of changing to the west or southward of west.

Frost occurs frequently during the spring, but never on the coast. It is only observed in the interior of the mainland or on the large island of Chiloé.

Water-spouts are very rare on the land, in fact almost unknown. They are, however, often seen along the coast of Valdivia and Chiloé when a calm sets in after a squall from the NW., and also when the wind is about to change to the W.

The prevailing winds on the coast of Valdivia and in the archipelagoes of the south are those from the NW. and SW. quadrants. Northeasterly winds are damp, and precede bad weather; whereas, those from the southward and the eastward are dry, and bring good weather. They are not strong, however, and of short duration. The obstacle presented by the Andes very seldom allows them to reach the coast with any force. There are cases on record, however, when they have blown with considerable violence. *Winds.*

The following table shows the results obtained at Ancud during 1863 and 1867, the numbers representing the per centum of time they have blown during each month:

Month.	N.	NW.	W.	SW.	S.	SE.	E.	NE.	Calm.
January	13.0	23.0	9.5	40.5	5.5	1.0	0.0	4.0	3.5
February	10.5	18.5	14.0	32.5	19.5	1.0	0.0	0.0	4.0
March	19.5	10.0	33.5	18.5	12.5	0.0	0.0	0.0	6.0
April	19.5	22.0	13.5	14.5	6.0	2.5	5.5	6.0	10.5
May	34.5	21.0	8.5	1.5	5.0	6.0	8.0	15.5	0.0
June	15.5	40.5	10.0	7.5	5.5	2.0	3.5	8.5	7.0
July	15.0	28.5	10.0	12.5	7.5	13.5	1.5	6.5	5.0
August	17.5	29.5	16.0	3.5	3.0	4.6	1.0	16.0	9.5
September	23.5	9.5	13.5	8.5	21.0	13.0	0.0	3.0	18.0
October	14.5	6.5	25.0	30.5	8.0	3.0	1.0	3.0	8.5
November	20.0	20.5	12.5	25.5	1.5	1.0	0.0	1.0	18.0
December	19.5	30.5	7.0	24.5	0.0	1.0	0.0	1.0	16.5
Percentage for one year.	16.9	21.7	14.4	18.2	7.9	4.0	1.7	5.4	8.9

From this table it will be seen that the NW. wind is the most prevalent; next comes the SW., the N., and last the W. wind. The wind from the NE. and SE. quadrants might be called transient; they occur less frequently than the calms.

The winds from the S. to the W. prevail in the summer, and those from the N. to the W. during the winter.

Changes of the wind. From the beginning of spring, that is, the end of March, until September, the winds rise in the NE., blowing gently at first, but freshening as they veer to the northward; they degenerate rapidly to a gale; especially when they pass quickly to NW.; the atmosphere becomes damp and disagreeable. Sometimes the wind oscillates at first between N. and NE. From NW. the wind passes to SW., sometimes by degrees, at others with a sudden shift; very often it changes forward and backward between NW. and SW., and backs to N. for several days; but during such change it never goes back to the eastward of N. When it goes back from SW. or WSW. to NW., bad weather must be expected. In this case it generally comes again from the SW. at sunset, and the change is then so violent and unexpected as to endanger the masts of vessels.

Before the shifts from NW. to SW., a clearing in the sky is generally observed between W. and SW., or a simple rent in the clouds to the SW. This is a sure sign, and mariners should be warned by it. This phenomenon resembles the bull's-eye of the cyclonists, and was called by the Spaniards *el ojo de torro*. When the wind from the NW. is very fresh, and it rains very hard at the same time, a sudden change may be expected to the W. and SW. This usually takes place about sunset. It may happen that the shift occurs during a heavy rain, or so quickly after its announcement that there is no time for preparation; but this is rare.

From SW. the wind passes to S. without any diminution of its force; then to SE., where it also often retains its strength, especially on the coast during the months of June, July, and August; it dies out between SE. and E., and rises again in the NE., after a calm of more or less duration. During these changes the barometer stands high.

The winds from the NE. are very seldom fresh, and do not increase much until after passing the N. or NNW. points. When the winds blow from the N. or NE., the vessels of the archipelago of Chiloé do not leave their anchorages, as the mariners know that the strongest squalls and heaviest rains come from those points.

From September to March the winds from the SW. quadrant predominate. They sometimes blow heavily, but with a clear sky and fine atmosphere. When the wind blows from the S. and SE., it becomes cold, and the weather fine. The wind dies out at SE.; and after a calm, with a clear sky, the wind springs up light from the NE.; it then becomes cloudy, and rain commences to fall heavily. This generally lasts for two days, sometimes longer, until the winds change slowly against the sun to the southward of W., after which the fine weather sets in again.

During the spring and summer, lightning is an indication of strong wind, and it often precedes a sudden shift. During the winter lightning is also common, and indicates changes of the wind.

When the change of weather takes place during the night, sparks of electricity are seen in the atmosphere toward the horizon, and reports similar to those of a rifle are heard. Thunder is rarely heard at sea, but is frequently over the thick forests of Chiloé and on the continent. Certain physical, thermometric, and barometric phenomena accompany the changes of the wind. After calms of short duration, which are accompanied by a high barometer and clear sky, the atmosphere between N. and E. becomes very clear, and a bluish or light-green tint is observed in the sky in those directions. The mercury then commences to fall slowly, and the temperature rises a little, according to the season.

These phenomena together are a sure indication of a blow from the N. to W. in a few hours. A dark band rises to the northward, the sky becomes covered with a thick mantle of clouds, and the northeast breeze freshens as it changes toward the N. At this point the wind has almost reached its full force, as it only increases a little when passing the NW. point. It generally remains variable between WNW. and NW. for a little while before it shifts gradually or suddenly toward the W., WSW., and SW. During the greatest force of the squall it does not rain, but sometimes drizzles lightly. The atmosphere is very hazy, and the clouds pass from N. to S. As soon as the rain commences, it becomes necessary to watch the horizon in the W. and SW., in order not to be surprised by a sudden shift of wind.

As long as the barometer falls, the weather remains bad, and the wind continues steady from the NW. If, however, the barometer remains at a stand during the time that the wind is strongest, double attention must be paid, as the change of wind is near. Sometimes this is announced by a sudden rise of the mercury, but this latter phenomenon ordinarily happens simultaneously with the shift of wind.

With NE. and NW. winds the barometer sometimes falls as low as 28.27 during the strongest part of the gale. In an ordinary gale the barometer does not fall below 28.78. In case the shift of wind is only momentary, the barometer rises a little, then remains fixed for a short time, and afterward, when the wind has jumped back to NW., it again falls rapidly.

It happens sometimes that during a fresh southwest wind, the barometer falls slowly. In that case it is certain that the wind will shift in a squall to NW. and NNW., and that it will rain heavily when the wind is about to jump back to SW. or W. During these squalls the temperature is notably lower than during those which begin at the NE. and N., from which circumstance it can be supposed that they have a different origin from the former. These blows are fresh and short, and generally die out in the W. or S., toward sunset. The duration of the gales commencing from NE. is uncertain; ordinarily they do not last longer than twenty-four hours, but they have continued two, three, four, and even ten days.

In May, 1867, a northerly gale commenced on the 14th, and blew variably from N. to NW. until the 27th. It changed to the W. on the 31st, and the breeze remained strong from that quarter for several days. In July, 1868, a gale from the NW. lasted more than five days.

It is, therefore, evident that the gales on this part of the coast of Chile follow the same laws as those of high southern latitudes. The sudden shifts which occur from NNW., NW., and WNW. to W., WSW. and SW., prove that these gales come generally from the westward, or some point farther to the N. If the wind, after shifting to SW., jumps back to NW., the vessel must have drifted to the center of a cyclone of small velocity, or two cyclones must be following each other closely; but in all cases navigators should never lose sight of the fact that they are in a revolving gale, in which the wind turns with the sun, and varies to the observer in the opposite direction.

Some anomalies in these shifts of wind are cited. On the 14th July, 1866, the Peruvian and Chilian squadron experienced a gale which shifted from NE. to SE., and thence to SW.; the barometer fell to 29.44. This explained by the theory of storms, is that the center of the gale passed to the northward and right of the vessels of the squadron. It is known that such inverse gales in high latitudes are always accompanied by heavy squalls.

PART II—FROM CONCEPCION TO COPIAPÓ.

Proceeding to the northward the intensity of the rains diminishes and the harvests depend upon the humidity of the year. The southerly winds which predominate during nine months of the year are dry, and fatal to the cultivation exposed to their influence; when the year is unusually moist the grass is abundant, and the cereals yield forty and even seventy fold, but when it is dry the cattle even have to be taken to the Cordillera. In certain localities, as in the vicinity of Valparaiso, the nightly dews moisten the ground and favor vegetation. *Baromet-r.*

The barometer in summer varies between 29.92 inches and 30.32 inches, in the fall between 29.53 inches and 30.04 inches; in the winter it falls as low as 29.14 inches. As in *Barometer.*
2 C

the preceding region it falls with north and rises with south winds.

In summer the temperature is moderate and rarely exceeds 77°, and does not fall below 68°. In the autumn the mean is between 57° and 59°, and in the winter between 53° and 55°.

The phenomena which we have described in Part I, also takes place in this section, but with less force. In the bays the breeze is fresh during the day and very light or calm during the night. In the spring the weather is generally fine, and the winds generally vary between SSE. and WSW. They sometimes alternate with winds between NNW. and W. In the summer the breezes are from S. to SW., and are accompanied by fine and dry weather. In the fall the north wind is most frequent, but after a calm it frequently blows from NW.; it is during the winter, from the end of May to September, that rain and northerly gales are most frequent; when the south wind sets in the weather becomes fine. Thunder and lightning are very rare.

It can be generally stated that there are two sets of winds on this coast; both are well defined but are of unequal duration. The winds from the N. and NE. are frequent in the winter, from the end of May to the middle of September. From September to the end of May, however, the southerly winds predominate and they often blow with great violence. These winds sometimes blow from a little to the westward and sometimes from a little to the eastward of south.

During the three months of the bad season, frequent calms, and light and variable breezes from the W. to SW., are experienced; these alternate with gales from the north, which cause heavy rains, not only on the coast, but also a considerable distance to seaward.

During the fine season, while the south winds prevail, the sky is generally clear, and there is but little rain even in the summer; however, there are at times strong northerly winds, accompanied by heavy rains, that last for two or three days. These exceptions are more rare to the northward of Coquimbo than to the southward.

During this season of the year, the south wind sets in about 11 a. m., sometimes later; it freshens rapidly, and often renders the anchoring of sailing-vessels difficult. It is called the sea-breeze, although it actually blows along the coast; it generally dies out toward sunset, though sometimes it lasts until midnight. At sea, between 25° and 35° south, this breeze sometimes obliges vessels to take two reefs in their top-sails; it is less strong than near the land, but it does not die out during the night. It will be observed that as the land is approached the breeze is stronger during the day and lighter during the night; or, when very near the land, that it falls calm during the night, then a light land-breeze springs up from the eastward; this comes from the Cordillera, and only lasts a few hours. It is almost always accompanied by a clear sky. If during this breeze the sky becomes cloudy, it is a sure sign that the breeze at sea is very light, or that it is calm. It is also an indication of rain. A cloudy sky in the winter is a sign of an approaching wind accompanied by rain.

The shifts of wind generally take place against the sun, as on the southern portion of the coast; the most violent coming from a point between north and west, it is well then for vessels to seek shelter behind the rocks or land which make out to the westward. During the summer it is best to anchor near the land, so as not to be driven to sea by the squalls from the southward. In winter it is best to anchor farther out, on account of the northerly winds.

In the region under discussion the approach of north winds is very well marked, the sky is overcast, and there is little or no wind unless it may be from the eastward; the swell sets from the northward and the tide is stronger than usual; the distant land is remarkably distinct and elevated by refraction, and the barometer falls. All northerly winds are by no means gales, and sometimes a whole winter may pass without a northerly wind which merits that name; at other times they are of frequent occurrence. The northerly gales never last longer than one day and one night, and ordinarily not so long. They shift to the westward with the weather still cloudy and with the strength of the wind undiminished, and moderate as it hauls to the southward. It is very difficult to fix the northern limit of these winds; it is certain, however, that they are much less dangerous at Coquimbo, although strong northerly winds and heavy seas have been known at Huasco (28° 21' S.) and at Copiapó,

(27° 20′ S.)   The English ship Conway experienced a heavy northerly gale in 25° S. latitude and 90° W. longitude.

At Valparaiso (33° S.) these winds do not set in until the middle of May and they continue until September, decreasing continually in intensity. They have occurred, however, in April.

The foregoing remarks apply to a belt extending from 180 to 240 miles to seaward.

Currents.

The waters of the South Pacific ocean, in a belt of which the mean latitude is about 50°, run in an oblique direction NE. toward the west coast of South America; thence as a cold river along Chiloé to the Galápagos islands, while another branch, separating opposite the gulf of Peñas, forms the Cape Horn current.

The northern branch of this current is called the " Humboldt current, or the current of Perú." It runs to the NE. to about the latitude of the island Mocha and then continues to the northward.

This direction (NE.) drifts vessels toward the coast, and when they are but a short distance from the land they should take this current into account, as it has caused several shipwrecks. Its change of direction from the NE. to N. takes place about 180 to 240 miles from the coast.

From the island of Mocha the current follows the trend of the coast, that is a direction between N. and NNE.; its direction varies a little with the wind; its diameter being about 120 miles abreast of Valparaiso, and it increases gradually. This current, aided by the wind, carries enormous masses of sand along the coast of Araucania and as far north as Valparaiso. This is deposited at all capes projecting to the westward and in all the bays and harbors penetrating to the eastward.

The temperature of the water in the Humboldt current is 39° 20′ in 47° S. latitude, and this increases farther to the northward. In the latitude of Valparaiso it is 52° 52′; in that of Coquimbo, 57° 02′; of Cobija and of Arica, 64° 58′. The observations of Dupetit–Thouars have proven, however, that these numbers are means, as the temperature of the current follows the variations of that of the surface. The velocity of this current varies greatly on the coast of Chile and Perú; it is generally greater near the land than at sea. Between Valparaiso and Cobija it runs at the rate of 26 miles per day; 15 miles per day is about the average velocity. The current is sometimes altogether arrested and sometimes reversed. Lartigue, and afterward Fitz-Roy, observed the current to run south in different latitudes, after strong northerly wind; but, according to Fitz-Roy, this does not prove anything definitely. The Russian admiral Litke observed no currents to the northward until north of Concepcion, and then very feeble, although he kept near the shore, the winds being light and variable from NE. to NW. During one day he had the current running to the southward. Between cape Pilar (54° S.) and Valparaiso, (33° S.,) Admiral Fleuriot de Langle observed a NE. current in the spring; a WSW. current during the summer; E. ½ N. during the fall, and WNW. during the winter. From this we conclude that the southerly current is rare, the northerly ones much more frequent, and that the latter run alternately to the eastward and westward of this point. The last-mentioned observer speaks of strong and frequent currents to the westward during the summer.

To the northward of Chiloé the rise and fall of the tide is *Tides.* never more than 6 feet, the general limit being 4.9 feet. It is not, however, the same for the archipelago of Chiloé and Chonos. When the flood-tide sets in it runs around both the north and south end of Chiloé at the same time. The former runs with great force, through the straits of Chacao, into the gulf of Ancud, where it produces violent eddies. After running around the gulf of Reloncavi, and around the archipelago of Calbuco, its general course is S. During this time the other branch, in running to the northward, passes through the channel which separates the Guaitecas from Chiloé. These two tides meet in the latitude of the Desertores and Chaulinec islands and counteract each other. The ebb-tide runs to the northward, in the gulf of Ancud, and to the southward in the gulf of Corcovado.

It will be readily understood that at certain points near the limit of the two floods there are very high tides. At the Carelmapu islands, opposite San Carlos de Ancud, the rise and fall of the spring-tides amounts to 10 feet; in the straits of Chacao it is 14 feet; in the cove of Oscuro, 22 feet, and at Huildad, 16 feet. If, in addition to this, the influence of the wind is considered, the irregularity of the tides on the E. coast of Chiloé is not astonishing.

Magnetic variation. The variation is easterly on the whole coast of Chile. From observations at various dates and from various sources, the variation diminishes to the north, on the average, about 20 minutes for every degree of latitude, the decrease gradually diminishing northward. The absolute value of the variation, however, increases about $1\frac{1}{2}$ minutes annually. The curves of variation trend from the coast in a WNW. direction, but only for a short distance, to about the 83d meridian, where they bend down in a WSW. direction.

Coasting routes. The northerly and southwesterly winds, which are the most prevalent on the coast of Chile, determine the course to be taken in all coasting voyages. Valparaiso, being a central position, will be taken as the point of departure and of arrival.

From cape Pilar or the gulf of Peñas to Valparaiso. Vessels bound to Valparaiso, or a point to the northward of it, from cape Horn or cape Pilar, should get well to the westward, and, when the wind shifts, be able to steer directly north without being encumbered by the land. This run is generally short, vessels frequently making from cape Pilar to Valparaiso in eight days; from the gulf of Peñas in six; twelve days being a medium passage from the strait of Magellan to Valparaiso. Vessels from the gulf of Peñas to Valparaiso should follow the same rule; they should get well off the land, as far as the meridian of 80°.

From Valparaiso to the southward. The southerly winds render this a hard and tedious passage; it is frequently necessary to carry two reefs in the topsails, and to stand on the port-tack, unless a northerly gale should allow of a direct course. These, however, continue generally but a short time. The stretch on the port-tack must be regulated by the season and the point to be reached.

From Valparaiso to the bay of Concepcion. If bound to Maule or Talcahuano, during the fine season, the stretch on the port-tack should extend but half the distance between Valparaiso and Juan Fernandez, and then stand in for the land. By this course the chances of a north wind are reserved, and the sea is smoother nearer the land. Very often the topgallant sails can be carried near the coast when the topsails have to be double-reefed farther out on account of the heavy sea.

To within a certain distance of the land the winds vary

between SSE. and SSW., which must be looked to when beating to the southward; in that case vessels should run on the starboard tack with SW. winds, and on the port tack with SE. winds. The first blow, generally, during the day, the last during the night. Near the coast, account must be taken of the currents and tides. According to Admiral Cosnier, this is the way in which the coasters of the country navigate.

Admiral Fleuriot de Llangle recommends the use of the thermometer, instead of sounding, on this part of the coast. According to his observations, the air has always a much higher temperature than the water, about 5°, and the temperature of the water falls on approaching the shore.

A vessel bound south of the bay of Concepcion, to Chiloé or Valdivia for example, must stand on the port tack, especially in the bad season, until off Juan Fernandez or Mas-a-Fuera, which are points easily recognized; then tack, but keep close to the wind until in the latitude, or even south, of the place of destination. *From Valparaiso to Valdivia and Chiloé.*

This passage, which requires about twenty days, is tedious and laborious. It is advisable to stand out from Valparaiso on the port tack for about 600 miles, or until the 86° or 88° meridian is reached; thus, a vessel would sometimes pass to northward of Juan Fernandez and Mas-a-Fuera. In these latitudes, (30° or 35°,) with the winds generally from the . southward, it is best not to try to make southing without westing; it is advisable, therefore, to stand to the SW. on leaving Valparaiso, at the start, even should the wind be from the northward. After reaching the 86° or 88° meridian, it is best to stand as far as 50° without losing any westing, if possible, and then steer a direct course for the straits of Magellan or cape Horn. If bound to cape Pilar, vessels will make the land too far south if they do not take into account the influence of the current, which, from the gulf of Peñas, sets to the southward along the coast. *From Valparaiso to cape Pilar or cape Horn.*

There is no difficulty in passages to the northward, as they are generally made with a fair wind. *Routes from Valparaiso to the northward.*

If bound to Arica, or a port to the southward of it, it is best to remain in sight of the land as much as possible, but not so near as to risk its vicinity cutting off the breeze; the influence of the coast-current is then felt and the danger of missing a port, owing to fogs, is avoided.

If bound to a port to the northward of Arica, as Pisco, Callao, or Paita, it is best to stand a little out to sea, so as to have the wind more steady and to avoid the bight of Arica, which is subject to calms.

Routes to the westward.

Under this head are comprised the routes to California, China, Polynesia, Australia, and India, and the return from those countries to Valparaiso. A few general remarks, only, will be made on these.

From Valparaiso to California and return.

On leaving Valparaiso for California it is best to stand NW. until in latitude 15° south and longitude 95° W., where the Maury route from cape Horn to California is intersected; Maury's route should then be followed. On the return trip from California, as soon as well clear of the land, steer to the south, crossing the NE. trades and equatorial calms, and taking the SE. trades good full; when upon taking the westerly winds of the Southern ocean and reaching 35° south latitude, which parallel is to the southward of Valparaiso, a direct course can be made to that port during the winter; but make the land a little to the southward during the season of the southerly winds. If the westerly winds are not well defined in latitude 35° it is best to stand farther to the southward.

From Valparaiso to China and return.

The passage from Valparaiso to China is made with the help of both trades. It is only necessary to stand far enough to the north or south, according to the season of the monsoons, keeping far enough to the northward during the NE. monsoon, and far enough to the southward during the SW. monsoon to reach the point of destination. For a return voyage it is best to stand to the eastward of the archipelago of Japan until in 30° north latitude. From there the westerly winds will take a vessel near the Sandwich Islands; from this point the route from California to Valparaiso is taken. This route is very laborious during the NE. monsoon, and many vessels take a route to the westward of Australia.

From Valparaiso to Australia, (Sydney.)

In this passage a vessel has simply to keep in the SE. trades, avoid getting too far north, and make southing little by little, in passing to the southward of New Caledonia.

For the return trip a vessel should pass to the southward

of New Zealand, as that route approaches a great circle, and make the longitude on the parallel of 45°. The northing is made little by little by keeping in the general current and afterward in the Humboldt current.

There are no special directions for the passage from Valparaiso to Polynesia, as it presents no difficulties. To return to Chile it is necessary to make use of the westerly winds which blow to the southward of 35°.

The voyage from Valparaiso to India can be made in two ways, around cape Horn, or through the trade-wind belts of the Pacific. It is best during some seasons, especially those of the SW. monsoons in the China seas and Indian ocean, to take the route around cape Horn. But this is always rough, and a good look-out for ice must be kept on the passage between cape Horn and the cape of Good Hope. In either case it will be necessary to study the winds of the different seasons. *From Valparaiso to India.*

There are several lines of steamers, having their terminus at Valparaiso, which make regular trips to Liverpool, by way of the strait of Magellan on the one side, and to Panama and the intermediate ports on the other, connecting with steamers at Aspinwall to the West Indies, the United States and Europe; in addition to which, there are several smaller steamers engaged in the coasting trade of Chile. *Steam navigation.*

Compared with its population, Chile is one of the greatest commercial countries of the world. It is also the port of entry for the south of Bolivia and for the northwest provinces of the Argentine republic. Its commercial importance shows yearly progress, attributable to its agricultural and mineral wealth, as well as to its industrious and peaceable population. *Manufactures and commerce.*

The manufactures of Chile under the patronage of the government are increasing. Among them are hempen cloths, cordage, soap, tallow, leather, charcoal, flour, brandy, the coarser works in gold, silver, copper, and iron, earthen jars, and ponchos; *charquí*, or beef dried in the sun, is also produced in considerable quantities. The greater part of the foreign trade is with Great Britain. The articles mostly exported are lead, copper, silver, wheat, flour, barley, hides, and wool; and in the imports are included nearly every variety of manufactures and foreign products.

During the thirty years preceding 1874, the value of the imports and exports of Chile has increased from, imports, 8,596,674 pesos—(the peso being, at the average rate of exchange, 96.8 cents)—exports, 6,087,023 pesos—to, imports, 38,418,000 pesos—exports, 36,541,000 pesos. The total value of the commerce of Chile, including her coasting-trade, for the year 1874, was 120,795,000 pesos. Of this, the imports from the United States amounted to 2,150,000 pesos, the exports to the United States 553,000 pesos.

Railroads. Chile was one of the first of the South American states to encourage the construction of lines of railroads. At present, there are completed and in course of construction about 805 miles of road; 580 miles of which is the property of the government, 225 being that of companies. A road across the Andes between Chile and the Argentine Confederation is in contemplation. The length of the road would be about 1,023 miles.

Telegraph. In 1873 there were in Chile 3,043 miles of telegraph. The Trans-Andine telegraph, connecting Santiago and Buenos Ayres, was opened to the public in 1872.

CHAPTER II.

Variation, 1876.—Cape Tres Montes, 21° 25′ E.; Chonos and Guaitecas archipelago, 20° 19′ E.; increasing about 1′ 30″ annually.

Cape Tres Montes, the southern extremity of the peninsula of Tres Montes and Taytao, is a remarkable promontory 2,000 feet above the level of the sea; is readily recognized, and is a comparatively safe point to make in bad weather, with the wind blowing on shore. There is no sensible current, and with the wind on shore, or, when overtaken by the night, a vessel has sea-room to lay-by until daylight.

Peninsula of Taytao.
Cape Tres Montes.

Cape Raper is 14 miles to the NNW. of cape Tres Montes, in latitude 46° 49′. From it the rocks and breakers project 1½ miles to seaward. This cape is one-third higher than cape Tres Montes. The side toward the sea is barren, and ends almost perpendicularly. The three sand-beaches between these two capes cannot be approached, owing to the constant heavy swell and breakers.

Cape Raper.

The coast from cape Tres Montes to capes Raper and Gallegos is free from outlying dangers; the water is deep, and the land has an elevation of from 2,000 to 4,000 feet.

The tides have seldom a greater rise than 6 feet.

Tides.

Cape Gallegos is a wall-sided promontory which rises perpendicularly from the water, and is barren toward the sea. To the eastward of it is San Andres bay, which cannot be used by large vessels; small vessels can find a good anchorage in Christmas cove, which is not quite half a mile broad, but is protected against all winds excepting those from the north; these winds may seem dangerous at first, but it will be found that they are more so in appearance than in reality, for they do not become violent until well to the westward of north, and the sea has not time to rise before that. The Beagle passed several days quietly in this anchorage, while a strong gale from the westward was blowing outside. It has from 11 to 19 fathoms of water, sandy bottom.

San Andres bay.
Christmas cove.

Cone Inlet.

Cone inlet is a long and narrow cove about 1 mile to the southward and eastward of Christmas cove; it extends into the shore to the foot of a remarkable conical mountain, 1,600 feet high; it is perfectly sheltered and has a depth of water of 16 fathoms; it is a passable stopping-place for a steamer, inside of the island at its entrance, where she can moor with stern lines on shore; it is not, however, a suitable anchorage for a sailing-vessel, being very narrow and difficult of access. There is no swell inside Cone inlet even with NW. winds; there is a natural basin on the NE. side.

In the northern part of the bay of San Andres, at the bottom of a deep indentation, there is a large basin, called *Useless cove*, as no vessel can enter it. Cape Pringle is the northern point of the bay of San Andres. Between this cape and Rescue point, 18 miles to the northward, the land falls considerably. There may be good anchorage in Stewart bay and Cliff cove, but they have not as yet been explored.

Port San Estevan.

Port San Estevan is 20 miles NE. of cape Pringle; a good anchorage with 10 fathoms water is found behind the scattered rocks off point Rescue; good water can be obtained in the river at the head of this arm of the sea, and also in the brooks near the anchorage. Dark hill, which rises 2,150 feet above the sea, is an excellent landmark for this port. Vessels can pass 200 yards outside of the rocks of Rescue point, and anchor on the west shore under the shelter of the reefs.

Hellyer rocks.

Hellyer Rocks is a dangerous group of rocks, hardly showing above the water, on which the sea breaks; they are N. 1° E. from point Rescue and 6 miles from the nearest land, Duende island.

Cape Taytao.

From point Rescue to cape Taytao the coast is broken and abrupt. The latter cape is N. 10° E., 25 miles from the entrance to San Estevan. There are undoubtedly anchorages to leeward of Usborne islands, behind mount Alexander, in Cornish cove, or more to the northward in Burns inlet, but they have not been as yet examined.

Cape Taytao is one of the most remarkable promontories on this coast. It resembles an island with a pointed summit; is 3,000 feet high, rugged, barren, and steep, and is

Approaches to Anna Pink Bay.

Coast from Rees Island to Cape Tres Montes.

surrounded by several rocks, but none of these are more than one mile from shore.

On rounding this promontory Anna Pink bay is to the east. The Anna Pink, a transport of Lord Anson, sought shelter in one of the coves of this bay against a westerly gale; she anchored behind the island Yuche-Mo, but dragged across the bay, and after having slipped or cut her cables, she reached Port Refuge in the SE. portion of the bay, where she repaired damages in security.

Anna Pink bay.

On the west shore of the bay of Anna Pink is Yuche-Mo island, 460 feet high. On this island were found the ruins of a large log hut and a number of goats, by the officers of the Beagle. It is probable that the Anna Pink did not anchor close enough to this island, as there is good bottom on the east side in 15 to 20 fathoms of water, with Penguin island bearing N. 21° E., and the summit of Yuche-Mo S. 66° W.

Yuche-Mo island.

Canaveral cove is situated on the southern side of the entrance to Port Refuge; though small, it is very convenient for refitting and repairing.

Canaveral cove.

Patch cove, to the westward of Gallegos island and WSW. from the entrance to Port Refuge, is so small that it cannot be used by vessels of more than 200 tons.

Port Refuge is bounded by high mountains, and to the ESE. sheltered by Yuche-Mo; it is a long, narrow, and deep inlet, in shape of a cornucopia; the bottom is rocky, with the exception of Lobato cove, where it is sand, in 27 fathoms of water.

Port Refuge.

About 6 miles NE. of Yuche-Mo are the Inchin islands. Near them, to the northward, are the Tenquehuen, Menchuan, and Puyo islands, among which there are probably several good anchorages, with the fresh water, wood, wild herbs, fish, and everything that is found on this coast. The western extremity of Menchuan island is low, and a cluster of rocks extends off the NW. point about two miles; it must therefore be given a wide berth.

The archipelago of Chonos and of Guaitecas consists of a multitude of islands, extending from the Pulluche channel on the south, which is the northern boundary of the peninsula of Tatayo, to the gulf of Corcovado on the north. These islands, lying in groups of from twenty to fifty, are

The Chonos and Guaitecas archipelago.

of various forms and sizes, some having a circumference of 70 miles, others of only a few yards. The largest is Santa Magdalena, separated by the Yates channel from the cordillera of the coast. It contains mountains and volcanoes covered with snow, and was, until lately, supposed to be a portion of the cordillera. Melchor, Traiguen, and Rivero are among the largest of these islands; they are traversed by the three principal channels, which run nearly east and west; Ninnalaca channel to the north of Melchor island; Darwin channel to the northward of Rivero and Traiguen islands; and Pulluche channel, the southern boundary of the archipelago. These connect with the Moraleda channel on the eastern side of the archipelago, which runs nearly north and south, and, with the channels Errazuriz and Costa, which are to the westward and eastward of the island Traiguen, extends through the whole length of the archipelago. The channels mentioned, as also several of the minor channels, have been examined and partially surveyed, more recently by captain Simpson and officers of the Chilian navy. On the coast of Taytao, as also throughout the archipelago, there are numerous bays and coves where vessels of heavy tonnage can find refuge. The depth of water varies, but in general the water is deep well up to the land. Banks and rocks under water, when in a depth less than 12 to 14 feet, are, as a general rule, buoyed with the growth of kelp and sea-weed. In deeper water the strength of the current does not permit this growth. The general aspect of these islands, which seem to rise suddenly out of the water, is abrupt, mountainous, and rugged. To the south some resemble a sugar-loaf; to the north they are in form of plateaux or elevated tablelands. They vary in height between 2,000 and 4,000 feet. The peaks of Cuptana, 2,950 feet, are constantly covered with snow, and in their vicinity are mineral waters. It is supposed that the mountains of these islands contain minerals, especially iron. Most of these islands seem to have the same geological formation as the cordillera, being of volcanic origin. Basalt, quartz with black sand, pumice-stone, outcrops of lava, with caves and grottoes covered on the inside with many-shaped stalactites. Some of the westernmost islands, as Huafo, Ypun, and Huamblin, evince a Neptunian origin ; they are formed of sediment, and have gradually risen from the bottom of the sea. Nearly all these islands are off-shoots of the great cordillera. The channels and estuaries are submarine valleys, lower than the pampas of the east.

The plains, which are a short distance from their shores, have a rich vegetation.* It is hard to penetrate the thickets, but if one succeeds in reaching the summit of one of the high islands, there is a splendid view of seventy or eighty different islands, separated by numerous channels. Nearly all of them have lagunes of fresh water in their interior, which abound with fish and are dotted with islets.

There is no fixed population on these islands. It is supposed that they were inhabited during the last century. At the end of the last century, the influence of the Jesuits drove the Chonos Indians from the archipelago ; transporting them to Chiloé under the pretext of conversion, they were allowed to die in misery. Some few escaped by the isthmus of d'Ofqui and went to Patagonia.

In some localities burial-places have been found, from which skulls were taken which resemble those of the race of Payas of Chiloé. They were more generally found in caves closed by branches. Some mummies, in oval cases made of the bark of the cypress, have also been found ; but all these have been removed or destroyed.

During the spring and summer the wood-cutters visit these islands to cut wood and hunt the seal. They discovered in the caves, which are not reached by the high tides, large deposits of guano, which is now being exported.

In 1866 three thousand laborers, placed on these islands to cut wood, did great harm by burning the woods, and by hunting the seal-calves almost to extermination.

The aquatic animals, such as the seal, sea-cat, &c., are abundant; the sardine and the robalo (*perca labrax*) are also found. The seal is diminishing. The larger animals of this species, the sea-elephant and sea-leopard, have entirely disappeared since the beginning of this century.

The fauna of these islands is quite rich. On Inchin and Tenquehuen are a large number of small goats which seem to be of European origin, somewhat degenerated. On Huafo there are wild dogs about 2 feet high, with short hair and ears. On the opposite coast, on the continent, the puma, fox, guanaco, and large deer are found. The birds of these islands are geese, chickens, hawks, cat-birds, thrushes, and goldfinches. There are no reptiles or venomous animals.

The principal value of these islands consists in their forests. The trees attain a good size, suitable for construction. For railroad-sleepers and spars are the alerce ; the cypress, which abounds on elevated localities; the oak, which attains a circumference of 13 feet and the height of 50 feet; the mañiu ; the tepu ; the laurel ; the teniu ; the ciruellio, which is suitable for cabinet purposes; and the hazel. In 1871, three hundred thousand railroad-sleepers were cut from these forests.

Climate.

The climate of these islands is severe, especially in the winter; the changes in the temperature are very abrupt; showers mixed with hail, accompanied by furious squalls, occur frequently, as do also heavy snow-storms, which render the navigation of the narrow channels dangerous. Earthquakes are rare in spite of the volcanic formation of the islands. In the summer there are some beautiful days, and the sun even dries up the brooks, causing a want of fresh water; but this is rare.

It is estimated that from 156 to 192 inches of rain fall annually. On an average day it is rare to see the horizon farther than five miles, the atmosphere is so thick and saturated. The climate is not unhealthy; local diseases are rare, and epidemics unknown.

Barometer and thermometer.

From observations made during the spring and summer of 1865–1866 and 1866–1867, by M. Veisthow, the mean height of the barometer was 29.6; that of the thermometer 52° 34'. The general averages for the three following months were:

	Barometer.	Thermometer.
December	29.63	52° 16'
January	29.63	55° 22'
February	29.00	53° 60'

From the averages obtained at Melinka, which is in 43° 50', it will be seen that both the barometer and thermometer are lower than in the corresponding northern latitudes.

Captain Simpson verified these observations in the summer of 1871. During the day the temperature varied between 50° and 53°, and during the night between 46° and 50°.

During the time of these observations the minimum pressure observed was, on October 7, 1865, at 6 p. m., 28.91 inches. The maximum pressure was observed the 14th of October, 1865, at noon, 29.91 inches, during clear weather and a SW. wind. The thermometer during that period stood highest on January 1, 1867, when it was 64°.40 at 6 p. m., during a calm and clear sky, as on December 30, 1865, from noon to 6 p. m. during a calm and a light SW. wind, with clouds at the horizon. It was at its minimum, 41°.00, the 13th and 14th October, 1865, at 6 a. m., with generally clear and calm weather, with occasional hail-storms.

By inspection of the table it will be seen that the results Winds. obtained for this part of the coast are nearly identical with those for the region comprised between Chiloé and Valdivia.

During the summer and spring the winds from the NE. and the SE. quadrant are very rare; the winds generally blowing from the N., NW., W., SW., and S. The winds generally die out during the evening, especially those from the N. and S.; the winds from the SW., on the contrary, freshen, or at least keep their force, in the evening. This confirms what was mentioned in the first chapter, namely, that the shifts to the SW. generally take place about sunset. During the mornings the NW. winds predominate. Generally speaking, the winds from the N., NW., W., SW., and S. are about equally divided. Calms are also as common during the mornings and at noon as any of these winds; during the evening they are about twice as frequent; in other words, the calms occur one day out of six in the forenoon, one day out of seven during the day, and one day out of three during the evening.

The NW. and SW. winds deposit all the evaporation of the Pacific ocean on the Cordillera, which causes the frequent rains. In the estuaries, as that of the Rio Aysen, it rains less, though 11 inches has fallen there during one night. The calms and the NE. or NW. winds bring the heaviest rains. There are only passing showers with the W. or SW. winds, but it invariably rains in torrents during SE. winds.

3 c

The barometer does not stand as high in this locality as in the latitude of Valparaiso; for instance, in all this hemisphere the barometer falls before the winds from the N. and NW. and through their duration, and rises during the calms or when the wind goes to the SW. or S. In consulting the barometer, however, its height must not be taken into consideration as much as its greater or less variation from its original reading.

During the voyage of the Chacabuco, in March and April. 1870, Captain Simpson observed a minimum barometer of 27.02 inches during a NW. gale, and a maximum of 30.4 inches during a SE. wind with clear weather. The heaviest gales blow from the NNE. to NW., often flying in a furious gust to the W., SW., and S. before the barometer rises and the thermometer falls. The hygrometer is also very useful. A great degree of saturation denotes a north wind and a small one south winds. Vessels at anchor must keep a good look-out for the shifts from NW. to SW.

Sometimes the wind after shifting suddenly from NW. to W. returns to NW., as it does more to the northward, when it blows with more violence than before; in fact, there are two gales following close to each other.

During the winter, the gales are accompanied by thunderstorms, and the traces of the passage of the lightning are frequent on the declivities of the mountains.

Winds observed at Melinka during the spring and summer of 1865–1866 and 1866–1867.

Month.		N.	NE.	E.	SE.	S.	SSW.	SW.	WSW.	W.	WNW.	NW.	NNW.	Calm.	Number of days.
October, 1865	6 a.m.	2	2			2	1	4				4	1		1
	noon	3				1	1	5	1	2		3		3	
	6 p.m.	1						8	1			4		2	
November, 1865	6 a.m.	2							1		1	1		2	
	noon					1			2			1		2	
	6 p.m.	1						1				3			
December, 1865	8 a.m.	5	1	1		2	2	3		3		6	1	5	31
	noon	5	1			3	2	3		2	2	7		2	
	6 p.m.	6				3	2	3		2		7		6	
December, 1866	8 a.m.	3	1	3		10		3	2	4		3	1	1	31
	noon	3			1	9		5		4		3			
	6 p.m.	2				4		11	1	2		2			
January, 1866	6 a.m.	5	1			6		4	4	3		6			31
	noon	2				4		6	5	5		7		9	
	6 p.m.	1				3	1	10	4	4		5		7	
January, 1867	8 a.m.	8	1		1			1		4		1		6	21
	noon	6	3			5		5	1	2		2		9	
	6 p.m.		1			1	1	6	2	3				7	
February, 1866	6 a.m.		1	1		2		9	1	4		4		2	21
	noon	3	3			6		3		3		3		4	
	6 p.m.	3	1			1		3	1	5				9	
February, 1867	8 a.m.	7		1		3		5	1	5				5	
	noon	7				4		4		3				3	
	6 p.m.	3				5		2		3				15	
March, 1866	8 a.m.	2				3		3		2		16		6	
	noon	4				1		5		1		16		12	
	6 p.m.	3	1			3	1	1				11		3	31
March, 1867	6 a.m.									2				11	
	noon							1		1					
	6 p.m.	1	1							1					
General average	7 a.m.	32	3	4	2	34	1	32	9	29	1	41	9	34	24
	noon	3	6	1	3	29	3	36	11	23	2	43	1	29	27
	6 p.m.	20	3		2	19	2	43	9	21		33		72	24

* The last 8 days of the month. | First 6 days of month. ‡ First 5 days of month.

The following general instructions will be an assistance in the navigation of the channels of the archipelago.

1°. Every dangerous submarine rock is marked by sea-weed, or is visible from aloft; strong currents, however, submerge the sea-weed.

2°. The sea-weed, in smooth bays, commences generally in seven fathoms of water, where the bottom is of large rocks; in three fathoms, where the bottom is of small stones or pebbles; it does not grow on sandy or shell bottom, or on leeward points which are washed by a heavy sea.

3°. In the channels transverse to the coast the flood-tide runs to the eastward, the ebb to the westward, and those parallel to the coast trending to the northward and southward; the flood in general runs to the north and the ebb to the south. In the large channels bordering the cordillera, the tides increase in velocity as the cordillera is approached.

4°. The winds generally incline to the west; on this account vessels should keep on that side of the channel where there are winds, good anchorages, and easy of exit.

5°. It is recommended to the navigator coming from seaward to anchor or tie up as soon as possible after entering the channel, preferring the northern shore, and send boats to seek those acquainted with the channels; these are readily found during the summer; too much confidence should not, however, be placed in them; they are generally unacquainted with the soundings, and are only useful in indicating the channel. With experienced lookout aloft, and the usual precautions, there is much less risk in this navigation than is usually supposed. To a novice in these waters the principal difficulty is the great depth of water surrounding the anchorages; but the general correspondence between the height of the land and the depth of the soundings in its vicinity will rarely deceive, especially in the bays and coves where the beach is seen.

Outside this archipelago the currents are weak, inside they attain great velocity, which varies much with the direction of the wind, the age of the moon, and the tidal hour.

The length of the continent, at Melinka and in the open ports, the difference between the extreme stages of the tide is at the syzygy about 8 feet; but in the contracted parts of the channels it is often 20 feet, and has a velocity of 8 miles per hour. According to recent observations by Captain Simpson, the tides during the summer are stronger during the night than in the day; in winter this phenomenon is reversed.

This arm of the sea, the Pulluche or Wickham channel, Pulluche or Wickham channel. separates the peninsula of Tatayo from the Chonos archipelago, and is the southernmost entrance to the archipelago; it opens to the eastward of the bay of Anna Pink, between Yenche-mo and the Skyring islands on the south, and the Inchin and Tenquehuen islands on the north; it connects with the several channels running to the northward through the archipelago.

The Wickham entrance is shown distinctly on the chart. After passing the islands Black and Bister, which leave to starboard, steer for Clemente island, borrow toward it and steer to the ESE.; on arriving to the east of Clemente island be careful not to take the Williams channel, which runs to the NE. and has a group of islands at its entrance; two miles farther on, after passing between the island of Guerrero and the small island of Ricardo, is the entrance of the Pulluche channel; after doubling a rock, on the summit of which are several dead trees, an anchorage will be found from whence pilots can be sought. Farther within the channel the tides run with considerable velocity; with a favorable current it is desirable to keep in midchannel.

After passing the Utarupa channel and the SW. point of Chacabuco channel. Humos island, the Pulluche takes the name of Chacabuco channel, which it keeps to the Costa channel.

On the north shore of the Chacabuco channel, near its Port Archy. intersection with the Errazuriz channel, at the SE. end of Humos island, is Port Archy. The Janequeo anchored here with point Archy bearing N. 80° E., and Observatory Cay S. 75° E.; the anchorage is exposed to SW. winds, but is somewhat protected by a line of shoals having on them from $1\frac{1}{2}$ to $4\frac{1}{2}$ fathoms of water; the two small islets called Observatory Cays are a part of these. The Janequeo was anchored in 11 fathoms of water, bottom sand and stone, about 600 yards from the shore; it is better to anchor in the western part of the cove in 10 fathoms, about the same distance from the shore, it being more sheltered.

Wood and water can be obtained. About 2¼ miles to the northward of point Archy, in Errazuriz channel, is a cove formed by Ramahuel islet and Humos island; near the land on the south side of the island vessels are sheltered against winds from N. to SSW.

Tides.

It is high water, full and change, at Port Archy at 1ʰ 30ᵐ; rise 10 feet; velocity 2½ knots; direction of the flood NE., ebb SW.

Bay of San Ramon and port San Miguel.

The bay of San Ramon opens after passing Errazuriz channel, and is formed by the islands Rojas and Traiguen. The diameter of this bay is about 4 miles; it offers a good anchorage in all its parts. In its western corner is port San Miguel where vessels can ride to one anchor in 17 to 20 fathoms of water, bottom sand.

To the northward of the outside entrance of Pulluche channel, between the islands Iuchin and Tenquehuen, are the islands Clemente, Garrido, and Isquiliac, all of which have steep and barren coasts about 3,000 feet in height.

Errazuriz channel.

To the westward of the island Traiguen, and connecting Pulluche and Darwin channels, is the Errazuriz channel. About a third of its length from the northward it intersects the Vicuña channel, which, running to the westward, communicates with the Utarupa channel. At the intersection of these two channels is a bank of shells forming a beach.

Utarupa and Williams channels.

Farther to the westward are successively the Utarupa (large channel) and Williams channels, which run from south to north, and also connect the Pulluche and Darwin channels. Opposite these, on the opposite side of the Pulluche channel, are two estuaries which run into Taytao to the foot of the cordillera of the coast. Care must be taken not to confound these estuaries with the Ortuzar channel, which ends opposite Errazuriz channel.

Utarupa channel is 6 or 7 miles wide, but its eastern shore is so full of islets and banks that its navigable part is reduced to 2 or 3 miles. The western shore is clear and deep.

Williams channel is tortuous, and near its middle is but 164 feet wide. In this strait the tide runs with great velocity, and, though safe and deep, it cannot be recommended for large vessels.

Darwin bay.

The name of Darwin bay has been given to the large

opening between the islands Tenquehuen and Vallenar. In its center is a dangerous islet, called Analao.

Darwin or Agüea channel enters the archipelago of Darwin or Agüea channel. Chonos in latitude 45° 22′ 30″ S., and longitude 74° 29′ 00″ W., between the islands Garrido and Isquiliac. It is perfectly safe for the largest vessels, and it presents no serious difficulties.

The Darwin channel is considered the best, not only because it is less tortuous than any of the others, but because it has good anchorages at both extremities, Vallenar toward the ocean, and Port Lagunas or Español at its eastern end. It is shorter than Pelluche channel. It is impossible to mistake the entrance of this channel, either from the ocean or from the Moraleda channel. Coming from seaward no great mistake can be made in its distance from Tres Montes, and when approaching the land the Vallenar islands will be distinguished, and the entrance itself will be clearly recognized from its wild aspect and high cliffs, free from rocks and islets.

The coast trends about E. 2° N. for about 6 miles, then the channel inclines a little more to the northward. Here a white, rocky, and well-detached islet will be seen, which must be left to starboard, and when passed a deep branch of the sea, running to the north, and containing several islands, is opened. After passing this opening, the coast runs about ESE., until another island is doubled, when the principal channel running to the northward will be seen. It is about 29 miles from the entrance. On each side the channel is skirted with forest trees, and in all probability smoke from some vessel will be discovered in this locality, when a pilot should be sought, if one has not been already obtained. Before entering Moraleda channel the Darwin channel contracts and joins the former about 4 miles to the southward of Port Lagunas, at which point it is not more than 1 mile wide. Some stores can be found at Port Lagunas. On leaving this channel, hug the coast to port until it seems to incline to WSW., then stand to the N. or starboard shore until a small white island appears to be in mid-channel; leave this island to port, and keep in mid-channel; the swell of the ocean will soon be felt, and the open sea come in sight. This heavy swell is the greatest inconvenience when leaving this channel, and vessels should not attempt to run out unless wind and tide are favorable.

If on coming from the sea it is desirable to anchor, a good place will be found on the south side, about 3 miles inside the entrance of Williams channel; the anchorage is in 15 fathoms of water, near two points formed of heaps of stones, where the tide is little felt; it is called port Yates. Mayne mentions that all the western islands of this archipelago are marked too far to the westward.

Pichirupa channel.   Pichirupa is the only channel running from S. to N. which connects the Agüea and the Ninualac channels; it runs to the eastward of the island of Victoria, and joins Ninualac channel between the island Kent on the east and the island Melchor to the west. This channel is a labyrinth of shoals and islets.

Vallenar road.   North of Darwin bay and between Isquiliac and the Vallenar islands is the road of Vallenar; it is well marked by Isquiliac mountain, which is very steep, 3,200 feet high, and ends in three peaks; it is an excellent anchorage, and can be easily entered and left. The best anchorage is in 14 fathoms, with rocky and sandy bottom, near a small islet at the SE. extremity of Three Finger island. The Beagle took refuge here during a SW. gale.

Tides.   It is high water, full and change, at Vallenar road at 12^h 18^m, rise 5 feet.

Huamblin or Socorro island.   Huamblin or Socorro island is about $9\frac{1}{2}$ miles long, and is 30 miles NW. of the Vallenar group; it is from 400 to 790 feet high, and comparatively flat. It is heavily wooded, and its shores are generally sloping and covered with verdure; here and there are remarkable ravines, which are in strong contrast with the somber color of the vegetation. There is anchorage under this island, to the eastward of it, in from 8 to 15 fathoms, $1\frac{1}{4}$ miles from the shore. It is high water, full and change, at 12^h.

Ypun or Narborough island.   Ypun or Narborough island resembles Huamblin island, and is entirely different from the neighboring islands, which are high, steep, and generally barren on the sea-side, while Huamblin and Ypun are comparatively low, flat, and fertile. Already valuable, these two islands will probably soon become more so from the abundance of vegetables and provisions which can be raised on them. They can both be easily approached and left. The sea, which breaks constantly on the advanced rocky points, is a sufficient mark to avoid them. The peculiar aspect of Ypun and Huamblin seems to indicate that they are of a Neptunian origin, in opposition to the volcanic origin of the others.

There is a good anchorage under Ypun, in 12 to 16 fathoms of water, bottom sand and clay. The harbor of Scotchwell, in the SE. part of the island, not only offers good shelter against storms, but is also a sure and convenient harbor for watering, repairing, or taking in wood. It has to be approached from the northward. The approach from the southward has been examined and no hidden dangers found, but it is very narrow, and some submarine rocks may have escaped notice. A great many seal are found on Ypun and Huamblin.

The bight to the eastward of Huamblin, between the islands Vallenar and Ypun, is called Adventure bay. Bordering this bay to the eastward are the islands Stokes, the summit of which, mount Philip, is 2,765 feet high; Rowlett; Williams, 2,525 feet; James, with Sullivan peaks, 435 feet in height; Kent, Dring, and Lemu. Among these islands there is no good or accessible anchorage. In the middle of this bay are the islands Paz, 1,050 feet in height, and Liebre. They are remarkable for their conical form, and offer no shelter. *Adventure bay.*

Ninualac channel commences in the middle of Adventure bay. It is one of the principal ones leading into the archipelago from seaward. According to Mayne, it is 10 miles too far to the northward on the French and English charts. Its entrance, according to Simpson, is in latitude 45° 03' S. The channel opens between the islands James and Kent. James island, which is surrounded by three peaks called Sullivan, is the Chirconlahuen marked on the charts of the Jesuits in 1766. The peaks can be seen a long distance; there is a small inlet at their foot on the south side. *Ninualac channel.*

To the westward of the inlet, nearer the open sea, is port Concha. It is formed to the westward by a small detached island, and to the NW. by a contracted channel. This port is narrow for ships, and in its northern part there is but little water; it terminates in a beach of white sand and low, swampy country. The tide is very strong around the island. *Port Concha.*

After leaving port Concha there is no anchorage for a considerable distance on the north coast of the Ninualac channel; but vessels can, according to the pilots, anchor on the side of Kent island, in a small estuary at the entrance of the Pichirupa channel. Six miles from the entrance of Ninualac channel is an island which divides it into two equal parts. It stops the sea, but increases the current on either side. The island is clean, there being over 50 fathoms of water within 250 yards of it. The same is the case near the land on the sides of the channel. The island must not be approached nearer, however, as a shoal surrounds it which is covered by only 2 to 3 fathoms of water. Opposite this, on the north side of the channel, is a spacious bay, in which the water is too deep for anchorage. Boats only can land on the small inlets.

Estuary of Cisnes.

Half-way between the ocean and Moraleda channel, outside of a small inlet called Gatos, is the first anchorage on the north coast of Ninualac channel. Three miles before reaching this, opposite a white spot on the north coast, there is a rock in the center of the channel which is only covered during very high tides. There is very little seaweed on it; vessels can pass safely within 250 yards of it; at that distance there are 12 fathoms of water.

One-half mile to the eastward of Gatos inlet is Cisnes estuary. This is a good anchorage, in 12 to 15 fathoms of water, on the NW. side of its entrance, about 250 yards from the land, but it is not well to get too near the visible reef. On entering this estuary, which is five miles long, hug the eastern point where there is 8 or 9 fathoms of water, which depth continues to its end; near the reef the depth is much less. A great many vessels could winter in this estuary.

Guaitecas islands.

The group of islands between the Ypun and the Guaitecas does not offer any easily accessible anchorage, though many coves for small vessels can undoubtedly be found.

On Midhurst, the southernmost of the Guaitecas, there is a peak 2,079 feet high. There is another of the same height on Chaiffers island, called Mount Mayne; it is on the eastern side and 8 miles from the peak of Midhurst. The four islands to the northward of Midhurst are high and from 4 to 7 miles long, at nearly equal distances apart, and remarkably parallel to each other in an east and west direction.

The northern and most important of the Guaitecas islands is Guaiteca Grande; its elevation above the sea-level is 1,900 feet; it contains Port Low, one of the best harbors on the coast, capable of containing a number of large vessels. It is situated on the north coast east of the island of Huacanec. Port Low.

Coming from the westward the Guaitecas islands appear like a chain of hills. At their NE. extremity is seen a remarkably-formed island, the summit of which is shaped like a table, its SW. part sloping and terminating in the low land; this is Guaiteca Grande. At a distance of 20 miles this table and chain of hills is visible, the chain of hills appearing to be in the middle of the group of islands. On the left facing to the SE. a high mountain with but one peak, which appears entirely isolated, will be seen. Farther to the left is the table already mentioned beyond which there appears an opening, the low land on the west resembling several small islands. Directions.

On approaching Port Low it is necessary to give a wide berth to several rocks on the N. and NW. coast of Guaiteca; and account must be taken of the tide, which is felt beyond Huacanec and which produces eddies outside of point Chayalime. When approaching from the SW. the rocks most to seaward are N. 31º E. about 2 miles from point Patgui. On these the sea always breaks. In order not to be drifted too close to these rocks it is well to keep to the northward of an east and west line drawn from the islets, to the north of Huacanec. When the east point of Huacanec bears south Port Low can be entered. Besides its outer roadstead Port Low has another inside harbor formed by Guaiteca Grande and Huacanec island; it is very secure and has 9 fathoms of water, sandy and muddy bottom.

The ordinary provisions can be found at Port Low; also, excellent water, wood, shell fish, (including oysters,) and wild herbs. The otter and sea-calf hunters from Chiloé planted potatoes here during the last few years; a few of these can be procured. Provisions.

It is high water, full and change, at Port Low at $12^h.40^m.$, with an average rise of 8 feet. Tides.

Melinka.

Melinka, also called Puerto Arena by the fishermen, was created for convenience in cutting the wood of the archipelago, in hunting the sea-calf, and for storing the guano. It is in latitude 43° 53′ S., and longitude 73° 47′ 03″ W. The entrance is between Guaiteca Grande and the round island Canelo; the course is nearly SSW. until some houses in the north of the bay are opened, when steer directly for them, anchoring in from 6 to 15 fathoms of water, pebble bottom, with the point of Guaiteca bearing E. by N. This harbor is entirely sheltered, and appears to be perfectly safe. In a cove there are traces of numerous Indian habitations. Captain Simpson gives the following description of Melinka. Melinka is a spacious harbor, open on the east to the Huafo channel; it is exposed to the NNE. wind, which often blows with great violence; the anchorage is also subject to strong eddies, produced by the currents; the western portion is sheltered against the prevailing winds, but the bottom is rocky, and it is also agitated by strong currents, it being the continuation of the channel opening into the Pacific. The anchorage in the small cove, formed by a visible reef and the point, is really good, with excellent holding-ground of sand in 8 to 12 fathoms of water; it is, however, narrow. The northern entrance to this port has no hidden dangers, and vessels can pass within 27 yards of both sides of Westoff island; the smallest of the entrances is preferable when the wind permits; it is quite clear, and has no less than 9 fathoms in midchannel; S. 47° W. leads directly from Queytao island to Port Melinka. Canelo island can be passed on both sides, and in a calm or fog vessels can anchor in 18 to 25 fathoms within ½ mile to the north of Westoff island.

Tides.

It is high water, full and change, at Melinka at 12^h 45^m, rise 7 feet; the flood and ebb set in and out to the westward.

Morale la channel.

The Moraleda channel trends N. by E. and S. by W., and separates the Chonos archipelago from the continent; the depths in this channel vary between 75 and 175 fathoms from its northern limit to the point where the channel is narrowed by the islands fronting the Playas Largas, ten miles to the northward of Port Lagunas, the bottom being mud and dark marl, with vestiges of shells. To the south-ward of this point, toward the islands of Traiguen, across the narrows between Lagunas and the entrance to the estu-ary of Aysen, the depths decrease suddenly to 38 fathoms, mud and sand. This is the principal channel for vessels going S. for wood; its breadth is 5 to 6 miles. Following it to the southward, the anchorages are—

On leaving Melinka to enter the Moraleda channel, Locos island, which is the northeasternmost point of the archi-pelago, is left to the westward. The first anchorage is Port Ballena, in latitude 44° 16′ S., longitude 73° 33′ W. It is situated opposite to the volcano of Melimoyo and to Refu-gio or Huatimo island. This port, on the island Muilchey, is an excellent harbor, sheltered from all winds, and the sea never rises in it. Its entrance is free from danger, all the rocks being visible; the shore can be passed within 100 yards. Oysters are found here. Port Ballena.

To the S. and E. of Port Ballena, the Quinchel group are the most prominent. The southeasternmost and most salient, Gorro island, is very remarkable, having the shape of a cap, from which its name. N. 45° E. from this island, about 3 miles distant, is Chacabuco rock, which is about 264 feet N. and S. and about 132 feet E. and W.; and in a spot of about 24 feet diameter there is but 3 feet of water at one-third tide. On the borders of the rock the soundings are from 2 to 3 fathoms, which increase to a depth of 25 and 30 fathoms at a distance of 125 feet. At low water the rocky bed is vis-ible, and it most always breaks. There is but little sea-weed upon it. To clear it, pass ¼ mile to the E. of el Gorro del Quinchel. Quinchel islands.

The fishermen bound from Malinka to the Moraleda channel pass inside of these islands, as it is shorter. About 4 miles southeasterly from Chacabuco rock are two high rocks, the one black the other white.

On the western shore, under the southeast point of Fran-cisco island, protected to the southward and to the west-ward by small islands, is Port Nassau, with a depth of 19 to 31 fathoms on a sandy bottom. The bottom of the western portion of the anchorage is rock. Having passed the Quin-chel islands, the mouth of the channel running to the NW. is seen. This is the one before mentioned as used by the Port Nassau.

fishermen, leading to Melinka. The pilots say that it is free from hidden dangers; but it is too narrow for large vessels.

Port Letreros or Tuhuenahueuec. Passing this entrance and the points before mentioned, where the Moraleda channel narrows, and opposite to point Calqueman, the western extremity of Magdalena island, is the small island of Letreros, which is low, and projects in front of one much larger and higher. To the westward of these is a large island with a mountain 2,766 feet in height, the only one on whose summit the snow lies during the summer—from which its name, Cerro Nevado. To the northward of Letreros island is the anchorage of Letreros, to the southward of it Port Cuptana.

Port Cuptana. In clear weather, before arriving at these anchorages, a quadrilateral of rocks projecting to the eastward, called el Enjambre, will be seen. Some are barren, the others surmounted by trees. When abreast of them, the remarkable mountain of Tangbac, or Americano, will be in sight. This marks the southern part of the Ninualec channel.

By keeping ½ mile to the eastward of Enjambre there is no known danger. Passing outside of Cayo Blanco and the group of islets extending from it to the southward and westward, port Frances, or Espineira, lies west, the best anchorage being as near as the soundings will admit to Transito island. A good anchorage is about ¼ mile from the island, in about 15 fathoms; bottom mud; outside the small island, which here protects the anchorage to the northward. Inside the entrance, and a little to the southward of midchannel, is Janequeo rock, usually submerged. It lies about ¼ mile from the northeast point of the island, which forms the southern protection of the anchorage. By keeping well over to the northern side, near the islet, the rock is avoided. Port Frances is in latitude 44° 46′ 35″ S., longitude 73° 39′ W. From its entrance, the extinct volcano of Motalat, 5,445 feet high, bears ENE. It is high water, full and change, at port Frances at $2^h 30^m$; rise, 9 feet.

Ninualaca channel. Seven miles to the southward of the entrance to port Frances is the eastern entrance to Ninualaca channel; in the southern part of the mouth of this channel is a submerged rock bearing N. 20° E. from the small island Pajel and N. 38° E. from Silachilu islet; there are 3 feet of water in it at low water, it is flat and marked by very little sea-weed. The eastern part of the Ninualaca channel is full of submerged rocks.

To the southward of the entrance to Ninualaca channel, with its opening on the southern end of the island of Tangbac, is port Tangbac or Americano, in latitude 45° 2′ S., longitude 73° 43′ W.; it takes its name from mount Tangbac, at whose foot it is; this mountain is 1,968 feet high and remarkable in appearance. Passing to the southward of the islets off the southeastern point of the island, large vessels should anchor at the foot of the cerro Americano in from 15 to 19 fathoms of water, bottom sand; on the opposite side of the channel the bottom is rock. There is anchorage here for but two or three vessels, as the bank is very steep. Farther in there is a well-protected basin, called La Darsena, with from 6 to 8 fathoms of water, but across its entrance there is but 15 feet. There are some apple-trees here, and fish and shell-fish are abundant. The establishment of the port is 2ʰ 40ᵐ; the rise and fall of the tide 10 feet.

Port Tangbac, or Americano.

Opposite to the eastern entrance to the Ninualaca channel, on the eastern shore of the Moraleda channel, is the mouth of the Puyuguapi channel, which runs to the eastward and then to the northward, making the circuit of Magdalena island, the largest of this archipelago, which it separates from the cordillera of the main-land; to the NE. of the island it turns to the westward, taking the name of the Jacaf channel, and opens again into the Moraleda channel opposite to the Quinchel islands.

Cay or Puyuguapi channel.

On the eastern side of the Moraleda channel there are no known anchorages for large vessels between the island Refujio and Aysen estuary, although there are several small coves and sand-beaches. Besides which, on all the exterior of this coast the prevailing winds blow on shore.

Fourteen miles to the south of Tangbac, on the west side of the Moraleda channel and under the island San Melchor, one of the largest of this archipelago, is port Lagunas, in latitude 45° 19′ S., longitude 73° 43′ W., one mile to the north of an island, named on some charts Barba; as it is not inclosed by high mountains to the NW., the heavy squalls which are generally felt in the estuaries of the cordillera, and always at the foot of high mountains, are not experienced here. It is surrounded by land to the SE., and

Port Lagunas.

the easterly winds, which never amount to a gale, are the
only ones which can enter it. The anchorage is in 14 fathoms, sand and shell; farther in, within 100 yards of the
mouth of a small brook, the water shoals rapidly to 6 feet,
the surrounding country is low, and seems to be favorable
for cultivation, though most of it is covered by impenetrable thickets. Besides the island of Barba, there are some
islets to the southward of the harbor. There is a small
settlement for wood-cutters. The rise and fall of the tide
is 6.8 feet.

Estuary of Ay-
sen.
Port Lagunas is situated in face of a great opening into
the cordillera, called the estuary of Aysen. Here the
Moraleda channel ends and splits into several branches, of
which Darwin channel to the SW., and San Rafael or Costa
to the S. are the principal ones. The entrance to the estuary of Aysen is full of small wooded and mountainous
islands, of which Chaculay, Churreciie, and Meninea are
the largest. After passing Lagunas the opening of the
estuary will be seen to the southward of Churreciie
island. It is full of islets, 4 miles wide, and is 27 fathoms deep. The channel running NE. contracts between Colorada island and the continent. The estuary is
free of islands, and has a width of 3 miles as far as Port
Perez, which is in latitude 45° 15′ S., and 73° 22′ W. It
lies in the NW. angle of a spacious bay comprised between
the northern part of the estuary Aysen and the five islets
called las Cinco Hermanas; the anchorage is in from 20 to
30 fathoms, sandy bottom; vessels should moor upon indications of bad weather, as the anchorage is exposed to
strong winds though protected from the sea.

To the northward, in the middle of a plain covered by
vegetation, is a small volcano which was in eruption about
twenty years since. Seven miles NE. by E. is mount Maca,
9,700 feet high, of conical shape, its summit covered by perpetual snow. The bay is bounded on the west by high
peaked mountains, and on the east by a higher chain covered with snow. At the foot of the latter are hot springs,
called el Baño, with a temperature of 187°.

Opposite the island Colorada, on the southern side of the
estuary, is a narrow bay 3 miles long, with 8 to 13 fathoms
of water; it is entered by a cut in the mountain between a

low, dangerous point and a mountain 2,700 feet high. On the same coast, opposite Port Perez, is the estuary Manco.

From the ensenada de Baño the estuary runs SE., with Head of the es tuary of Aysen. an average breadth of 2 to 3 miles; its shores are surrounded by mountains covered with perpetual snow and furrowed by deep ravines; 8 miles from el Baño, behind a small point, is a good anchorage in 25 fathoms, ½ mile from the shore, which is covered by heavy forest; the two points which form this indentation are not very clean. In the middle of the channel the water is very deep, varying from 75 to 95 fathoms.

On the SW. side, 15 miles from Port Perez, is a small, crooked point, forming a small cove on either side; to the west of it are 10 fathoms, to the east from 19 to 24; the shores are wooded. At this point the estuary enlarges, running to the eastward; 4 miles from this it forms a funnel, running to the NE. for 4 miles, leaving Carmen island to the south and Barelda to the north. The passage to the southward of Carmen is ¼ mile wide, and has 70 fathoms of water. On the side of the continent is a small peninsula, behind which is situated port Chacabuco; this harbor is 1 Port Chacha- buco. mile wide, and is perfectly sheltered; the anchoring-ground is about ⅓ mile from its end, in 18 to 25 fathoms, bottom mud. It is necessary to moor, as furious gusts sweep down through the mountain gorges, throwing the spray to a height of 12 or 15 feet. Two miles farther to the NE. is another cove, but it has not sufficient water for large vessels.

To the northward of Barelda island the water shoals Rio Aysen. rapidly from 35 to 3½ fathoms; a short mile from this bar is the mouth of the river Aysen proper, having but 11 feet at low tide. This river runs between mountains of less elevation than those around the estuary. The river has 3½ fathoms of water to within 3 miles of its mouth; at this point is the partially obstructed confluence of the two branches of the river; the northern one is again divided into several branches, of which one is the outlet of a deep lake situated about 3 miles from the confluence. The principal branch is the eastern, which is about 200 feet wide; about 4 miles from the junction are rapids where the current runs ten knots. After that it becomes again naviga-

4 C

ble, and 1 mile from the rapids it divides again into two parts; the southern branch is called the Rio Blanco, the northern or principal one runs for 15 miles between mountains and forest; it then becomes a mountain torrent, the water of which is blackened by a mixture of snow and water, and its bed is obstructed by fallen trees.

By following it for 55 miles, captain Simpson, in 1871 and 1872, reached Patagonia. The pass is in 45° 25' S. latitude.

The surrounding country is entirely wild. In some places the soil is very deep. The river is difficult to navigate, the currents, the sudden rises, and the *débris* of all kinds from . these virgin forests arrest the smallest boat at every moment. At its source, 45° 20' S. and 72° 20' W., the mountains are high without attaining the prodigious height of those to the north; but the Andes at this point are very difficult to cross.

The mean height of this section of the Cordillera is 4,840 feet; the perpetual snow commences at 5,246 feet, excepting in sheltered places, where it is found much lower.

At the mouth of the estuary of Aysen is the junction of the channels of Moraleda from the north, of Darwin from the southwest, and of San Rafael or Costa from the south, besides of many unimportant estuaries.

Traiguen island. Following the channel of San Rafael, Traiguen island is situated between Pulluche and Darwin channel, and forms the western side of the channel of Costa; it is one of the largest of the archipelago.

Costa channel. Costa channel is the northern part of the San Rafael channel. It was named by Captain Hudson, of the Janequeo. It runs between the continent and Traiguen island, is about 30 miles long by 1½ miles wide, and leads from the Pulluche channel to Port Lagunas. It is deep and without danger as far as the island Raimapu, near Pulluche channel; there the best channel runs between this island and the islet facing the northwest point of the estuary Quetralco, as the other between Raimapu and the more western islands is filled with dangerous shoals.

Confluence of Raimapu. Raimapu lies just north of the confluence of four branches of the sea; to the north the one just mentioned; to the south the estuary of Elefantes, or San Rafael continuation of the Costa channel, leading to the lagoon of San Rafael:

to the eastward the estuary of Quetralco, which cuts into the continent; and to the westward the Pulluche channel, leading to the open sea. When it blows fresh this point is dangerous, as the ebb coming from three estuaries rushes toward the Pulluche channel, occasioning violent currents and a rough sea.

The estuary Elefantes, the continuation of the San Ra- Estuary Elefan-
tes.
fael channel, extends from the confluence, 40 miles to the southward of the estuary Quetralco.

On the eastern shore of the estuary Elefantes, about 2 Rio de los Hue-
mules.
miles south of Quetralco, is the mouth of the river los Hu-
emules; its source is in the Cordillera of Patagonia. Dur-
ing 1871 and '72 it was explored carefully; large herds of deer having been seen on its banks, it was hoped that by tracking them the pass by which they crossed the Andes might be found.

In 1872 captain Simpson arrived at a pass in the Cordil-
lera in latitude 46° 06' S. This river empties by two mouths and forms two branches, which unite about 3 miles above; at the fork it is about 2,000 feet wide; it is subject to formi-
dable rises. One branch descends from an immense glacier 13,210 feet high, which is about 20 miles in the interior; it extends down the Cordillera on an inclined plane, 4 or 5 miles long by 1 mile wide. The waters of this river are very turbulent and very shoal.

Eight miles to the southward of the confluence of Raima- Road of los
Mogotes.
pu, on the side of Simpson island, or the west side of the channel, is the road of los Mogotes. It is formed to the southward by some small islands of rock, low, and crowned by trees and resembling champagne corks. The anchorage is 100 yards from the beach in 7½ fathoms, fine sandy bottom. To the southward of it the channel contracts from 3 miles width to 1 mile, between point Pescadores and Simp-
son island. At this point there are 10 fathoms of water. After passing it the middle of the channel must be taken again. The channel averages about 1½ miles wide its entire length.

Going south the entrances of the channels Liucura and Tuahuencayec are left to the right. The aspect of the channel is the same as farther north. On the east side the Cordillera, abrupt and covered with snow, rises from the level of the sea; on the west side the hills are not so high, but the ravines are as deep and are covered by an almost impenetrable vegetation.

The chart and directions do not show any danger in the middle of the estuary of Elefantes, excepting an uncovered shoal just before reaching the entrance of the lagoon.

Quesahuen point is about 40 miles from the confluence; it is on the continent, on the east side of the channel, and at the entrance to the lagoon Elefantes.

Elefantes la-goon. The termination of the San Rafael channel is Elefantes lagoon, which is a vast swamp covered by from 6 to 12 feet of water. The entrance is partially closed by a chain of small islands, and is divided by a tongue of sand, which is nearly level with the water. Vessels crossing the lagoon are obliged to double the island to the northward. All the southern side of this lagoon is inundated low land, of which the borders are submerged even at low tide, and are obstructed by dead oak trees. According to the oldest pilots and from the analysis of the decomposed woods, this land was sunk during the earthquake which took place in 1837. The tide rises and falls about 12 feet. This lagoon was some fifty years since frequented by species of seals, called the sea-elephant and sea-leopard, which gave about seven times as much oil as the ordinary species. They have been exterminated by the hunters.

Los Tempanos river and San Rafael lagoon. Pieces of ice have been found floating in Elefantes lagoon, and it was discovered that they were from an immense glacier, visible from the lagoon, and were brought by a river which empties into its SW. corner. This river, named de los Tempanos, can easily be entered with the flood-tide; it is about 330 feet wide and 7 fathoms deep; in the summer it is almost obstructed by floating ice; its banks are, at first, low and swampy, but afterward high. This river is the outlet of a large lagoon, whose southern extremity is, according to Simpson, in 46° 39′ S. This would be the northern portion of the isthmus of Ofqui, which unites the peninsula of Taytao with the continent and separates the gulf of Peñas from San Rafael channel.

One mile farther south the Cordillera opens, and a cataract is heard whose waters descend toward the gulf of Peñas. This is probably the San Tadeo, or one of its tributaries.

This second lagoon, the San Rafael of the missionaries, is nearly circular, and has a diameter of 8 or 9 miles. The great glacier of San Rafael projects into it. The latter extends from an immense sea of ice, situated in the Cordillera, which, running from N. to S., at a height of 3,000 feet, covers a large extent of the mountain-chain of the sea-coast. The glacier descends by a gorge more than 3,300 feet wide, and runs over 4½ miles into the lagoon, and at its extremity it is still more than 4 miles broad. It forms a kind of trapezium of 6½ miles of horizontal height; its perimeter is formed by precipices 328 feet high, and its surface is a chaos of peaks and crevices. Large blocks of ice, which detach themselves from it every moment, are found in the lagoon. The noise of the fall of these avalanches is heard in the lagoon Elefantes.

One mile to the northward of this glacier there was no bottom at 58 fathoms. The glacier therefore reaches to that depth; it is probable that it is the cause of the close of the passage to the gulf of Peñas, which formerly existed.

About 10 miles to the southward is another glacier, which originates in the same bed of the Cordillera as the former. These glaciers are the most distant from the pole of any that have been found at the level of the sea.

Returning to the estuary Elefantes, about 5 miles to the NNE. of point Quesahuen, is the entrance to the estuary San Francisco; 6 miles from its entrance is Esploradores bay, into which empties a deep river. *Estuary San Francisco.*

The first channel to the west, after leaving Elefantes lagoon, is the Tuahuencayec channel. It connects the Elefantes with the Barro or Aau estuary. On entering the Barro channel, two branches are formed by the island San José, and another smaller island to the westward of it. This channel runs between the peninsula of Sisquelau to the south and Nalcayec island to the north. To the northward of the former channel, and 6 miles to the southward of point Pescadores, is the entrance to Liucura channel, also connecting Elefantes and Barro estuaries, and running between Nalcayec and Simpson islands; its western entrance is narrow. *Tuahuencayec channel. Liucura canal.*

The Barro estuary extends from the Chacabuco channel, on the north, with which it connects by the narrow channel of Renjifo, to the southeastern entrance of the Tuahuenca- *Barro or Aau estuary.*

yec channel, where, taking a WSW. direction, it is named Puelma estuary. Two miles from the western point of the channel Tuahuencayec, on the south side of Puelma estuary, is the Thompson estuary; it runs to the southward and connects with the estuary Chasco by a narrow channel, in which there is from 8 to 16 fathoms of water. In Thompson estuary are two coves, one to the westward, called Port Tupper, in which the anchorage is not good; the other opening, to the southward, is Port Barcelo, which is frequented by woodcutters.

Thompson estuary.

Port Tupper.

Port Barcelo.

Puelma estuary. The Puelma estuary runs to the WSW. about 24 miles, and ends in a shallow basin, filled with rocks and shoals. This basin is at the foot of Dark hill, the western slope of which reaches to the gulf of St. Estevan. This end of the sound is surrounded by low land, interspersed with lagoons, which are overlooked by elevated peaks covered with snow. The mountains are of quartz and basalt, of easy access, and wooded to a great height; their soil is different from that of the Cordillera.

In Puelma estuary is an anchorage of the same name, where vessels can anchor in 7½ fathoms, bottom sand and rock. This port, which is situated on the north bank of the estero, is remarkable for a point of sand, which runs to the eastward, and for its position at the foot of mount Fonck, which is 2,600 to 2,950 feet high, is the most barren place of the neighborhood, and ends in a double peak. Behind it are other higher mountains, covered with snow. From the top of these a large lagoon may be seen, which is only separated from Puelma by a small line of hills, about 2 miles across; this lagoon contains fresh water and discharges into Puelma channel. To the north of Puelma estuary there are several estuaries running into the peninsula of Tatayo from the western shore of the Barro estuary; from the south they are Albano, Vidal, Silva, and Verdugo. The channel to the southward of Fitz-Roy island, between the Barro and Vidal estuaries, is called Esperanza channel; to the northward of this and east of Fitz-Roy island, these waters connect, through the Alejandro channel, with the Chacabuco channel by la Carrera del Diablo.

All these estuaries and channels are at the foot of high mountains, of a slate or granite formation; the former are covered with oak, the latter with cypress.

Variation from 20° 35' to 19° 04' easterly, in 1876, increasing annually about 1 30'.

The island Huafo is situated about 20 miles WNW. of Guaiteca Grande, and about the same distance to the SW. of Chiloé. It has no harbor, excepting for small craft. The highest part of the island is the northwest point, the weather point, which rises about 800 feet above the sea; reefs run out from it to northward and westward for about 3½ miles. This island is formed of hard clay, which can be cut with a knife, like chocolate; its center is low, but the ground rises again to the SE.; it is thickly wooded, and, while the natives were living in peace, sheep could be procured. Small and Sheep coves offer two landing points on the east coast. Huafo island.

The island of Chiloé is large and fertile, and has not the wild and desolate aspect of the other islands along the west coast of Patagonia; the characteristic of the island is lower, its contour more rounded, and its extended forests are of the thickest vegetation. The opposite coast of the continent, the Cordillera, is as rugged as that farther south; it is more elevated and more thickly wooded. On Chiloé there is no land more elevated than 2,600 feet, and the average elevation is about 500 feet; the small island San Pedro excepted, which has an elevation of 3,200 feet. Chiloé.

The island Chiloé, with its archipelago comprising some small islands situated in the gulf which separates it from the continent, is bounded on the S. and SE. by the gulf of Corcovado, on the NE. by the gulf of Ancud, and on the N. by the narrow channel called the strait of Chacao. Chiloé is 100 miles long from N. to S.; its greatest breadth being 38 miles at the Mamelles de Matalqui; near its center a deep indentation reduces its breadth to 14 miles. The whole island is a mass of rock, covered by earth clothed with forests, which furnish excellent timber, large quantities of which are exported. Some lignite is found on the island, but not in large enough quantities to work. The population of Chiloé and the neighboring islands, including the Chonos archipelago, which belongs to the province of Chiloé, is about 65,000, some 5,000 of whom are occupied in fishing and coasting.

There are numerous good ports on the many small islands between Chiloé and the continent, and along the east coast of the island itself. Small quantities of provisions can be obtained, principally potatoes.

Island and port of San Pedro. The island San Pedro, which is separated from the south-east extremity of Chiloé by a narrow channel, resembles from a distance a mountain composed of rounded masses. On approaching, it is seen to be wooded to the summit, though its height is 3,200 feet. The northern part of the channel separating the island of San Pedro from Chiloé is called the San Pedro passage, or Port San Pedro; that to the south the Huamlad passage. Port San Pedro is a long, narrow, but safe anchorage; it can be recognized by a white rock near the northeast point of the entrance. If the tide be low on entering or leaving, care must be taken to avoid a bank with 3 fathoms of water, which runs from the south shore two-thirds across the entrance. A promontory on the north shore of the anchorage, in line with the next point to the westward of it, bearing N. 81° W., leads between White rock and the shoal extending from the south side. The best anchorage is off a sandy head on the north shore, about 1 mile inside White rock, in from 7 to 10 fathoms, sand.

To the eastward of Huamlad passage, and to the southward of San Pedro island, are some clusters of rocks running out about 2 miles to the SSE., named the Huamblin rocks.

Canoitad rocks. The southern coast of Chiloé must be given a wide berth; it is filled with rocks and shoals, extending in places 8 miles off shore. The Canoitad rocks lie south, a little easterly of point Cogome; they are 43 feet high, and lie 4½ miles off shore. As the tide stream sets toward them they are dangerous in calms and at night. There is a deep channel 1 mile wide between them and the Caduhuapi rocks. These rocks are to the southward of point Olleta, and

Caduhuapi rocks.

extend from the coast about 3½ miles. The sea can generally be seen breaking on these rocks at a distance of about 7 or 8 miles.

Off the southeastern point of Chiloé is Quilan island, which is 4½ miles long, and has a long hill at either end. The roadstead to the eastward of the island is unsafe. The south coast of Chiloé, being, as before mentioned, full of rocks, must be avoided and passed at distances varying from 4 to 8 miles. There are many submarine rocks around the island Yemcouma, and to the eastward of the southern extremity of Quilan, extending 6 miles from Chiloé, the sea breaks over them with great violence. San Pedro and Quilan islands are not inhabited. Quilan island.

The southwest point of Chiloé, cape Quilan, is wooded. In its vicinity are cliffs of a light-yellowish color, about 300 feet high. Although there are many trees, the adjoining land is less wooded than the eastern and more sheltered parts of the island. The profile of the land is rounded, without rapid descents, and lies very often horizontal; it is an undulating country formed of hills and valleys. The cliffs at the edge of the sea are irregular, and do not extend very far. From cape Quilan to cape Pirulil, 35 miles to the northward, the coast maintains the same aspect. Between these there is no anchorage of any kind; there is hardly a place sufficiently sheltered for landing a whale-boat. Cape Quilan.

Five miles to the northward of cape Pirulil, bordered by a low beach which is constantly swept by a heavy surf, is Cucao bay. The mountains of Cucao, from 2,000 to 3,000 feet high, are at the same time the highest and most level parts of the island; they are wooded to the summit. Cucao bay.

Cape Matalqui is also a remarkable point, 34 miles from cape Pirulil. The heights which surmount it reach an elevation of 2,000 feet. Seen from the sea they present three summits; they are called the Mamelles of Matalqui; they are conical, can be easily recognized, and are an excellent landmark when approaching Chiloé from the SW. To the NNE. of cape Matalqui are the Matalqui rocks, extending about 1 mile off shore. Cape Matalqui.

There are no outlying dangers along this coast to the northward of cape Quilan; 5 or 6 miles from the land the water is everywhere deep.

Tides. It is high water, full and change, off cape Matalqui at 12^h; rise 6 feet. Between capes Matalqui and Cocotue is the inlet of Chepu, behind which low land alone is visible; this causes the heights of Cocotue and the mamelles of Matalqui to appear as islands from seaward.

Cocotue bay. The heights which end in cape Cocotue do not attain a greater elevation than 1,000 feet; they give the name to a bay to the northward, between them and the peninsula of Lacuy. This bay is dangerous and always exposed to the prevailing winds; in its center are two groups of islands and sunken rocks, while its shores north and south are rugged and full of visible and hidden dangers.

The east coast of the bay is low, foul, and sandy, and backed by low hills, at the foot of which are swamps and small lagoons. It offers no shelter to vessels; though at certain seasons of the year, during calms, some of the small coves to the north and south can be used by fishing-boats.

The swell from NW. causes a slight current which sets toward the shore, and which might become dangerous to vessels near the shore during calms or in light winds.

Peninsula of Lacuy. Joined to the main island by a low isthmus is the peninsula of Lacuy, surmounted by mount Centinela, about 300 feet high. This isthmus is the western limit of the inner harbor of Ancud. Cape Caucahuapi is the western extremity of the peninsula; in its vicinity are no outlying or hidden dangers. Capes Caucahuapi, Guabun, and Huachucucuy are promontories with bold steep cliffs; the latter is a high steep bluff. These capes are the first land seen when making the land near San Carlos de Ancud.

Huapacho shoal. In the NE. part of the circular and unexplored bay of Huachucucuy, which opens between this cape and that of Huapacho, is Huapacho shoal, a circular extent of rocky reefs ½ mile in diameter, whose center is 3 miles N. 81° E. of cape Huachucucuy and 1¼ mile N. 87° 30′ W. of cape Huapacho, with which it is connected under water. This reef must be avoided, especially during the night, as the land back of it is low and cannot be distinguished. After rounding Huachucucuy point a vessel should keep to the northeastward with Huapacho point to the southward of E. by S., until the light

Approaches to San Carlos de Ancud.

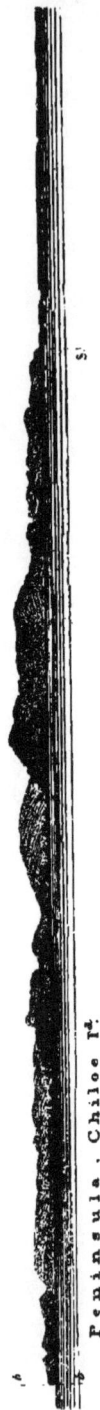

Main Land

L a c u y

Corona Hill
8.607E.

N.¼°E.

N.50°E.

Peninsula. Chiloe I⁴.

Approaches to Port Melinka from the north.

Entrance to Port Melinka
Nestoff I¹

bears S. 37° E., to avoid Huapacho shoal and Osorio rock, before keeping to the southeastward.

Bearing N. 68° W. of point Huapacho, distant 1,180 yards from the point, is Osorio rock with 2½ fathoms of water over it; between it and Huapacho shoal the depth varies from 6 to 10¾ fathoms; between it and Huapacho point the depth is but 5 fathoms; it should be avoided by sailing-vessels going out of Ancud or Chacao channel with light winds, as the current may set in the direction of the rock at the rate of 3 to 4 miles per hour, according to the state of the sea. *Osorio rock.*

Huapacho point forms the northern extreme of the Lacuy peninsula; it is a rounded cliff, barren on top and broken toward the sea; it is 4½ miles east of cape Huachucucuy. The low extremity of Huapacho is sometimes called Tenuy; it probably advances farther to seaward than it formerly did. Vessels passing must take into account the tides, as they run very strong in this vicinity and should give Huapacho point a berth of 1¼ miles. *Huapacho point.*

One and one-half miles to the SE. of cape Huapacho is Corona point; both are formed by a chain of small hills separated from the large peninsula by the low isthmus of Yuste. *Point Corona or Huapilacuy.*

On point Corona, 1½ miles S. 42° E. of point Huapacho, is a *fixed white light, flashing* every *two minutes;* it is dioptric, of the fourth order; the building is of wood and the tower of masonry, 32 feet high, painted white. The light is 197 feet above the sea and in clear weather visible 12 miles. It is on the following bearings: *Light: Lat. 41° 46′ 55″ S.; long. 73° 52′ 30″ W.*

West point of San Sebastian island.............. N. 39° E.
Center of the island Cochinos................... S. 48° E.
Cape Huachucucuy.......................... N. 82° W.
There is a signal mast on point Corona.

Point Aguy is a little less than 3 miles SSE. ¼ E. from the light-house. It is a rounded hillock surmounted by a fort. It is the eastern part of the peninsula of Lacuy, and the northern point of the roads of Ancud. *Point Aguy.*

A *fixed white light* visible between *two* and *three* miles, is exhibited from the cross-trees of the signal-mast, on the upper part of Aguy point; it is useful as a guide to the anchorage; near this light is a telegraph station. The harbor of Ancud is on the south side of the peninsula of Lacuy, *Harbor light: Lat. 41° 49′ 30″ S.; long. 73° 31′ 00″ W.*

between it and the main-land of Chiloé, its entrance between point Aguy and Cochinos island, is about 2 miles
wide; from this island the harbor extends about 8 miles to
the westward with an average width of 2 miles; it is di
vided into two parts, the roads, which extend between the
town and points of Aguy and Arena, and the harbor, which
is to the westward of point Arena; the bottom is generally
fine sand or mud and sand.

Roadstend.   Vessels which draw more than 12 feet cannot anchor off
the town of San Carlos, but on the west coast of the harbor
ENE. of cape Baracura, the eastern part of point Arena,
there is about 5 fathoms, and they will be sufficiently sheltered from the prevailing winds. To the NW. of the town,
about 1 mile from the shore, is a bank on which the Shearwater touched; since the survey of 1835 this bank has considerably increased; instead of 18 feet of water there are
now but 8.8 at mean low water spring-tides.

San Antonio
bank.   There is also another shoal, called San Antonio, with 12
feet of water on its western extremity, about 1 mile S. 31°
E. from point Aguy; a cylindrical-shaped buoy, painted red
and white, with a vane-spindle, has been placed near its
northern edge in 4 fathoms of water, the point Aguy bearing N. 22° W., point Baracura bearing S. 78° W., the custom-house at Ancud bearing S. 25° E. Both to the northward
and to the eastward this buoy should not be approached
nearer than 230 yards. The water is deeper to the westward
and the bank extends to the south from the buoy. With
westerly winds a vessel should not anchor off the mole of
Ancud.

Nunez bank.   On the eastern edge of the bank which extends between
Aguy and Baracura points, and ⅓ mile E. of Nunez bank,
is placed a red and white buoy in 3¼ fathoms of water,
marking the western side of the channel; from the buoy
point Chairura bears S. 87° W., point Baracura S. 53° W.,
and the mole of the custom-house of Ancud S. 31° E. This
buoy should be given a good berth and never be passed to the
westward. It is said that there is another shoal with 18
feet about 600 yards SW. of Arena point, but its existence
is doubtful, as the pilots and fishermen have often searched
for it without success. A spur of rocks with 12 feet water
extends about 400 yards to the eastward of point Baracura.

Since 1835 the whole port has filled from 3 to 9 feet, especially on its east side, in front of the town.

On a small point to SW. of the town is a government mole; the dock alongside of it is partially obstructed by sand and rocks; it is not well to try to land here after half ebb or before half flood, as the sea breaks between these two limits. Boats can land with safety behind the small mole, but during low tide they can only make fast outside, where there is always a sea with west winds. The anchorage off the town of San Carlos is not safe.

The harbor is always smooth, but more contracted, it be- _{The harbor.} ing but 1 mile wide. There is an inlet, called the Dique, on the side of the peninsula, and Quetralmahue cove at the inner extremity. The shores are well peopled, the fields are well cultivated, and a short distance in the interior there is wood valuable for construction.

Sand or stone ballast must be discharged on the part of the coast of Lechagua comprised between Morro de Poquillihue and the estuary Muñoz.

The Chilian government established a small coaling-station at Punta Arena, where it can be taken aboard easily, as the point is steep and the sea always smooth.

Ancud (the Lapi of the Indians) was named by the Span- _{Description.} iards, successively, Port Ingles and San Carlos, (1767;) the Chilians gave it its present name in 1834. Its foundation really dates from 1768, when the capital of Chiloé was transferred to it from Chacao, in the strait. From a maritime point of view the location of the town is bad; the choice of the locality was that of a captain of dragoons, Don Carlos de Beranger y Renaud. The town is built on two small elevations, separated by a narrow gorge containing a brook which empties near the mole. The houses are small and of wood. The plaza, on which there is a flag-staff, is on the southern hill. The population of Ancud is from 5,000 to 6,000. Oysters can be procured here, as in all the ports of Chiloé, in great abundance; fowl and good potatoes are cheap, but provisions are more expensive than at Valparaiso.

There is a beach for hauling up vessels and a yard for construction and repairs; all labor and repairs, however, are very expensive; the constructors combine to raise the prices, instead of creating competition.

To the SE. of Ancud is mount Bellavista, 560 feet above the level of the sea. From this all the bay and the valley of the Rio Puleto is overlooked. In the war of independence of Chile, the last battle with the Spaniards was fought here.

Mount Caucaman, to the southward of Ancud, 734 feet high, serves the inhabitants as an infallible weather guide; when its summit is covered with clouds it announces a rain, when, in spite of threatening weather, it is free from clouds, rain is not to be soon expected.

Doña Sebastiana island. Four miles N. 70° E. from cape Huapacho is Doña Sebastiana island, 160 feet high. Between Sebastiana and cape Chocay is a sand-bank about ½ mile SE. of the eastern point of the island. A shoal, called Achilles bank, extends 4 or 5 miles to the westward. About 2 miles from the island there are 3½ fathoms on the bank at low water, and at 3 miles 5½ fathoms; at the edge of the bank are from 13 to 18 fathoms. The water over the bank is much agitated; it bubbles and boils during calm weather, and during gales it breaks in high short waves. This bank or reef extends to the westward in the line of Chocay point. Vessels should not approach this island or Carelmapu islets to the northward, as the tide produces violent eddies near them.

Carelmapu islets. The Carelmapu islets are a chain of rocks north of Sebastiana island, stretching SE. and NW. The northwest islet has an elevation of 141 feet. They should not be approached to the westward nearer than 4 miles. It is always better to keep closer to cape Huapacho than to Doña Sebastiana or Carelmapu islets.

Yngles bank. Yngles bank, which lies about 2 miles south of Sebastiana and east of Corona point, must also be carefully avoided. It is a very dangerous shoal, 3 miles in length, and very badly defined, especially to the east; the sea breaks on it with great violence. The least water is on its eastern part, called by the natives Arenillas rock. It has but 6 feet of water over it; bottom either sand, sandstone, or hard tosca. It is said that this bank has extended ¾ mile to the NW. since 1835. In order to pass to the northward of this bank, point Coronel, the northeastern point of the strait of Chacao, must be kept open to the northward of point San Gallen; to pass it to westward, the SW. islet of the Carelmapu must be kept to the westward of Sebastiana.

Cochinos island is surmounted by two peaks; it is about 1 mile from the heights of Guihuen, on which Ancud is situated. A bank covered by 5½ feet of water makes out 1 mile from its eastern point, and a small rock has been reported 384 yards NW. of it. C chinos Island

Point Mutico is 2 miles ESE. of Cochinos. There is a patch of rock about 1 mile NNW. from this point. All the bottom near it is very irregular and foul; masses of seaweed, which appear to cling to large stones as well as rocks, are sometimes seen on it. A small ledge of rocks, ¼ mile long, runs out from point Pecheura, which is 3 miles NE. of point Mutico. It is the tail of a bank which extends between it and point Mutico; it is connected with Yngles bank by depths of 2¼, 3, and 4 fathoms. Point Mutico.

At the extremity of this 2¼-fathom line of sounding is a rock, on which the Chilian frigate Esmeralda* touched lightly. It is about 4 miles W. of Puñon point, and on the line drawn between Cochinos and point Carelmapu. Vessels bound to San Carlos de Ancud with southerly winds should make the paps of Matalqui and the heights of Cocotue, from thence keeping along shore round cape Huechucucuy at a distance of 1 mile, steering N. 64° E., and keeping point Huapacho S. of ESE. in order to avoid Huapacho shoals, until the light bears S. 37° E., keeping fully 1 mile north of a line drawn between capes Huapacho and Huechucucuy, then run for this light until within ½ mile of the shore, when keep along it at this distance, rounding Corona and Aguy points. By conforming to these and the direction given above, Achilles and Yngles banks will be cleared. Esmeralda rock.

Between point Aguy and cape Baracura a vessel must not come nearer the coast than ½ mile, as a shoal, called Nunez bank, extends out about that distance. Half-way between these points, on its eastern edge, is placed a red and white buoy, as before mentioned. Merchant-vessels anchor in 4 fathoms, sand and mud, by keeping the light-

---

* The governor of Chiloé has reported that neither the Guillermo nor the Esmeralda rocks are situated in the positions given to them on the charts.

house over the western angle of the fort on point Aguy, and the south point of Cochinos island and point Matico in line ; the town will then bear ESE. This anchorage is not safe. A vessel bound to Ancud with northerly winds must make the land NW. of cape Huechucucuy in 41° 42' south lati- tude. The peninsula of Lacuy is about 328 feet high, and can be seen sufficiently far.

Tides.   It is high water, full and change, at San Carlos 12^h 14^m; rise 6 feet.

Strait of Cha-
cao.   Chacao strait lies east of Yngles bank ; it separates Chiloé from the continent to the north ; is about 11 miles long and from 1 to 2½ wide. In the channel the depths vary from 9 to 42 fathoms. On its north side is cape Chocay, the western extremity of this coast, with point Carelmapu 3 miles to the eastward. They are both peaked cliffs, off which the tide runs with great velocity. To the southward of point Carelmapu the water is not deep, there being but 3½ fathoms ½ mile distant, outside of which the water deep- ens. To the eastward of this point a shoal runs along the shore, extending off ½ mile. At the entrance of the strait

Topaze rock.   H. M. S. Topaze, drawing 22½ feet, touched at half-ebb. The shoal is 1 mile S. 29° W. of point Carelmapu, with cape Chocay bearing W. 33° N. At the moment the vessel touched no bottom was obtained with 10 fathoms of line from the channels.

Puñon or Page-
ñun point.   Puñon point, at the southern entrance of the strait of Chacao, is low and has a sandy beach two miles from this point. On a line drawn between it and Sebastiana island is a rocky head in 3 fathoms, called Guillermo rock. By keep- ing point San Gallan well open to the southward of point Coronel a vessel will pass to the northward of this rock.

West of the position assigned by the charts to Guillermo rock there is another rock with 2 fathoms of water on it at low water, bearing from Carelmapu point S. 24° W. Dis- tance, about 2 miles.

Periagua rocks.   About 1 mile WNW. of Puñon point is a rock which is uncovered at low water, and another lies directly to the west- ward of it.

The easternmost rock uncovers an hour or two before low water ; it is part of a reef which runs N. 64° E. and S. 64° W. for about 400 yards. The western extremity of the

reef rises in two points, which are 2 feet above the water at
low tide. Close to this rock, on the NW. and SE. sides, are
9 fathoms of water, whereas in the direction of the reef there
are only 4½ fathoms at a distance of 95 feet from it. Captain Vidal Gormaz gives the following description of this
rock: " It is black, and extends in an east and west direction. It is 65 feet long and 32 broad, and seems to be divided from north to south. It is encircled by a whirlpool
of ½ mile diameter; the current sets over it, and the fishermen say that it sucks in the boats. It should be passed at
a distance of at least 440 yards."

Four hundred yards to the westward of the former is the
other reef mentioned; it is about 200 yards long in an east
and west direction. On its lowest part, the eastern extremity, there are but 3 feet of water at low tide. On the
western extremity is another rock, which has 6 feet of water
over it. Both of these rocks are covered with sea-weed
which can only be seen at slack water, as the currents are
too strong at other times.

According to the last Chilian researches, these shoals are
connected by a chain of rocks of very irregular shape. During the strong flood very violent eddies are produced off
Puñon point, from the irregularity of the bottom. Between
the point and the chain of rocks shoal-water extends about
one-third of the distance.

About 1 mile W. 20° S. of the western Periagua rock and Prince of Wales rock.
1½ miles from point Puñon is a rock covered by 9 feet of
water. It is supposed that the steamer Prince of Wales
was lost on this rock. Its bearings are points Puñon and
Quintraquin in line, and the NW. extremity of the heights
of Guihuen shut in by the eastern extremity of Cochinos
island. This rock was searched for but not found. It cannot, however, be asserted that it does not exist.

To the eastward of Puñon point, between it and point Lacao bay.
Quintraquin, there is the island and shallow bay of Lacao.
In it the anchorage is not good, though a vessel may wait
here for the tide. Point Quintraquin is a barren cliff steep-
to; east of it there is a good anchorage in 7½ fathoms, sand
and shell, about ⅓ mile from the land.

Point San Gallan, also on the main-land of Chiloé, is Point San Gal-lan.
5

about 2 miles from point Quintraquin. It is steep, about 500 feet high, and covered with a remarkable clump of trees.

Point Coronel. The north coast opposite is clean, but low, excepting near point Coronel, the eastern limit of the strait, where the cliffs are 100 feet high; behind these the land rises to 200 feet and is thickly wooded. There is a timber-yard at this point.

Between point San Gallan and point Santa Teresa, on the continent, the strait is 1 mile wide, this being its narrowest part; ½ mile farther to the eastward the strait is divided into two narrow channels by the rocks Petucura and Seluian, either of which can be taken.

Petucura rock. Aside from the velocity of the tides, which may render steering difficult, there are no dangers in Chacao strait other than the Petucura and Seluian rocks. Petucura rocks, which uncover at half-tide, are in the narrowest part of the strait, and nearly in midchannel. Their southern part is situated at the intersection of two lines, one drawn from the extremity of point Coronel to the extremity of San Gallan, and the other from the summit of Santa Teresa to the summit of Chacao. This southern part, which has from 2 to 4½ fathoms on it at low tide, extends as far under water as the northern part, which uncovers, does above the water. The highest part of Petucura, having 2 fathoms of water on it at high tide, uncovers at low water for 69 feet east and west, and 10 feet north and south. From its summit, point Coronel bears S. 81º E.; point Tres Cruces bears S. 43º E.; point San Gallan bears S. 84º W.

Seluian rock. Seluian rock is more dangerous to large vessels than Petucura; it lies S. 65º E., ½ mile from it, with 2 fathoms of water over it at low tide. The water is deep around it, as around Petucura, except to the eastward, where a reef makes out ¼ mile. Seluian is at the intersection of a line drawn from point Remolinos to the summit of a hill near point Coronel with another from point Santa Teresa to the summit of a hill half-way between point Chacao and point Tres Cruces. The line which joins Seluian and Petucura is parallel to the rocky coast. The ebb and flood tides run by and over these Point Remolinos. rocks with great velocity. From this point Remolinos, which is small, rocky, and clear, situated about 1 mile SE.

of point San Gallan, the strait commences to enlarge. The stretch between these two points is free from dangers.

Chacao bay is on the south side of the eastern entrance to the strait, 1¼ miles from point Remolinos, and opens to northward of cape Chacao. An excellent anchorage will be found ½ mile north of the cape in 9 fathoms of water, bottom mud and sand. About 1 mile to the eastward of Chacao head there are some rocks, which extend from the shore about ½ mile to the northward. *Chacao head and bay.*

When the Spaniards established themselves on Chiloé they made Chacao their headquarters and moored their vessels in this bay, but the strong currents of the strait, and the rocks which divide it, rendered this anchorage unsuitable for their indifferent vessels, hence the foundation of San Carlos.

The small promontory Tres Cruces is the eastern limit of the strait on the Chiloé side; it is 1¼ miles from point Coronel, there being a depth of 55 fathoms of water between them. *Cape Tres Cruces.*

It is high water, full and change, in Chacao strait at 0ʰ 50ᵐ; springs rise 16 feet and neaps 7. *Tides.*

Through the entire strait of Chacao the current increases in proportion as the distance between the shores diminishes. From 2 miles off cape Huechucucuy the current attains a velocity of 3 or 4 knots, between Sebastiana and Huapacho, which is about their average velocity in the gulf of San Carlos; off point Puñon they run from 4 to 5 knots, and between points San Gallan and Santa Teresa from 5 to 7 knots. The flood generally sets from W. to E., following the sinuosities of the channel; it runs ESE. between Sebastiana and the light-house, and inclines to SE. on leaving the strait. The Petucura and Seluian rocks divide the current without changing its direction. The strong eddies in the gulf of San Carlos render this navigation still more dangerous. *Currents.*

On leaving San Carlos there are two routes to the strait of Chacao, one by passing to the northward and westward of Yngles bank; the other to southward and eastward; the latter route is not advisable, though the mail-boats and large coasters, called periaguas, use it constantly, but the patrons of these vessels cannot be trusted as pilots in vessels draw- *Directions between the straits of Chacao and San Carlos.*

ing more than 6 feet. During low water there is only 4 fathoms between Yngles bank and the Playa de Huicha, which joins Mutico and Pecheura points; the accidents to the Esmeralda and Prince of Wales also lead to the belief that there may be other unknown rocks, which cannot be easily found on account of the strong currents. The deepest water is on the line connecting the paps of Huechupulli with the low part of point Carelmapu; this line passes through the middle of the Cochinos, whereas point Puñon remains a little to starboard, but as long as the hydrography is uncertain it is best to take the other passage. In the route to the northward and westward of Yngles bank, round point Agüy, ½ mile distant; when it bears W. steer N. by W. to get on the bearing already given for clearing Yngles point. The Corona light must be passed at a distance of not more than one mile, and if the vessel be set to the southward by the current, a course to the westward of N. by W. must be taken. When Corona light bears W. Agüy point should bear S., then steer for the western cliff of Sebastiana until east of point Huapacho, when head to the eastward for the channel, keeping 1½ miles from point Carelmapu, to avoid Topaze rock; as soon as that point bears N. the middle of the strait should be taken, and the Petucura and Seluian rocks can be passed on either side. Not less than 7½ fathoms will be found on this route. From point Carelmapu to San Carlos, after gaining a position a short mile to the southward of point Carelmapu, steer for point Huapacho. When Agüy point bears south, run for it, passing it within ½ mile, and when south of the western cliff off Sebastiana round the point, as before directed for the anchorage.

The state of the tide and wind are the principal things to be taken into consideration when going through this strait; there is anchorage on the south side in Lacao bay, between Puñon and Quintraquin points, close to the land on the east side of Quintraquin point and in Chacao bay. Though the tide runs with great velocity through this strait, it sweeps on either side and not over the Petucura and Seluian rocks; the strait is not so formidable as considered by the Chilian coasters; but as the westerly swell is heavy and constant and the current so strong, the passage should not be attempted without a pilot and a favorable wind and tide.

The gulf of Ancud opens to the eastward of the strait of Gulf of Ancud. Chacao. It is bounded by the continent, the islands of the continent, and Chiloé to the west, north, and east, and by the islands Desertores and Chaulinec to the south. The strong currents at the entrance to the strait decrease in velocity toward the southern part of the gulf, except in the contracted passages, and there they only attain to $\frac{1}{2}$ knot per hour. The rise and fall of the tide reaches its maximum in the gulfs of Ancud and Corcovado.

After entering the gulf of Ancud, Chilen bluff will be Chilen bluff. seen 5 miles to the southward of point Tres Cruces. It is a low point of shingle; at its extremity there was a remarkable tree in 1835; $\frac{1}{2}$ mile in the interior the island rises suddenly to about 150 feet; about $2\frac{1}{2}$ miles to the NW. of it there is another hill of the same elevation which ends in a rocky point, between which and Chilen bluff is a good cove for boats, called Manao bay.

Vessels can anchor NE. of the rocky point, $\frac{1}{2}$ mile from the land, in 15 fathoms. Between point Tres Cruces and Chilen bluff the tides near the land have a velocity of from $1\frac{1}{2}$ to 2 miles per hour. Between this bluff and point Tique, to the NNE., on the continent near the entrance of the strait, the eddies and tide-rips are very strong when the wind is fresh from the southward. They are called Raya de Tique. Off Chilen bluff the tide is but $\frac{1}{2}$ knot per hour.

Two miles SW. of Chilen bluff is Linao cove, which. though Linao cove. small, offers a good anchorage. A reef of rocks runs out about 1 mile from cape Huapi-Linao, the southern limit of Linao cove. This cape is a large isolated hill, connected with the land by a low neck, which, seen from the east, is a good landmark for entering. The soundings in the bay are from 4 to 6 fathoms. It is about $\frac{2}{3}$ mile broad at the entrance, and 2 miles long.

Maypú bank is a shoal with $4\frac{1}{2}$ fathoms of water, reported Maypú bank. $5\frac{1}{2}$ miles E. from cape Huapi-Linao, and 4 miles S. 60° E. of Chilen bluff.

Queuiao point is a shingle beach between Huapi-Linao Point Queniao. and point Hueniao; it is dry at low water to about 1 mile from the land. The small village of Lliuco is situated 4 miles SE. of Huapi-Linao. Between these points the shore, formed of high wooded cliffs, is about 200 feet high. To

the SE. of the village the land is low as far as point
Queniao, when it again rises to 200 feet. The point had a
remarkable isolated tree on its extremity in 1835. The
water is shoal 1 mile from this point.

Strait of Cauca- Point Queniao is at the NE. entrance of Caucahue strait,
hue.
which is formed by Chiloé and the island Caucahue.
There is 8 fathoms at the NE. entrance, 22 fathoms off
Quenche, and 50 fathoms at the SE. entrance; bottom
mud and rock.

Port Huite or About 1½ miles SW. of Queniao is a sand-bank, in 2 fath-
Oscuro cove.
oms of water, about ¼ mile from the shore; the water then
deepens suddenly from 7½ to 12 fathoms near the rocky
point which forms the small but useful bay of Oscuro, or
Port Huite. This cove, a little inside the strait of Caucahue, can often be of great advantage, as the tide rises and
falls from 16 to 19 feet, as it is always calm and the depths
are good to the beach; the entrance is about 600 yards
wide, and the point of the shoal is steep-to; there are at
least 7 fathoms within 165 feet of low-water mark, and
from 12 to 15 fathoms in the middle of the cove; bottom
mud and sand. The west side of the entrance to the strait
is formed by a rocky point, with stones projecting out about
100 yards; vessels should keep over to the opposite side of
the entrance, under cape Lobos, a peaked hill of 250 feet
elevation, the north point of Caucahue island, which is a
promontory covered by trees. Behind this promontory the
ground falls suddenly and rises again at a short distance.
A vessel can be laid ashore, or hove down and repaired
with facility and security in this cove. It is one of the few
places on the west coast of America which offers these
advantages.

Tides. It is high water, full and change, at Oscuro cove at $0^b 55^m$;
springs rise 20 feet. The flood tide is very strong during
the springs, and sets to the northward.

Quenche. In this strait there is another good place for stranding
vessels, called Quenche. Its entrance is about 3½ miles to
the southward and eastward of Queniao point, at the narrowest part of the strait, the channel being contracted by
two sandy points. On passing these care should be taken
to keep to the Caucahue side until the beach or stranding
place of Quenche, which is the continuation of the western

point, is entirely open; the anchor can then be dropped in 14 to 19 fathoms, muddy bottom. This varadero is about 230 feet long; during spring-tides the rise and fall is from 23 to 26 feet; the fall is sufficient to beach vessels drawing 12 feet. The beach is of fine sand and gravel, with an inclination of 5°. An hour after the beginning of the ebb it is dry through its entire length. The upper part of the beach is triangular, steep, and formed of small stones; in its upper angle a small stream empties from a gorge running WSW. This stream traverses a natural mole, the bottom of which is also of sand and gravel, and 8 feet above high-water springs; inside of this dam vessels of 300 tons can be hove down, entering the stream at high water. There are two rocks about 100 yards from its entrance.

To the eastward of this varadero the land, as at that point, is flat and about 6 feet above the level of high water. Three hundred and twenty-eight yards inland it rises suddenly, and presents rugged mountain ravines, thickly wooded or naked, and inhabited by the native Indians. To the westward the land is high from the water's edge, and is crowned by magnificent trees. It was proposed to substitute the varadero of Quenche for that of Tenglo (Puerto Montt) or of Huito, (Calbuco,) but the locality was found to be wholly barren of resources, and having only a few huts, inhabited by about 250 Indians. It is said that the E. and SE. winds, though rare, produce a sea which is troublesome on entering or leaving this harbor. At this point is the greatest rise and fall of the tide of any place in the gulf, which is without doubt due to the fact that it is on the line of junction of the two tide-waves which enter, one by the gulf of Corcovado, and the other by the strait of Chacao. The currents are weak, but the beach is not so well sheltered as that of Tenglo or Huito.

The southern extremity of the SE. entrance to the strait of Caucahue is point Chogon; it is a hill 200 feet high, situated about 1 mile to the southward of point Quintergen, the southern extremity of the Caucahue island. The latter point is low and stony; a bank makes out from it about ¼ mile. The next point south on Chiloé is Quicavi hill. Between it and point Chogon the coast recedes a little; in the middle of this indentation the river Colu empties; it is

Point Chogon.

Quicavi bluff.

very small, and is only accessible to boats. There is a shoal about ¼ mile in extent off point Quicavi; at its extremity is a rock which uncovers at half-tide; it is said that there is another rock nearly in midchannel which uncovers at low water, but the officers of the Beagle could not find it. There is a tide-rip between Chauques island and Quicavi hill. The safe part of the channel is hardly 1½ miles wide; in order to pass through it, point Huachuque, on the western-most of the Chauques islands, must be kept in line with the northern extremity of Meulin island, the coast of which has a cleft at this point of a remarkable form. This direction is to be followed until point Quicavi shows out well; coming from the north or south the same bearing answers. There is not less than 9 fathoms.

Quicavi lagoon.   About 1 mile to the southward of Quicavi point is Qui-cavi lagoon; boats entering after the first quarter of the flood can remain afloat in some parts of it even at low tide. This lagoon can be recognized by a narrow border of pebbles, on which there is a grove of trees, and from which a reef runs out to the eastward of Quicavi hill. Behind it the land, thickly wooded, rises to an elevation of 250 feet. Here and there are some few spots under cultivation. There is a safe but contracted anchorage off the lagoon; from it the entrance of the lagoon bears N. 15° W., and the point which forms it to northward, N. 39° E. The entrance, how-ever, must not be approached too closely, for almost along-side of the beach no bottom is found with 20 fathoms. The rise and fall of tide is 20 feet; in the lagoon, the chan-nel leading to it has that depth, and at low water is almost dry.

Cháuquis isl-and.   The Cháuquis group consists of four islands, and is 3 miles to the eastward of Quicavi bluff. They are divided by a chan-nel which first runs NE. and SW., then N. and S., and which is 1½ miles wide at its narrowest part. In its center are 48 fathoms of water, and often no bottom at 55 fathoms. The SW. island is the biggest, having an elevation of about 350 feet, and it forms a chain which runs NW. and SE. On the NE. island is a round hill, less elevated than that of the western islands; the rest of the island is much lower. There are some cultivated spots on these islands, but they seem to be little inhabited by the Indians.

A small islet, connected with the coast by a reef of rocks, lies off the SE. point of the western Cháuquis; ¼ mile from this islet there is no bottom with 28 fathoms; but in the channel between the east and west Cháuquis the water shoals suddenly to 2 fathoms on a reef which runs out from the rocky point near the islet. To the northward of this reef is the entrance to a narrow boat-channel separating the two western islands. Reefs make out from the NW. points of the NE. and NW. islands; the latter extends 1½ miles from its extremity, in 10 fathoms of water. The western hill of Meulin is a little open of the low point, under the hill of the SW. island. To the southward of this island is Dugoab reef and a small island, named Tac. A large bank, dry at low-water springs, is marked on the charts 10 miles north of the Cháuquis.

Four miles to the northward of the NW. point of the NW. island of Cháuquis, and the same distance from point Quicavi, is Pulmun reef. It extends NW. and SE., and uncovers in two places ¼ mile apart; the sea breaks on it. *Pulmun reef.*

From Quicavi hill to point Tenoun, the coast of Chile is flat for ½ mile back of the beach. A reef makes out from point Tenoun, which is dry at low water for more than ¼ mile from the shore. For ½ mile from the beach the water is shoal, when it suddenly deepens to 9 fathoms. This reef does not make out exactly from the extremity of the point, but from a hill situated a little to the northward. The point is low and thickly wooded for ¼ mile, when it suddenly rises to 200 feet. The southern side of the point is steep-to, and at less than ¼ mile from it the depth exceeds 19 fathoms. One hundred yards from the beach there are 7 fathoms of water. Vessels can, if necessary, anchor in a moderate depth near Tenoun point, 100 yards from the shore, off a small village marked by a church. Vessels are here sheltered from N. and NW. winds. A bank, dry at low water, is reported to lie between this point and Linlin island. *Tenoun point.*

The flood, which comes from the southward, doubles point Tenoun and sets across to the Cháuquis islands. The ebb, which sets to the SW., passes around and close to the point, and during spring-tides has a velocity of 2 knots. *Tides.*

Ports Calen and Quetalco. Five miles to the westward of point Tenoun is port Calen, which is sheltered from the prevailing winds. Two miles farther is Quetalco, which has capacity for a large number of vessels, with a moderate depth throughout. It is perfectly sheltered from the N. and W. winds, which are so frequent in these localities during the entire year. Vessels can await in these ports until a storm subsides; or, if bound to the southward through Dalcahue channel, await high water.

Linlin island. Linlin island, 4 miles to the SW. of point Tenoun, at the entrance of the Quinchao channel, is low in the center, and rises gradually to a round hill, which terminates in a higher elevation at its north and south extremities. One mile from the NW. side of this island there are 27 fathoms; bottom sand and mud. After passing Linlin, no bottom is found at 164 fathoms in the middle of Quinchao channel. To the southward of Linlin there is a small island called Linna.

Cahuache, Meulin, Quenac, and Tenquetil Island Cahuache, Meulin, Quenac, and Tenquetil islands are situated SE. of Linlin, half-way between the Quinchao and Cháuquis islands. There is a round hill on Cahuache 246 feet high, from which all the surrounding islands can be seen. The northern part of the island is low; the southern falls abruptly to the shore. Off its NE. point is Tenquetil island, which is connected with it by a reef, on which there is but water enough at low tide to float a boat.

Cahuache is entirely cleared and well cultivated, with many apple-trees around the houses. The population consists of Indians and a few Spaniards.

Meulin, to the NW. of Cahuache, is undulating. One and one-half miles NW. of Meulin is a chain of rocks, on which there is but 2 fathoms of water; this chain is separated from the island by a narrow channel 1½ miles long.

Quenac, to the southward of Meulin, extends 3½ miles in an east and west direction. A shoal in 3 fathoms of water, covered with sea-weed, lengthens its SW. point; it is necessary to pass 1½ miles from the point. There is a passage between Quenac and Cahuache.

Tiquia reef. Tiquia reef is placed 2 or 3 miles to the eastward of Cahuache; it is about 3 miles in length from NW. to SE., and ½ mile wide; it uncovers at low water. It is reported that there is a passage between this reef and Cahuache through which a brig passed.

The Chelin and Quehuy islands are situated between the SE. point of the Quinchao and Lemuy islands; a vessel can anchor temporarily to the southward of Quehuy. The NE. extremity of Quehuy, called Imel, is connected with it by a narrow neck. One mile to the SE. of Imel is a bank of pebbles, which is dry at low water and narrows the channel between Imel and Chaulinec. A French brig was lost on this bank. Chelin and Que- huy islands.

Chaulinec, which is to the eastward of Quehuy, is a hilly island 6 miles long with 1½ miles average breadth; it is terminated to the eastward by a beach of pebbles. To the northward of Chaulinec, and to the eastward of the SE. extremity of Quinchao, there are two small islands, called Alau and Apiau. Reefs run out from the northern extremities of both, that of the latter for 2½ miles. At the SW. extremity of Alau, between it and Chaulinec, at the entrance of the channel, is a small cove or harbor which affords good anchorage for vessels of small draught. The point is steep-to, and the channel on that side clear. Vessels can anchor temporarily to the SE. of Chaulinec in 17 fathoms, sandy bottom; this anchorage is open and exposed to strong currents. Chaulinec, Alau, and Apiau islands.

In the narrow channel which separates the islands of Chaulinec and Chiut, there is a bank or reef reported about 2 miles SSW. from the island of Chiut; it extends NNE. and SSW. about 160 yards, with a breadth of about 32 yards; it is probable that some of the rocky heads which form this bank uncover at low water. Bank of Chiut.

The group Desertores is situated ESE. of Chaulinec. In midchannel between them and the latter island there is 95 fathoms of water over a bottom of coral and broken shells. Talcan, the largest, is 9 miles long and 4 miles broad, and has a deep inlet in its SE. extremity. The smaller islands, Chulin, Chiut, Nihuel, Ymerquiña, and Nayahue, do not offer any shelter to vessels excepting at the northern extremity of the latter island, which is divided by a channel having from 2 to 9½ fathoms, but which can only be used by coasters. There are some few rocks ½ mile outside of the SE. point of Nayahue. Desertores islands.

There are many scattered rocks off the S. and SW. coasts of Talcan about 1 mile from the shore. Off the north point Talcan.

a shoal extends $1\frac{1}{2}$ miles; it has on it from 4 to 6 fathoms of water; 2 miles from this point is a rock which rises about 10 feet above the water, and is much frequented by seals. A good lookout must be kept for these rocks by vessels seeking an anchorage.

Driver rock.   Driver rock is reported to be $2\frac{1}{2}$ miles NW. of Talcan, at the entrance of the channel between Chiut and Chulin, even with the water.

Talcan inlet.   Talcan inlet is at the SE. extremity of Talcan island; its depth varies from 7 to 12 fathoms, and for two miles from the entrance the shore can be approached within 200 yards; the land on both sides rises gradually to about 200 feet and is covered with thick woods. At the end of the inlet the land is low and the shore flat and muddy; there are some few huts along the shore. It is visited by the inhabitants of the other islands during the fishing-season. The current runs about 4 knots during spring-tides.

Immediately inside of the entrance, between the exterior points of the island, is a bay containing several banks of broken stones and sea-weed; and about $\frac{1}{2}$ mile from there, in the line of the point, is a reef of rocks which uncovers at low water. To the northward is a small channel which leads to the bay, having in it from 7 to 9 fathoms up to the entrance of the harbor, which is nearly closed by broken stones; the maximum depth 200 yards from the entrance is 3 fathoms.

In the middle of the channel, between the continent and the SE. point of Talcan, there are 85 fathoms; the channel is hardly 2 miles long.

Solitaria islets.   The Solitaria are small islets surrounded by a reef; they lie $5\frac{1}{2}$ miles S. 73° W. from the western point of Nayahue, nearly in the middle of the channel between Chiloé and the Desertores, in latitude 42° 45' 30" S., longitude 73° 10' 50" W.

Quinchao island.   Returning to the islands which border Chiloé, Quinchao island is the largest of the group of islands situated in the bay which opens south of Tenaun point. It is about 18 miles long in a S. 48° E. and N. 48° W. direction; the northern part of Quinchao channel turns gradually to the westward as far as the NW. point of the island; there it turns suddenly SW. into the Dalcahue channel, and is not more than 1 mile wide. The small village Dalcahue is situated on Chiloé; to southward of it is a saw-mill.

About ½ mile to the eastward of the turn in the channel Dalcahue chan- nel. there is from 4 to 9 fathoms, muddy bottom; there is no water on the north coast, and vessels must not approach nearer than ⅓ of the width of the channel. The shoal which commences here runs around the bay off Dalcahue and in front of the saw-mill of this village; it takes up half of the channel. In the center of the channel is a bank which un- covers at low-water springs, leaving a passage on either side of it; at low tide there are 10 feet of water in the one on the side of Dalcahue, whereas there are 20 in that on the side of Quinchao, it being more narrow. The shoalest part of the Dalcahue channel is between the point of the pass and the saw-mill, which is a little farther south than Dalca- hue; it has only 10 feet at low and 24 feet at high water. The Dalcahue channel can therefore be cleared by vessels drawing less than 23 feet. A sailing-vessel should never attempt this channel with a head wind. If bound to the northward she should wait between points Huenao and Que- lirquehui, or in the bay of Curaco, and never attempt to double point Quelirquehui before the flood has run at least 3 hours. If bound to the southward she should wait for a favorable time at the entrance of the channel in the ports of Calen or Quetalco in the Quinchao channel.

During springs the tide runs nearly 4 knots; the ebb sets Tides. to the southward and the flood to northward. The rise and fall is about 15 feet.

Dalcahue channel opens to the southward into a large Relan cove and reef. bay formed by the two coasts; on that of Chiloé is the small cove and village of Relan; at the entrance of the cove there are 18 fathoms of water. The coast is steep to as far as the E. point; there a shoal of pebble and larger stones commences, which uncovers at low water to about ¼ or ½ mile from the land. This shoal reaches to a low, stony beach, where it terminates in a ridge which extends more than 1 mile to the SE. On the NE. side of this ridge the water is shoal for some distance, but on the SW. side there are 3 fathoms close to the shore. The ebb sets in with Tides. great force across Relan reef to the SE. in the direction of the channel between Lemuy and Chelin islands; between

---

Lemuy and the continent the currents are scarcely felt, the little there is sets to the eastward. During springs the water rises 18 feet above the level of mean low water.

Castro inlet.   From Relan reef to Castro inlet the channel is from 2 to 3 miles wide; in the middle of the channel there are 42 fathoms of water; muddy bottom; it leads to the entrance of Castro inlet. The eastern point of the inlet is low and stony, but it can be passed within ¼ mile in 12 fathoms. The eastern side of the entrance is formed by Lentinao islet, which is united to Chiloé by a sand spit which is dry at low water; from the exterior of this islet a stony point runs out to the eastward 200 yards; the southern portion of this point is steep-to.

To the southward of Lentinao islet is the small harbor Quinched, where a vessel bound to Castro can await a favorable time to run up, in case she found the wind baffling in the first two reaches; this is generally the case with north winds, though they may be strong outside; and no anchorage can be found in either of these reaches until too near the shore for safety.

One-half mile above the second reach of Castro inlet the east coast can be approached within 100 yards, but the other side is flat and shoal to within ½ mile of the beach, and the soundings decrease too rapidly to admit of navigating by the lead.

Going up or down, the eastern shore should be kept; not leaving it more than two-thirds the width of the channel.

The eastern coast is formed by steep and thickly-wooded cliffs, about 150 feet high. The western coast rises gradually to a height of from 400 to 500 feet; 5 miles from the shore is a chain of mountains of almost equal height, thickly wooded, having an elevation of about 1,000 feet.

About 2 miles below Castro is a small cove, on the east shore, in which a vessel can anchor in case of necessity; there are 20 fathoms between the points, but it shoals suddenly as soon as the line connecting these points is passed. Point Castro is a level plateau, about 93 feet above the sea-level, which separates the small harbor in the north from the river Gamboa, to the south; it is terminated by a low, pebbly point, steep-to on the north side, but on its southern side a bank commences, which follows the western shore of this part of the inlet. The small harbor of Castro, Castro harbor. to the northward of point Castro, is $\frac{1}{2}$ mile long and $\frac{1}{4}$ mile wide; between the points are 7 fathoms, but this depth decreases gradually to 3 fathoms, $\frac{1}{4}$ mile inside. The best anchorage is very near the south point; the northern side is shoal 100 yards from shore. In steering for the harbor a vessel must keep along the east shore until off the port, and then steer a course perpendicular to the entrance. By this, the bank to the southward of point Castro, which extends $\frac{1}{2}$ mile out, will be avoided.

The town of Castro is situated near the exterior part of Description. point Castro. It consists of several small streets with wooden houses. There were two churches in a very poor condition in 1834. Castro has, in fact, been almost abandoned. The few inhabitants are poor. San Carlos d'Ancud and Castro, 38 miles apart in a straight line, are connected by a road through the forests; this road is 50 feet wide; in its center is a path of 5 feet width, formed by logs placed transversely; it can only be used in dry weather. The earthquake of 1837 has changed the surroundings of Castro.

The greatest velocity of the current during springs does Tides. not exceed $1\frac{1}{2}$ knots: it is very weak during the neaps.

The village of Quinched is about 3 miles to the westward Village of Quinched. of the entrance to Castro inlet. The country is well cultivated and thickly peopled for 3 or 4 miles on either side of the inlet.

After passing Quinched the soil is only cultivated in spots; the remainder is wooded. About 1 mile south of the small harbor of Quinched the channel is $1\frac{1}{4}$ miles wide, and its depth in the middle is 41 fathoms.

Lemuy island is the southern boundary of the channel Lemuy island. leading to Castro inlet. On the north coast of this island Poqueldon is the principal village, located on the east side of a small narrow cove, about 150 feet above the water.

The cove is not deep enough to offer any shelter; there is an anchorage, however, $\frac{1}{4}$ mile from the east point, with from 4 to 5 fathoms of water. The village consists of about 20 houses built around a square; a church, the finest and largest between San Carlos and Castro, forming one side. Poqueldon, although smaller, seems to be more prosperous and in better condition than Castro.

At the NW. point of Lemuy is a cove, near which the shore is rocky and steep-to; at its entrance there is no bottom with 20 fathoms, but half-way in there is good anchorage for a small vessel in 7 to $9\frac{1}{2}$ fathoms, muddy bottom. It is difficult to land, as at high tide the water rises to the trees, and at low water it leaves a muddy beach. Some lignite has been discovered at Lemuy.

Détif promon-
tory.

'Détif headland, the southern extremity of the Lemuy island, terminates on the western side in a cliff 150 feet high, which is the eastern limit of the island. The cliff is surmounted by a round hill 250 feet high, from which the land descends gradually to a low tongue about $\frac{1}{2}$ mile in length; it then rises again to the eastward to the original height. A shoal of pebbles extends out about $\frac{3}{4}$ mile from the point; it is steep-to on the west side, but extends to the eastward to the next point about $1\frac{1}{2}$ miles; there are, at 100 yards from the extremity of the bank, $6\frac{1}{2}$ fathoms, and at 400 yards no bottom with 30 fathoms; the SE. extremity of the promontory is called point Détif.

Apobon   point
and reef.

From Apobon point, the northern extremity of Détif headland, a reef extends $3\frac{1}{2}$ miles to the eastward; there is a dry rock at the exterior extremity of the bank, and the latter uncovers at low water for about $\frac{1}{4}$ mile on each side of the rock; a vessel should never attempt to cross this reef, although there are 10 feet at low water between the rock and the land, as the current sets over it with great violence and very irregularly.

Yal point.

A bayou 5 miles long and very deep is formed by Lemuy island and Chiloé, which terminates at point Yal, where the channel is very narrow; the shores then separate, forming Yal bay. Outside of point Yal are two low shingle islands, which are connected by a spit, covered at low water. The passage between this double island and Lemuy is $1\frac{1}{2}$ miles wide and 12 fathoms deep; vessels can, therefore, work to windward in it. On the opposite side, that is between point Yal and the double island, the passage is narrow, but deep enough for all vessels.

Off the NE. extremity of the exterior island a dangerous ridge extends about $\frac{1}{4}$ mile in 2 fathoms of water; its extreme point seems to extend about $\frac{1}{4}$ mile farther out.

One mile south of point Yal is a steep hill, which forms tne southern point of Yal cove, and a little in the interior from this point is a flat hillock covered with trees.

The entrance to Yal cove is to the northward of the steep hill. Both points of this small harbor are steep-to, and there is no bottom with 20 fathoms in the middle of the channel between them, but half-way in good anchorage will be found in 5 to 12 fathoms, muddy bottom.

The steep hill forms also the north point of Yal bay. Between the points of this bay, which is 2 miles wide, there is no bottom with 64 fathoms; there is no anchorage in the outer bay of Yal until within ¼ mile of the hill, where there is 23 fathoms; the depth thence decreases to low-water mark. It is not a good anchorage, however, and vessels should only use it in cases of necessity. On the west side there is a hillock resembling that to the eastward, but less elevated; when coming from the south both will be recognized and they form an excellent landmark for the cove. The land behind them is low and thickly wooded. Point Tebao, the southern extremity of the bay, is low but steep-to, there being 9½ fathoms within 100 yards of the beach; a little to the southward the coast is flat and the shoal water extends out about ¼ mile.

The gulf of Corcovado is limited on the north by the Chaulinec and the Desertores islands, and extends to the southward between the continent, Chiloé and the Guaitecas islands. The currents, which are feeble in its northern part, increase near the entrance to the gulf, between Chiloé and the Guaitecas.

Point Ahoni is 5 miles SE. by E. from Yal bay. Near a little brook to the eastward of this point a bank commences with 3 fathoms, which borders the coast to the SE. for several miles. To the southward the coast is rocky and rises in cliffs to 150 feet. Point Lelbun is about 4 miles from Ahoni; off it the shoal just mentioned extends for 1½ miles; it deepens to 6½ fathoms, and is covered with masses of sea-weed. During springs the ebb-tide runs about 2 knots to the SE.

The low and rocky cape Aytay is about 3 miles to the southward of point Lelbun; about 2 miles from the land are several protruding rocks; they are part of a reef which

6 c

makes out from the cape; soundings show 5 fathoms 200 yards outside of the exterior rock. From there the edge of the large shoal already mentioned runs toward a sandy point covered by a grove of trees, situated 2 miles north of point Quelan; after passing this point the depth is 11 fathoms 1 mile from the land.

Point Quelan.   Point Quelan is a long and narrow spit, very low, and covered by trees, excepting about ½ mile from its end, where there is nothing but sand. This spit is about 600 feet long. On the SE. side the beach has a gradual slope, there being but 2 fathoms of water ¼ mile from the shore. Off the point the soundings show 6½ fathoms at less than 100 yards. The small island Acuy, 3 miles to the eastward of this point, is low at its SW. extremity, and terminates on the northeast in a cliff 200 feet high, off which are some detached rocks, extending out about 2 miles. The island is, in fact, surrounded by a bank of rocks and pebbles, covered with seaweed, a part of which uncovers at low water, and extends about 1 mile outside of the W. point. The channel between point Quelan and Tranque island is about 1 mile wide; the ebb sets to the westward, and, during neaps, has a velocity of 2 knots.

Tranque island.   Tranque island is 13 miles long and 3 wide; it is situated to the southward of point Quelan, and shelters Quelan cove and Compu inlet. A chain of hills traverses this island from WNW. to ESE., which is about 300 feet high at the most elevated point, near the NW. extremity; from thence it gradually declines to the SE., and terminates in a low point, called Centinella.

Numancia roc   During the war of Spain with the South American republics, the frigates Numancia and Blanca discovered the Numancia rock, 2½ miles N. 68° E. of point Centinella. It is black, about 65 feet long, and the sea washes over it. The existence of this rock has been questioned. H. M. S. Nassau subsequently passed the position several times without seeing it; on the other hand, it is reported to have been known to the sealers for many years. On rounding or passing point Centinella a good lookout should be kept for it.

The northern coast of Tranque has a gradual slope and is well wooded. There is a small bay at the NW. point of the island, where the channel suddenly turns to the southward;

off the bay there is an islet, inside of which there are 20 fathoms of water. The bay can be used by vessels awaiting the change of tide.

On doubling point Quelan, keeping along the outside of the ridge, is Quelan cove. Its entrance is about ½ mile wide, but the shores on either side should not be approached nearer than 200 yards; at that distance there are 3 fathoms, and, in mid-channel, 13 fathoms of water. The cove is about ¾ of a mile long, and has about the same width, but along the western shore, ¼ mile out, there is but little water; in all other parts there is good anchorage in 4½ to 8 fathoms. In its NW. angle is a narrow inlet, which can only be used by boats. There are several houses surrounded by cleared land; the population is Indian. In 1835 the surrounding country was very thickly populated. To the westward the land rises suddenly to 200 feet, and is thickly wooded.

Quelan cove and channel.

On the north coast, about 1 mile to the westward of the cove, is a small bay called Quelan bay; in it there is good anchorage in from 9 to 12 fathoms; off its entrance is a small island. It can be approached within 200 yards; inside of that the water shoals rapidly from 9 to 3 fathoms.

Quelan bay.

In the Tranque channel the flood runs along and around the points, then across, in the direction of the north shore, outside of the small island; inside of it there is very little current. In the narrowest part of the channel, during neaps, it runs at least 4 knots along the points of the rocks. Its direction is always to the northward.

Tides.

On the Chiloé shore, abreast of the NW. extremity of Tranque island, is an inlet called Compu. Neither this inlet nor that of Chadmo, to the SE., has been examined. About 1 mile from the turn in the channel there is a small cove on the Chiloé shore, which only appears practicable for boats. In the southern part of Tranque there are deep bays between the points on its southern shore.

Compu inlet.

Point Cuello is on the coast of Chiloé, at the entrance of the channel, SE. of Tranque, which makes another elbow at this point. About 1 mile SE. of this point a reef of rocks extends from NW. to SE. for about ½ mile; part of this reef is uncovered during low-water spring-tides. The portion with least water is marked by sea-weed, but the shoal extends about 200 yards on each side of it; inside, at the

Point Cuello

distance of $\frac{1}{4}$ mile, there is $3\frac{3}{4}$ fathoms, which deepens to 12 fathoms $\frac{1}{4}$ mile from the shore.

The small island Chaulin lies $4\frac{1}{2}$ miles from point Cuello, off the entrance of Huildad inlet.

Huildad inlet. Huildad inlet is S. 15° E. of point Cuello; its entrance is but 460 feet wide; inside of the ridge, which makes out from the north shore, it widens to about $\frac{1}{3}$ mile; 1 mile from the entrance it contracts again to 400 yards, when it opens into a large basin 1 to 2 miles wide and 4 miles deep. In the outer harbor is a good anchorage in 5 to 8 fathoms; the shores are steep-to, excepting along the indentation, which is behind the ridge of pebbles, and where there is shoal water 300 yards from the beach. In the strait between the two harbors there is $4\frac{1}{2}$ fathoms 40 yards from either shore and 20 fathoms in mid-channel. On the south side is a church with several houses. The other buildings are scattered along the shores of the harbor, but principally on the southern shore; the land around them is cleared. The ground rises in a gentle slope for about 1 mile, when it joins a chain of mountains about 300 feet high. If it is desirable to await a change in the wind or weather at Huildad, the outer harbor is preferable, as the NW. squalls are very heavy in the inner one, while the outer is entirely sheltered.

Tides. At the entrance of the inlet the ebb tides run nearly 4 knots during the springs, but slacken considerably inside; in the strait the tides are almost as strong as at the entrance.

Huildad shoal. Huildad shoal lies to the southward of the entrance to the inlet; between it and point Chayhuao it extends 1 mile from the land, and is almost entirely covered by sea-weed. The tides on its outer edge run about $1\frac{1}{2}$ knots during springs. This shoal is terminated to the southward by a long stony reef, which commences $\frac{1}{3}$ mile from point Chayhuao, and extends to the SE. Several of these rocks uncover at low water 1 mile from the point, and during spring-tides the entire reef to the outer rocks is dry. There is a passage between its southern extremity and the NE. extremity of Caylin island, in which there is deep water close to the reef. In this channel the flood sets to the eastward across the reef at least 3 knots at springs; after passing the reef it meets the outside tide coming from the southward.

Tides.

Between point Chayhuao and San Pedro passage is a deep bay, bordered by Caylin, Laytec, and Colita islands. To the NW. of the latter, on Chiloé, is the small cove Yalad.

Caylin is 5 miles long NW. and SE., and about 3 miles wide; its northern coast is steep-to; in the channel which separates it from Chiloé there is no bottom with 40 fathoms. After doubling a low point of pebbles on its northern shore there is an inlet which runs in to the land about 4 miles SE., and terminates in three small coves. It is not a good anchorage, as there are from 22 to 30 fathoms to within less than ¼ mile of the head of the bay, where it suddenly shoals to 11 fathoms, and 200 yards inside; it is dry at low water. The SE. coast of the island is composed of cliffs about 100 feet high; at their foot is a beach of pebbles. A reef with 4 fathoms at its edge extends 1½ miles from the beach.

In 1834 an Indian village was found here, containing about forty houses and 250 inhabitants, who raised sheep and poultry.

Laytec is 6 miles in length from NW. to SE., and 3 miles in width; it is separated from Caylin by a channel 2 miles wide, at the southern entrance of which there are 19 fathoms of water. Off its SE. point are some rocks, but no dangers have been found farther out than ½ mile, where there are 4 fathoms. Shoals with three fathoms extend 3 or 4 miles from the SE. part of the island.

Colita is low and thickly wooded; it is about 4 miles long and 1½ wide. The channel between it and Chiloé is very narrow, and probably offers no passage for vessels. Behind this island the land rises gradually from the beach, and forms a chain of hills more than 1,000 feet high. The passage between Colita and Laytec is 1½ miles wide. The tide runs about 1 knot in the channels to the northward of these islands.

Five miles south of Colita is point Yatec, the northern point of the entrance to San Pedro, of which a description has been given.

CHAPTER IV.

COAST OF THE CONTINENT OPPOSITE CHILOÉ.

Variation from 19° 03′ to 19° 38′ easterly, in 1876; increasing annually about 1′ 30″.

Taking up the description of the coast of the continent opposite to Chiloé and bordering the gulfs of Ancud and Corcovado—

Parua or Parga bay. To the eastward of point Coronel is Parga bay, entirely open to the southward, but sheltered against NW. winds; it should therefore be used only by vessels waiting for a favorable opportunity to pass through the strait of Chacao. Its shores are steep-to, and thickly wooded; 1½ miles from the beach there are 35 fathoms of water; but 3 miles from point Coronel, near point Tique, the breakers extend ¼ mile from the beach, and the shoal continues along the shore to the southern part of point Auque. At that point a dangerous spit, ¾ mile long, makes to the SE. into the Abtao channel. The dangerous eddies between points Chilen and Tique, called the Raya de Tique, have been already mentioned.

Abtao island. Abtao island is 7 miles to the eastward of point Coronel; it forms a narrow ridge, about 3 miles long and 766 yards extreme width. The northwest part is the highest; it ends in Pilquen hill, 155 feet above the sea. This is the only spot on the island where fresh water can be found; from it the ground falls, but rises again on the southern part of the island. In the SW. the island terminates in a long, low, and very narrow tongue of land, named point Quilque. From the NW. point some rocks make out about 400 yards, close to which there are 12 fathoms, and, ¼ mile to the northeast, 50 fathoms; banks make out from the E. and SE. extremities of Abato, called Huenuhuapi and Nahuel-huapi; the latter projects about 1 mile, and is covered with weeds; at its edge there are 5 fathoms. The bank is prolonged, parallel to the shore, to point Quilque, and from thence it extends ⅔ mile to the westward. On the channel side the water is clear 200 yards to the. north of point Quilque.

The channel which separates Abtao from the continent is narrow, but there is sufficient room for any vessel to work. At its western entrance attention must be given to the bank which surrounds point Quilque, and also to that which makes out from the continent, and is uncovered at all tides. On entering the channel from the strait, run for the hill which overlooks Abtao to the south, leaving point Quilque a little to starboard; the clear part of the channel is not more than $\frac{1}{8}$ mile wide. Abtao must be kept distant not more than 400 nor less than 200 yards; coasting along its inner side, there is from $9\frac{3}{4}$ to 20 fathoms of water. The continent must not be approached under any pretext until the northern part of point Auque, the eastern extremity of the continent, bears west. Between this point and point Lilihue, on the northern part of Abtao, the channel is only $\frac{1}{4}$ mile wide, but it is clean on either side. At its end, between the hamlet Challahué and Pilquen hill, it is a little less than $\frac{1}{2}$ mile wide.

Vessels can anchor in this channel at two points: first, at the entrance, a little inside of and to the northward of point Quilque, in $13\frac{3}{4}$ fathoms, sandy bottom close to the island; second, at port Abtao, off the entrance, in $14\frac{1}{2}$ fathoms 400 yards from the shore, point Challabué bearing N. This bay can hold a large number of vessels. In its NW. corner there is a watering-place; point Auque is surmounted by a look-out tower. The shores of the island uncover at low water 110 to 166 yards. Small vessels can be beached.

From Abtao channel the coast runs for about 3 miles W. by N., and then again to the westward, forming a deep bay; at its end, between points Peuque and Curaco, it has a width of 1 mile, and vessels can anchor in $25\frac{3}{4}$ fathoms, muddy bottom, and be sheltered against all except easterly winds. This cove is called Collihué. From the end of the cove the coast trends about E. by S. for 7 miles to point San Antonio, at the western entrance of San Antonio channel, which separates the island Quihua from the continent.

Lagartija island, a little more than $1\frac{1}{2}$ miles NE. of point Huenuhuapi, is a rounded peak with a gentle slope to the NW., and high steep banks on the rest of its contour; it is a little more than 400 feet in diameter and about 55 high.

The upper part is wooded, and the NW. part abounds in strawberries and potatoes. It is surrounded by a bed of pebbles, which is covered at high tide, excepting at the N. point, where a narrow ridge remains dry; on this side, the island can be approached within 400 yards. Off the SE. extremity of Lagartija is a shoal full of rocks and covered with sea-weed, which extends along the SE. and S. shore of the island at a distance of 2 miles. The outer rocks are called Cola bank, South bank, and Medio bank; they are divided by deep channels from the main bank, and are covered by $1\frac{1}{2}$ and $2\frac{1}{2}$ fathoms at mean tide.

The main shoal terminates at the NE. extremity of the island; between it and the main land are 33 fathoms, bottom sand and rock. There are no dangers.

Buoys.

Between Abtao and the banks of Corva, the tide runs $1\frac{1}{2}$ to 3 knots. The channel known as the Lagartija channel is about 1,310 yards wide, and has been marked by cylindrical iron buoys, painted red.

The first buoy is on the SE. extremity of the reef which extends from the island of Abtao, in 3 fathoms water at low tide, with the following bearings:

Point Nahuelhuahi, on Abtao, bears N. 39° W.; and the middle of the island Carva, N. 42° E.

The second buoy is placed on Medio bank, in $2\frac{1}{2}$ fathoms water at low tide, with the following bearings:

Point Huenuhuapi, or eastern extremity of Abtao, bears N. 68° W.; the middle of Carva, N. 34° E.

According to the pilots, there is a rock, with 2 fathoms of water over it, on the bank del Medio of Lagartija channel, 164 yards S. 19° W. of the buoy which marks the bank. The depth in the channel varies from 30 to 55 fathoms, and the bottom is of sand and rock.

Lami banks.

The western edge of the Lami bank is a little less than 2 miles from Lagartija, but the distance between the banks off this island and Lami is only 1,090 yards. These banks, several of which are always dry, are 2 miles long in an east and west direction, and 1 mile north and south. They are intersected by deep channels, through which the tide runs with a velocity of 1 to 2 knots; some portion of them is covered by sea-weed. It is said that they contain an excellent spring of fresh water, which the coasters prefer to that found on the neighboring islands. The dry portions have the following names: The western, *Ghal;* the middle, *Cailin;* and the eastern, *Quihua.* They are formed of sand and rock.

The passage between Lami and Lagartija banks is clean, having from 10 to 30 fathoms.

The channel which separates Lami banks from Quihua island is a little less than 1 mile wide, with about 26 fathoms in mid-channel.

Quihua is a large and rugged island, separated from the continent by a long, sinuous, and narrow branch of the sea, called San Antonio. It is deep at the entrance, having about $14\frac{1}{2}$ fathoms; it runs first to the northward, then SE., when its breadth diminishes; its shores approach each other, and the width at Quetrue is but 328 yards; at this point the channel is for $\frac{1}{4}$ mile dry at nearly all tides. From the estuary of Quetrue it takes the name San Rafael, and empties into Caicaen channel, where it is only 2 feet deep and 328 yards wide, and is only navigable for boats during high water. The tides enter at both ends and lose themselves in the numerous estuaries, the principal one of which is Quetrue, into which that of Tilao empties. It is supposed that the tides have formed this channel by gradually eating away the neck by which Quihua was connected with the continent. Similar changes have taken place at San Augustin, on the coast of Contao, and south of the island Huar. *Quihua island.*

The portion south of Quihua is clean; the beach uncovers about 500 feet. To the east of its southern point, called Chuyehua, there is a large single rock of about 328 cubic feet, called San Pedro. It is at the extremity of the low-water mark, and is an object of superstition to the natives; near it there are about 8 fathoms of water.

Tabon island, the southernmost of all the group belonging to Reloncavi sound, is $5\frac{1}{2}$ miles long from east to west, and about 19 miles in circumference. It is very irregular, and formed of a series of detached peaks, connected by small ridges of pebbles, several of which are covered at high water, when three separate islands are formed, named, from the *Tabon island.*

east, Lin, Ilto, and Polmallelhue. The latter is so narrow
that the high tides of the equinox again divide it into two
islands. Tabon is completely cleared of wood, and only
apple-trees are seen around the houses. Lin is 12 feet high,
and Ilto 157. The coves El Ded, Lin, and Ilto are formed
by them; the latter affords an excellent anchorage for vessels of any tonnage.

The south coast is clear at a very short distance from the
shore, but the northern one is very flat and dangerous. One-
half mile NW. of point El Ded is Borudahue bank, formed
of rocks and pebbles; between it and the island there is
only a passage for boats. Farther to the westward, about
½ mile WNW. of point Cholchollen, is a bank of the same
name. Three-quarters of a mile NW. of this point is Corvio
bank, and Culenhue bank between the latter two. All these
banks uncover at low water except Culenhue, which is only
dry at neaps; they are all covered with excellent shell-fish.

A rock, about 1,965 feet in length, called Polmallelhue, makes out to the westward of the western point of the
island; it is only uncovered at low-water springs.

The beaches of the island uncover for nearly 1,000 feet.
Between Tabon and Chidhuapi, there is no bottom with 40
fathoms line.

The passage between Lami and Corvio banks is 1¼ miles
wide. There are 20 fathoms of water, pebbly bottom; the
tides run from ½ to 1 knot per hour.

Tabon has about 350 inhabitants, who are occupied in
agriculture and cutting timber at Comau and Hualayhue.
Fire wood is very scarce.

Amnistia or Shearwater bank. About 3½ miles to the southward of Tabon is a bank of
rocks and stones, covered by 11 feet of water. It was re-
ported by H. B. M. S. Shearwater, and has a diameter of
about 1 mile.

About 10 miles to the southward of the east point of Tabon,
and 15 miles east of Queniao, the charts show a large bank
of about 4 miles diameter, which uncovers at low water. It
is probable that this bank has been confounded with the
latter.

Quenu island. Quenu is a small island 110 feet high, of the same charac-
ter as Tabon, situated to the NW. of the latter. A reef of

rocks makes out to the westward of point Pinto, the western
point of the island.   It extends WSW. for ⅓ mile, and terminates in three rocks, which only uncover at low water;
the bank uncovers almost entirely.   Eight hundred and seventy yards from this point there are only 3 fathoms of water.
Between this bank and Lami banks the channel is 1 mile
wide, with 35 fathoms of water; muddy bottom.

The passage between Quenu and Calbuco islands is clear.
It is ⅓ mile wide, with 20 fathoms in mid-channel; sandy
bottom.   Vessels passing through it must keep in mid-channel, steering due east.   Point Martin, the northern point of
Quenu, is cleaner than the southern points of Calbuco, but
it must on no account be brought to bear to the northward
of east.

The eastern and southeastern points of Quenu are foul to
a distance of 330 yards; from there the water is very deep.
The coasts are level, excepting in the SW. and in the middle
of the north coast, where there are some steep banks.   On
the southern point, Rumen, there is a small chapel.

Quenu is a small, picturesque, and fertile island 3,000 yards
long from NW. to SE., and 1,300 wide.   It is almost divided
by the unimportant estuary Puchivilo.   It produces potatoes, apples, grain, strawberries, and cattle; but the island
is poor, as it is badly cultivated.   It is entirely cleared of
wood; fire-wood is imported, and dear.

Calbuco is a small, highly cultivated island extending 3   Calbuco island.
miles WSW. and ENE., with an average breadth of 1 mile.
It has the same aspect as the preceding ones.   In connection with the island Quihua and the continent, it forms
Caicaen channel, which has 20 fathoms of water at the entrance, but shoals toward the outlet.   At that point, between
the town of Calbuco, point Yahuecha, and the continent, it is
only ¼ mile wide, with a maximum depth of 3 fathoms.   The
banks off the town and the point reduce the navigable portion to 225 yards in width.   The northeast shore of the
island, called Pecuta, where the town is built, uncovers for a
distance of 200 yards.   Close outside of this low-water mark
there are 6 fathoms of water, and beyond the depths increase
to 17 and 24 fathoms.   This island, with Quenu, Chidhuapi,
Chaullin, and Puluqui, forms another and a much deeper
channel, where the soundings are not less than 20 fathoms

in mid-channel, and which, after passing Quenu, is 1¼ miles wide.

Anchorage.   Vessels anchor ⅛ mile N. 60° E. from the town of Calbuco, with the northern part of point Yahuecha bearing N. 55° W., with the entrance of Huito inlet nearly shut in, in 15 fathoms of water; bottom fine sand mixed with mud. Ballast is thrown overboard about 1 mile to the northward of the anchorage.

Town of Calbuco or El Fuerte.   The town of Calbuco, founded in 1602, is one of the principal ones of the province of Llanquihue. It has deteriorated considerably since the foundation of the colony at port Montt. It is about 26 to 30 feet above the sea, and built on very uneven ground. Its comparatively large church has been struck by lightning, and is now in ruins. A lieutenant of the reserve from port Montt fills the office of subdelegate of the marine. The city was first founded at Rosario, on the N. coast of the Huito estuary, about 1¾ miles from its present position; but the frequent attacks of the natives forced the colonists to abandon it and establish themselves on the island Calbuco.

Chidhuapi island.   The island Chidhuapi is low and almost entirely under cultivation. Between its southern extremity and Tabon there is no bottom with 40 fathoms line; between its western part and Quenu the average depth is 30 fathoms; a bank runs out about ½ mile from its SE. point and obstructs half of the channel which separates it from Puluqui, which is but ⅓ mile in width. Along Puluqui island, at a distance of 200 yards, there is not less than 7 fathoms. The island is from 80 to 100 feet high and is divided into two parts called Lacao and Ahuenu; it is about 2½ miles long in a NW. and SE. direction; its maximum breadth being nearly 1 mile. The ground is very much cut up and the shores are alternatively low and high; there are isolated rocks 440 yards from its W. point. The island has been cleared of wood and no cattle are raised on it. The population consists of about 200 persons. There is a chapel on its SE. part.

Puluqui island.   Puluqui island is nearly 7 miles long from NW. to SE., and averages in breadth about 3 miles. It is the easternmost and largest island between the strait of Chacao and Reloncavi sound. On its eastern shore there are a few spots of cleared land, but on the other side, where the land is lower and more swampy, these are more numerous. From the south point, which is low, the ground rises in a gentle slope to a chain of hills about 230 feet high, which runs across the island in a N. and S. direction.

The south coast, between points Manao and Centinela, is shoal, to a distance of 550 yards, and entirely open to winds from east to west by the south. Manao, the SW. extremity, is about 2¼ miles from Centinela, the SE. extremity. A little to the westward of the latter the coast trends to the northward and forms the estuary of Poza de Llaicha, which is 4¼ miles long from N. to S. and communicates with the sea by a narrow channel which is dry at neap-tides. This estuary can only be used by small vessels or boats, although larger ones can enter it at high water; in the center it is very deep. To the westward of this inlet on the hill which overlooks the chapel of Llaicha is a rich graphite mine of excellent quality, though a little black.

From point Centinela the coast first trends N. to point Perhue, the eastern extremity of the island, thence NNW. to point San Ramon. This entire coast, 7 miles long, is called Pollollo; it is overlooked by steep and wooded cliffs from 180 to 220 feet high. It has no good anchorages. The beaches of large pebbles are dry at low water to a distance of 330 to 430 feet; beyond that they are steep-to.

Point San Ramon forms the NE. and N. extremity of the island; its northern part is separated from Tautil island by the pass of Tautil; near this point, which is surmounted by a chapel in ruins, is the lagoon San Ramon, whose level is 56 feet above the sea; it is 328 yards to southward of the chapel and 433 yards from the sea; it lies in a crescent-shaped hollow whose points turn to the NW. Its shores are swampy and covered with thick undergrowth, which runs up the hills from 100 to 165 feet, in the midst of which some cultivated spots and dwellings are scattered. Its outlet is to the SW. This lagoon is of recent formation; its water is fresh and deep, and shells are found in it.

From point Ramon, the western part of Puluqui forms the eastern shore of the Calbuco channel; in it there are four estuaries; the northern one is named Puluqui, the next Machil or Quniched, which opens between point Puluqui to the north and point Machil to the south. In its

center is a large bank, which uncovers about 4 feet at low-
water springs. This estuary is without importance; behind
it is the highest point of the island, an elevation of 230 feet.

Point Machil, on which there is a chapel, is the north
point of the estuary of Chauquiar, which can be entered by
vessels of all sizes. It has 14 fathoms of water at the entrance and then shoals toward the eastern extremity, where
there are 10 fathoms. The bottom is of small stones mixed
with mud; to the south it is limited by point Chechil, 135
feet high and entirely clear.

The estuary of Chopé opens between point Chechil and
point Chopé. The latter point is steep-to and about 100
feet high; this lagoon is the southernmost of the four; it is
about 178 yards wide at the entrance and extends a little
over 2 miles S. 37° E., contracting gradually in width; at its
end is the small chapel of Chopé, and a little to the eastward the small lagoon of Chipue; the hills which surround
the lagoon are strewn with houses and cultivation; this estuary is deep and without importance.

From point Chopé the coast of Puluqui runs SE. for 4
miles, to point Manao, forming the Chidhuapi channel.

Puluqui has about 2,100 inhabitants, which are occupied
in fishing, in cutting wood, and in cultivating potatoes,
wheat, flax, and grass. The coasts of this island are formed
of alluvial stones, and show large deposits from the sea.
The island is thickly-wooded, the inhabitants are scattered
over it, and there is no center of population. The agent of
the government lives at Chéchil.

Tides.

It is high water, full and change, at Perhue, at 12^h 56^m.

Chaullin island.

Chaullin is a small island situated opposite the estuary of
Huito, and about 1 mile to the NE. of the town of Calbuco;
its circumference is about ¾ mile; it is low, scantily wooded,
and abounds in shell-fish. The want of fresh water prevents
the cultivation of this fertile island.

Point Yahue-cha.

The hilly point Yahuecha lies between the town of Calbuco and the entrance of the estuary of Huito.

Huito estuary.

Huito estuary is about 4 miles long; its breadth does not
exceed ½ mile at any point; it has a semicircular shape,
with a funnel-shaped entrance; its depth, 17 fathoms at the
entrance, diminishes gradually to a narrows, where there is
but 3½ fathoms in mid-channel, 26 feet from the northern

shore. At this point where the southern shore makes out a
little, the Chilians barred one-half of the channel during
the war with Spain. The entire breadth is about ¼ mile;
the navigable part is 650 feet. This artificial bank has
increased; it is covered by cholgas, for which the inhabitants fish.

Farther in, the breadth and depth of the estuary increase again; ⅔ mile from the neck the breadth is 1½ miles,
and the depth 12 fathoms; bottom coarse sand; this is the
anchorage. Farther in again, the depth remains between
5½ and 6½ fathoms to near its end; the bottom is fine sand
and mud. Water can be taken in at the many springs, or
from the torrent at the bottom of the estuary. This estuary
is gloomy, thickly-wooded, and bordered by steep hills.
There are some few cultivated spots, near which huts are
built. Northward, and near the neck, is the Rosario hamlet,
consisting of a few small houses around a chapel.

The Chile-Peruvian squadron took refuge here during the
war with Spain, and built some fortifications on each side
of the entrance and at the narrows. In the year 1871,
some traces of the stone houses and fortifications remained,
though everything was covered with vegetation. Vessels
can anchor at the entrance to the inlet in 17 fathoms, bottom
fine sand.

It is high water, full and change, at 1ʰ 22ᵐ; rise from 16   Tides.
to 22 feet. The tide runs with a velocity of 1 mile to the
northward of point Yahuecha, and from 2 to 3 miles at the
narrows. During springs it reaches a velocity of 5 miles.
Off Calbuco the strength is about ¾ mile. The flood sets to
the northward and the ebb to the southward.

About 1 mile NE. of point Pelu, the northern point of   Rulo estuary.
Huito estuary, is point Metreucue, near which there is a
small dry rock; around this, Rulo estuary opens, having at
its entrance a width of 1½ miles, its depth being nearly 2
miles. It is surrounded by hills of moderate height, which
are cultivated. This bay is somewhat shoal; there are 5½
fathoms at the entrance and about 3 at its head; the interior lagoon is almost entirely dry. On entering the
estuary, vessels must keep close to point Metreucue to
avoid the bank of San Agustin, which makes out from the
NE. point of the bay. At this point there is a small hamlet,

with au unfinished chapel; the bank runs to the SW., is
dry for about 260 feet, and projects 230 feet farther under
water in the same direction; at its edge there are 3 fathoms
of water.

San Agustin. San Agustin is a small district containing 660 inhabitants. Behind it is a series of lagoons, and the ground
rises to au elevation of 230 feet.

Tides. It is high water, full and change, at 1ʰ 4ᵐ. At neaps the
rise and fall is 12 feet, and during springs 20.

Tautil island. The island of Tautil lies between points San Ramon and
San Agustin, the passage between the continent and the
northern extremity of Puluqui, where it forms two channels.

The channel on the side of the continent is narrow and
shoal, and remains dry on its southern portion from half-
tide. No vessel can pass through it. A tongue of sand
and pebbles unites this island with San Agustin. The remainder of the channel is formed by low and stony beaches,
and at its northern entrance there it a bank of detached
rocks, which is awash at low water and must be avoided by
boats which may attempt this passage.

The southern channel, which is between the island and
point San Ramon, is deep, and is used by steamers and
also by sailing-vessels with fair winds. This channel,
named El Paso de Tautil, is hardly 730 yards wide.

The strait of Tautil trends N. 53° E. and S. 53° W; it is about 1,000
Tautil. yards long, and has a mean breadth of 105 feet. The ground
is level and favorable for agriculture; it is about 72 feet
high; its western coast is cut up, and near it are many detached rocks; it contains drinkable water and several huts.

The coasts of the island are flat and rocky. The shores
on the eastern part uncover to a distance of 200 yards, and
on the NW. extremity 335 yards. To the SE., in the direction of San Ramon, or in the narrowest part of the channel,
a bank makes out about 328 yards to the middle of the
channel. At its edge are 3 fathoms at low tide; farther
out the depth increases to 5 and 5½ fathoms. Vessels passing through this channel must keep closer to Puluqui than
to Tautil.

To the NE of the northern extremity of Tautil a reef
makes out 483 yards; at its edge there is 3 fathoms at low
water. The reconnoissance of Captain Gormaz proved that

the banks of Tautil extend out a less distance than the charts indicate.

In taking this strait a vessel should pass one-third of the breadth of the channel nearer to point San Ramon than to Tautil, and steer for the Caicura islets, at the entrance of the estuary of Reloncavi, if they are visible; otherwise the southern part of Huar island must be kept open. This course must be kept until points Huatral and Huelmo are opened, when a course can be taken for the Huar channel.

If coming from the northward, a vessel should be kept in the middle of Huar channel, having Tautil 1 mile distant. When point San Ramon is on with point Metreucue, take this bearing as a course and the strait will be passed in 5 fathoms water. It is high water, full and change, in the strait at $1^h 7^m$. Rise and fall varies between 10 and 22 feet. The velocity is between $1\frac{1}{2}$ and 3 miles. The current, which follows the direction of the channel, changes about 25 minutes before slack water. In this archipelago there is considerable difference between the tides by day and those of the nights, amounting to nearly 2 feet. This phenomenon was observed during summer, and it is believed that it is reversed during the winter. Mr. Simpson observed the same in the archipelago of Chonos.

The gulf of Reloncavi has three entrances. The first is the strait of Tautil just described; the second is between the island Queullin and Nao islet; the principal one, however, is the strait between Puluqui and Queullin, which is 2 miles wide. The gulf of Reloncavi extends 20 miles to the northward, and is about 12 miles across from east to west, and very deep. In the strait between Puluqui and Queullin there is no bottom with 60 fathoms, and no bottom in the gulf with 120 fathoms, except in the vicinity of the islands and banks. Between Huar and the Caicura islets the depth varies between 140 and 160 fathoms. Generally speaking, vessels anchor under the lee of the islands or along the coast on either side according to the wind. The tide runs from 1 to $2\frac{1}{2}$ knots in the strait between Puluqui and Queullin, according to the age of the moon; in the narrowest part there are some eddies.

Huar island, on the west side of the gulf, is over 300 feet high, and is separated from the continent by the Huar pass-

(marginal note: Tides.)

(marginal note: Gulf of Reloncavi.)

(marginal note: Huar island.)

7 c

age, which is 1 mile wide. Its NW. point is formed by the small island Malliña, united to it by a narrow bank of pebbles, which uncovers at one-third ebb-tide. Malliña rises gradually toward the northward, where it attains its maximum height of 72 feet and terminates in a cliff. It is inhabited and cultivated. From its NE. part a bank makes out which is dry at low water. It abounds in shell-fish and is the rendezvous of the fishermen of the vicinity.

On the other side of Malliña is the roadstead of Quetrulauquen, which is open to northward. At the end of this estuary there is a nearly circular lagoon of about 547 yards diameter; this communicates with the gulf by the estuary of Quetru, into which it empties; this outlet is encumbered with fishing-stakes. This lagoon is separated from one of the estuaries of the south by a narrow tongue of land 433 yards wide; the northern part of Huar is a well-defined peninsula.

Point Cuervos, 120 feet high, forms the NE. part of the island; behind it, 217 yards from the beach, is a small lagoon, which runs from E. to W. 557 yards, with a mean breadth of 213 yards. It is 26 feet above the level of the sea, and the water is fresh.

Point Alfaro, on which there is a small chapel, is 128 feet high, and the western cliffs of the island, called Del Pedragal, are 250 feet high.

Huar is naturally divided in three parts by the estuaries of Chipue, Chanqui, Chenquohue or Chucahua; it forms the three districts of Quetrulauquen, of Alfaro, and of Chucahua.

Huar was colonized in 1610, and has more than 1,000 inhabitants; Chucahua seems to be the principal place. The inhabitants are occupied in fishing for shell-fish, which abound on the coasts. They cut on the island the laurel, the luma, and on the opposite coast the alerce and the cypress; they cultivate flax, corn, and grass. The earthquake of 1837 lowered the ground of the island.

It has been asserted that there is a reef to the NE. of Huar island which is dry during the low water of the equinoctial springs; it is not marked on the charts; M. Gormaz found, 2½ miles NE. of Huar, 74 fathoms of water, muddy bottom. This officer does not believe in the existence of this reef. During clear weather the isolated mountains of Osorno and Calbuco can be easily distinguished from Huar.

Janequeo bank is said to be situated 4½ miles E. ½ S. from the SE. point of Huar. La Covadonga searched for it without success; in the position indicated 122 fathoms were found.

The Pucari shoal is 1 mile S. 15 E. of point Blanca, the southeastern part of Huar; its center is 9 feet above low-water springs. The western part of this bank is formed of small, rounded pebbles, and the eastern of large pebbles with scattering rocks of moderate size, which do not uncover at ordinary low water. The bank is 547 yards long, NW. by W. and SE. by E., with a variable breadth of from 163 to 217 yards. This bank is covered with shell-fish, which are caught by the inhabitants of Huar.

Captain Gormaz thinks, with the people of this place, that the Rosario bank, which is supposed to lie 1 mile SSE. from Pucari, does not exist.

Point Huatral forms the entrance to Huar passage on the side of the continent; it is ½ mile NE. of Tautil. The coast between Huatral and Huelmo is rocky and shoal; the former is level at its foot and rises in shelves to the height of 260 feet; it appears to be the highest elevation in the vicinity.

Huelmo bay, to the northward of the point of the same name, is between it and Huelmo island; the largest vessel can enter it. The surrounding hills attain an elevation of 180 to 230 feet, and some streams of excellent water run from them. The shores are low and rocky; there are some houses, and to the west is the chapel of Huelmo.

Huelmo island is 3 miles N. 9° W. from point Huatral. It is 1,317 yards long, NW. and SE.; its mean breadth is 383 yards. At low water it is connected with the continent on the west side; sloops and large boats can pass through the channel on high water. The northern extremity is high and wooded; the south coast is more even; from it a chain of reefs extends to the eastward for ⅓ mile and ends in a dark-colored rock called Lobos, which is dry at half-tide. Between it and the coast is another rock above water, called Huelmo; it is of a whitish color, inclined to the westward, and is always uncovered; it resembles a boat under sail.

When after clearing the straits of Tautil the points Huatral and Huelmo are opened, vessels can enter the Huar passage between the island of that name and point Huatral. The channel is clear to the northward of Tautil.

Mallina. Vessels can anchor in the passage on the west coast of Huar island off the elevated promontory, which is Malliña island, about ⅓ mile from the coast, in 13 fathoms, bottom sand.

As all the bearings of these coasts are not well determined, it must be remembered that when Huar passage is entered from the northward the island Queullin should be seen.

Ilque or Ti-
quen bay. Point Capacho is directly to the northward of Huelmo island. The unexplored bay of Ilque is between Capacho and Ilque points; its shores are shoal 225 yards from the beach, especially on the west side, the rocks of which show as islets at high water and as large rocky banks during low water; it is surrounded by hills, from 200 to 215 feet high, on which are some houses; the chapel of Ilque is at the head of this bay.

Point Ilque. It is presumed that Ilque point was confounded with point Capacho on the old charts; it lies about 2 miles N. 15° W. of the latter; is not steep to, and rises to a height of 200 feet. A chain of rocks, uncovered at low water, extends from its eastern part for about ⅓ mile; a bank is reported to exist between this point and the island of Capirahuapi, on which, during the tides of the equinoxes and in calm weather, the bottom can be seen.

Anchorage of
Huenquellahue. To the northward of point Ilque a large bay opens, which has point Panitao for its northern limit; its shores are sloping and covered with bowlders; back of it are small hills 230 feet high; they form a table-land, wooded, inhabited and cultivated in spots; they limit to the east the extremity of the prolongation of the central valley of Chile, which ends on the north coast of the strait of Chacao. The opening of the bay is 3½ miles wide and nearly 2 miles long; in its SW. part is the anchorage of Huenquellahue.

Point Panitao, which is low at its extremity, rises rapidly to the height of 230 feet; it is steep-to, except to the north, where a bank makes out 1 mile, on which there is very little water. This point forms one side of the Maillen channel. To the southward is the chapel of Santo Domingo.

Maillen island. Maillen island is formed of regular hills, whose maximum height is 224 feet; on its western part, which is called point

Alta, or Pucheguin, Maillen is about 3 miles long from east
to west, and about 2¾ miles wide north and south. Southward of point Alta is a cove, called El Surgidero, which is
open to the westward and subject to eddies, but vessels can
anchor in it temporarily; to the SE. of this cove is Candelaria estuary, open to the SW., the principal one of the
island; at its head are two chapels and a stream of water,
the only one on Maillen; it is formed by the collection of
several small brooks, and is used to drive a mill; ½ mile to
the SE. of Candelaria is point del Banco, the southern point
of the island; next is point San Pedro, its eastern extremity; and, finally, to the north is the cove of Puqueldon,
from which a reef runs toward Tenglo; it is about ¾ mile
long, bottom coarse sand, mixed with shells and pebbles;
this reef is, however, outside of the line which the steamers
take when passing through the Maillen channel.

Maillen has been inhabited since 1868, and has at present
about 900 inhabitants. Its shores are formed by large
stones and rocks of greenstone; the soil is of alluvial
deposits, which cover the granite completely. Shell-fish
are extremely abundant, and form part of the nourishment
of the people, who go in the summer to cut wood in the
estuary of Reloncavi and at the foot of the Calbuco.

Capirahuapi island is situated to the southward of the
point del Banco, from which it is separated by a channel
730 yards wide, which cannot be used even by boats at
low water, as then the island is connected with Maillen by
banks of sand and stone; during high water, boats can pass
through without difficulty.

Capirahuapi is 1,000 yards long from NW. to SE.; its
western extremity is formed by a cliff 55 feet high; its
opposite extremity is a gentle slope, undulating, and covered
by prairies, on which numerous herds graze; it has two
fresh-water springs; its shores are formed of pebbles and
large stones. The NW. point, called Blanca, is level;
owes its name to the oyster-shells which cover the beach.
To the northward and south-eastward of the island the
water is shoal.

The channel between Maillen and point Panitao is clear,
and deep enough for all classes of vessels. The bank which
is indicated on the chart as extending 1 mile to the west-

ward of Maillen does not exist; the bottom is irregular, but the least depth, reduced to the lowest tide, is 5½ fathoms. Vessels should not approach too close to the shores; the least distance between which is at points Alta and Panitao, 1,073 yards. Panitao and Maillen banks are outside of the natural route of vessels.

Tides.

The strength of the tides in this channel varies from ½ to 1 knot per hour; the ebb is the strongest; the flood is sometimes insensible.

Bay of Chin-quiu.

The bay of Chinquiu is formed by point Panitao and the south coast of the island Tenglo; it is tolerably deep, but there is very little water ½ mile from the land; at its northern extremity is Chinquiu islet, which forms a small harbor in its western part, where a small, unimportant brook empties. In the center of the bay, near the continent, is Caullahuapi islet, which is connected with the coast by banks which are dry at low water.

Islands and channels of Tenglo.

Tenglo is the northern island of the gulf of Reloncavi; it is 3 miles to the northward of Maillen, is a little over 2 miles long NNE. and SSW., and has a mean breadth of 870 yards; it is formed of hills of moderate height, whose elevation increases to the north cape, which is 260 feet high; it is inhabited and cultivated.

The channel which it forms with the continent is narrow; in two places it is less than 1,000 yards; in its center, to the SE. of the digue, its maximum breadth does not exceed 559 yards. At the first turn, coming from the N., (that is to southward of the digue,) there are 1¾ fathoms at low water; this increases to 3 and 4 fathoms at the second turn. From thence to the southern entrance the coast of the island is stony; low points and covered rocks run out from it; the beaches are obstructed by fishing weirs, which are constructed of stones and of stakes interwoven with the branches of trees; they are regular reservoirs of fish. The coast of the continent is clear and deep. The shoal water and the sudden turns of the channel only allow of its navigation by small vessels.

A bank runs out to the SE. about ⅓ of a mile from the NE. of the island; at its edge are 2 fathoms, sandy bottom; a little to the south and SE. of this bank the depth is from 5 to 25 fathoms. The vessels which unload before going to the digue anchor here. The bank of the point uncovers for about 328 yards at low-water springs; the coast of the continent is also shoal for some 300 yards off shore, which must be kept in view when wishing to enter or leave the anchorage of the digue.

Just before reaching the NE. point of the island, there is a small digue; at its entrance there is 7 fathoms of water, muddy bottom, and close alongside of the shore there is 2 fathoms. Vessels should anchor to the northward of the digue, in 4 fathoms, about 150 yards from the shore. Vessels of moderate size must moor head and stern, as there is no room to swing. The anchorage is excellent, and sheltered from all directions. The bottom is a gently-inclined plane, formed of pebbles, muddy sand, and shells. The beach of the continent uncovers for nearly 166 yards.

Anchorage.

Tenglo is the port of Melipulli; the steamers of the P. S. N. Company always anchor there and repair their vessels; they are obliged, however, to haul in; a good dock could be built with little labor.

To the NE. of the passage of Tenglo is the important town Puerto Montt, the capital of the province of Llanquihue, and also the northern point of the gulf of Reloncavi. The shore is generally low, and uncovers from 166 to 266 yards at low-water springs : the northern shore is filled with isolated rocks; the western is formed of small stones, and has some outlying rocks in front of the N. point of Tenglo island. The anchorage is good and sheltered against SW. winds by Tenglo island, but entirely open to the southward. Off the town is a submerged plateau, on which vessels can anchor, about ⅓ mile from the city, in 17 fathoms, bottom fine black sand, with the NE. extremity of Tenglo bearing S. 64° W., the western extremity of the town N. 34° W., and the cemetery N. 76° E. This plateau is very steep-to, especially on the SE.; it must be approached with caution, as the beach uncovers for a long distance at low-water springs; the water then deepens suddenly to 7, 10, and 14 fathoms, and, if the anchor is dropped in less than 14 fathoms, vessels with a long scope of chain are apt to touch on the beach with a south wind.

Puerto Montt or Melipulli. *

* Melipulli, "four hills." The Chilians often use this name for Puerto Montt.

To be in good position on the bank, the low, pebbly beach at the entrance of the small harbor of Tenglo must be in line with the black house situated to the south of it.

When the wind is from the south boats cannot land on the beach, but it is said that the wind does not endanger vessels at anchor.

Light.

A small light is shown from the center of the town, visible about 3 miles.

Description.

The town of Puerto Montt, founded in 1853, is 15 feet above the water, and in front of some hills of a very remarkable formation, which shelter the town from the N. Its location was formerly swampy, and thickly wooded; now there are many fine houses surrounded by verandas; the streets are straight, and about 65 feet wide; the hills surrounding the town have been cleared, and are covered with vegetation. Puerto Montt has 2,500 inhabitants, of which 1,000 are Germans, who are the principal colonists of this part of Chile, and they agree well with the natives. Owing to the growing importance of Puerto Montt and its surrounding country, the territory has been erected into a province under the name of Llanquihue. A fine plaza, churches of different denominations, a library, schools, machinery for distilling liquors, breweries, tanneries, and machinery for making oil, indicate the prosperity which the industry and intelligence of the inhabitants has created. Fresh meat can be bought at low rates, as also vegetables and fresh water. There are two pilots. It has a ship-yard for repairing, and capacity for building vessels of 100 tons.

Tides.

It is high water, full and change, at Puerto Montt at 12ʰ 48ᵐ. Springs rise 18 to 20 feet, neaps 14 to 15 feet. These figures are averages; the tides are very irregular on account of the winds. The rise varies between 10 and 22 feet.

Lake Llanquihue.

Lake Llanquihue has a circumference of about 125 miles, and is 8.6 miles to the northward of Puerto Montt; it is separated from the gulf of Reloncavi by an undulating ridge 460 feet high. A road 9¾ miles long leads from Puerto Montt to Port Varas on the southern shore of the lake, the center of all the German settlements; the lake measures 19⅔ miles north and south and 24¼ east and west; it is 138 feet above that of the gulf of Reloncavi, and is of great depth. Cox states that no bottom was found with 197 fathoms. The winds are often so strong that navigation is interrupted. The N., S., and W. shores are low, wooded, and overlooked by hills covered with houses. The lake is covered with small vessels; a steamboat of 80 horse-power performs the postal service between Port Montt and Valdivia. To the N. of the lake a small colony, called Port Octai, has been founded; the steamer stops there; this port is difficult to enter but well sheltered. Several small rivers empty into the lake, which has but one outlet, the river Maullin, which leaves it on the SW. The eastern shore of this lake rises abruptly; to the NE. is the volcano Osorno, 7,265 feet high, and to the E. Calbuco, 5,570 feet high; they are separated by a swampy gorge, whose waters run into lake Todos los Santos; through this gorge mount Tronador, with its three peaks, and farther to the north the snow-covered peak Bonachemo can be seen. Port Varas, the southern extremity of lake Llanquihue, is in latitude 41° 19' 40'' S. and longitude 72° 56' 11'' W.

There are two routes between Chacao strait and Port Montt, one outside and the other inside of the islands. *Directions between the strait of Chacao and Port Montt.*

The one outside of the islands, to the south of Tabon and through the strait between Puluqui and Queullin, is very inconvenient on account of the fogs, squalls, and rains which shut in the shore. Amnistia bank south of Tabon, and the banks in the center of the gulf of Reloncavi, badly explored, and doubtfully located, are serious obstructions to this navigation. Sailing-vessels alone are forced to make this tedious passage.

The inside route is recommended as preferable, but a pilot should be taken either at Ancud or Puerto Montt. Steamers can pass between Abtao and the continent, through the Abtao channel. The buoys recently placed in the channel between Abtao and Lagartija render this channel preferable. They should then keep to the northward of Lagartija island and Lami bank, then through the opening between Quenu and Calbuco as far as El Fuerte; from thence along the NW. coast of Puluqui, to pass between this island and Tautil; then to the eastward of the Tenglo and Maillen islands through the Huar passage to Puerto Montt. The passage to the westward of Tenglo has an abrupt turn which a large vessel would find difficult to pass, but that of Maillen presents no inconvenience by keeping in midchannel between the shores.

M. Gormaz gives the following instructions for going from the strait of Chacao to Calbuco, passing between Abtao and Lagartija :

Leaving the middle of Chacao strait, steer S. 69° E. until Lagartija island and point Nahuelhuapi, on Abtao, bear, in line, N. 65° E., which will be when about ¾ mile from point Quilque, on the same island ; then steer S. 85° E. until Lagartija bears N. 29° E., when change the course to N. 15° E. until point Martin, the northern extremity of Quenu, bears S. 83° E., in line with point Blanca on Chidhuapi, and Pilquen hill on Abtao. On this course Lagartija will be passed within ¼ mile and in from 5½ to 27½ fathoms; sand and shells. Steer S. 83° E. for point Martin as far as the entrance of Quenu passage, which will be cleared by making N. 85° E. or E., keeping rather nearer the shore of Quenu than to that of Calbuco. When the small mountain on the SE. part of Calbuco bears WNW. Calbuco channel is entered by steering N. 34° E., which will lead to the anchorage, either off the town or in the estuary of Huito. After leaving Lagartija there is never less than 17 fathoms of water.

The passage between Corva and the Lami banks can also be used, but it is not to be recommended, unless it were buoyed, as the landmarks are not easily distinguished, and it lengthens the route.

In the paragraphs relating to these different passages all the details will be found ; it is always necessary to keep account of the tides when going through these straits.

Coast between port Montt and Reloncavi estuary. The coast between port Montt and the estuary of Reloncavi does not offer any convenient anchorages. Besides the great fall of the tide, and the rocks and cays with which it is studded, it is completely open to winds from NW. to S.

From the cemetery at Puerto Montt the coast trends S. 52° E. with many indentations. After the first turn is the Rio Chico de Pilluco, which comes from lagoons about 1¼ miles in the interior, and 460 feet above the sea.

To the southeastward of this river is point Peñas, 490 feet high, which overlooks all the surrounding hills. To the eastward is the unimportant river Pilluco; farther to the southward and eastward the coast falls, and forms point Coles or Coihuin. On all this coast there are cultivated spots, but the soil is poor.

From point Coles to point Cheuquemo the coast forms a large cove, about 4½ miles wide at the entrance; at its head the land is low, swampy, and inundated by the tides for a considerable distance. The cove is filled with the deposits from Coihuin river, which empties into it. The bank commences about 2¾ miles from port Montt, and unites with Coihuin bank off the mouth of the river. This bank extends in the center about 2 miles from the coast; its outer edge is steepto; on one side of a boat 2 fathoms will be obtained, and on the other 5½, and a little farther out 14½ fathoms. The upper part of the bank is of soft mud, and it only uncovers at low-water springs; at high water it is entirely covered; small vessels and boats pass over it; it connects with the coast at low water about 1¼ miles to the southward of point Cheuquemo. *Coihuin banks.*

During high water Coihuin river is accessible from the gulf through three openings; the estuary Tralhuempue to the west, that of Muermos in the center, and the mouth of the Coihuin to the east. To enter the river a pilot is necessary; it is navigable for 4½ miles, and is the outlet of lake Chapo, which is about 13¾ miles from the coast. *Coihuin river.*

From point Cheuquemo the wooded coast descends hardly perceptibly for about 3 miles, as far as Quillaipe island, which is nearly circular, with a diameter of about 666 yards; it is low, wooded, and separated from the continent by a narrow channel. *Pichi Quillaipe island.*

Point Quillaipe is a little more than 2 miles from this island. To the southward of it is a large cove, of the same name, which is dry for more than half of its extent, during low-water springs; before it are two rocks detached from the coast, the one 550 and the other 1,100 yards. *Point Quillaipe.*

From point Quillaipe the coast runs S. 37° E. to point Metri, which is high and wooded, forms several small coves, and for about 1 mile off shore is bordered by numerous islets. *Lenca cove.*

From point Metri the tide recedes from 340 to 440 yards, though ½ mile from the coast the depth is from 11½ to 14½ fathoms. Pajaros islet, from which Lenca cove commences, is a small island with a circumference of about 1,000 feet;

it is connected with the continent at low water; the cove extends to Chaica hill, situated at the foot of the mountains which form the northern entrance to the estuary of Reloncavi. At low-water springs it is almost entirely dry, leaving only two small channels through which the rivers Lenca and Chaica empty; there is a chapel on the shores of this cove. The entrance of the estuary of Reloncavi is 17 miles from Puerto Montt.

Twenty miles NNE. of this coast is the volcano of Osorno. Calbuco or Quillaipi is 11 miles to the right. Calbuco is separated from the estuary of Reloncavi by the Sierra de Rollizo and its branches.

Estuary or river Reloncavi.   The estuary or river Reloncavi was explored by Moraleda in 1795. The Chilian expedition of 1871 in the Chilian corvette la Covadonga, proved that his work was deficient in many particulars, and that great changes have taken place, especially at the head Ralun bay; the idea given of the topography of the surrounding country was entirely erroneous.

This estuary separates the Yate, a branch of the Andes, and its chain from the Sierra Rollizo and its spurs, which form an almost isolated group. Its entrance is in 41° 44′ S. and 72° 41′ W.; it runs east for less than 11 miles, then turns to NE. and N. by E. to the parallel of 41° 22′; its entire length is 30 miles.

The estuary of Reloncavi is a gorge in the Andes, filled by the waters of the Pacific, a phenomenon due to the sinking of that part of the Cordillera contemporary with the formation of the archipelago of Chiloé. Its mean width is 1½ miles, but this is reduced to ½ mile from point Relonhue, 5 miles south of Ralun. It has several small branches, which are formed by ravines in the mountains; they are almost unknown, and only frequented by wood-cutters.

The shores are rocky and steep, only presenting a gentle slope at the mouths of the rivers and at some few isolated points. The mountains which overlook them are steep and very thickly wooded, and have a mean elevation of 4,270 feet. Many of their summits, especially those of the interior, attain to a height of 4,650 feet, the limit of perpetual snow. Only mount Castello, 4,935 feet, and Yate, 6,950 feet, are above the average elevation of the Andes in this latitude.

The mountains on the south and east coast are more elevated than those on the north and west coast. The former are not higher than 4,590 feet, and are connected with Calbuco by low hills; at the height of 3,940 to 4,260 feet they are entirely barren; the snow does not adhere to their declivities; above this height their southern slopes are covered with snow; there are some glaciers in their ravines.

The depth of water in the estuary is great; at the entrance of the bay of Ralun, at the northern extremity, there are 33½ fathoms, muddy bottom. To the southward the depth increases; off Cochamo it is from 110 to 115 fathoms, between the bay of Yate and Marimeli islets 165 fathoms, and at the entrance of the estuary 250 fathoms is the maximum depth obtained. The bottom is always green mud, very fine and adhesive. In the gulf of Reloncavi the water is not deeper than 150 fathoms, and the nature of the bottom is the same. The gulf is not so deep as the estuary; there are no hidden dangers in the estuary.

For several years the inhabitants of the archipelago of Chiloé have visited this locality to cut wood, which is extensively exported, as in the archipelago of Chonos and Guaitecas. Some permanent buildings have been erected.

The winds generally follow the direction of the estuary. Winds. Combined with the tides they produce remarkable effects. With the winds from the western quadrants there is a heavy sea in the entrance as far as Sotomo; the currents of air contracted by the high mountains which surround the estuary drive into it with a violence often extraordinary, and follow the windings of the channel. The N. and NE. winds which come from the head do not usually pass the Marimeli islets, but those from S. to W. blow through the whole estuary and escape through the openings of the Petrohue and and Reloncavi rivers. With high winds from N. to W. the estuary is calm in the spring, but when the northerly winds blow in gales they follow all the turns of the channels. They form water-spouts and cause such violent squalls that they would capsize any vessel under sail. These effects are worse with the flood-tide.

With a SW. to W. wind, when the wind is blowing fresh outside, during the-ebb tide, the phenomenon called *curanto* takes place at the entrance; the crests of the waves are taken up and thrown with violence before the wind like a

heavy briny rain. As soon as the squall is over the whole
atmosphere cracks like ice when it is breaking up; no vessel could then carry sail without being capsized or dismasted.
These squalls are announced by a thick smoke, which the
people of that region call *el curanto*, which they dread, and
with reason. This phenomenon takes place even with an
ordinary SW. wind, though with less violence; even then
it is dangerous, especially for boats, which must run in
under bare poles or a close-reefed sail; it is necessary to
watch the state of the weather and of the tide.

In the winter, the N. winds, which alternate with the
SW. winds, are the most dangerous. It often happens that
there is a fresh wind in the estuary when it is calm in the
gulf of Reloncavi.

Directions.

In the mornings, there is frequently a calm or light E.
wind, which allows vessels to run in and out; but if the N.
wind blows outside, which can be easily ascertained by
watching the clouds, it is necessary to await a change in
one of the inlets on the northern shore of the estuary.

To enter with a N. wind, it is indispensable to take the
north, or Horno shore, and follow it as close as possible,
in order to avoid the swell and force of the wind, which
beats against the south coast; the swell is short and quick,
and breaks on board easily. With a south wind, on the
contrary, the south coast must be hugged for the same
reasons. It seems to be indispensable to have steam in this
estuary. The shores are perfectly clear, and the few remarks necessary will be made while continuing the description.

Tides.

The tides are very variable; they depend on the place
and age of the moon, but, above all, on the quantity of the
rains; the ebb is a little stronger than the flood. In the
rainy season, the flood is very weak, and the ebb, owing to
the torrents, acquires great force; it has a velocity of $2\frac{1}{2}$
miles, and often exceeds 4 and 5 miles between Marimeli
and the north shore. In summer, the flood runs about $1\frac{1}{2}$,
and the ebb 2, miles.

In the inlets the changes of the tides are of little importance; at Sotomo, the force is a little more than 1 mile; at
Cochamo, $\frac{2}{3}$, and at Ralun, $\frac{1}{4}$ mile. The local and atmos-

pheric conditions are always the principal causes of the variation in the currents. The rise is nearly 22 feet.

Caicura islets are situated at the entrance of the estuary, Caicura islets. and are divided into two groups; that of the north, called Piren, consists of two islets and some few cays; there is no inlet even for a boat, though they can land during good weather; the islets are covered with shell-fish. The southern one, called Caicura, gives the name to the entire group; it is a little more than ½ mile to the south of Piren. The principal island, Caicura, is rocky, steep to the south and west, well wooded, and cultivated in spots; its height in the center is 154 feet. To the eastward of it are four small, rugged islets, covered with vegetation, and separated by numerous rocks.

At the northern extremity of the principal island is an elongated creek, which can hold sloops and boats; it is open to the northward, has an excellent anchorage in soft mud, and is a sure shelter in heavy gales. This creek is an excellent resource for the vessels which frequent the entrance of the estuary, and as a stopping-place during calms and contrary winds. This island is of greenstone, with a slight covering of earth over a bed of shells; it has fresh water in a well at the summit of the island, which is only dry during very dry seasons; these are fortunately very rare.

The channel which forms the two groups is very deep; there are, however, three submerged rocks about ⅓ mile from its entrance, off the shore of Caicura, on which a vessel could readily run at low water; at ⅓ flood they are not dangerous for small vessels, but they should in all cases keep close to the cays of Piren. Large vessels should not try this passage.

At the N. entrance of the estuary is a hill, called the Morro del Morro del Horno; it is the beginning of a perpendicular Horno. shore bordered by mountains, which are all overlooked by the double peak of Cuitue, 4,460 feet. The shore is cut by the almost inaccessible small creeks, El Milagro, Cuitue with an islet, and El Cajon. The shore trends first ESE., then ENE. The sides of the mountains are covered with a thick forest of alerce.

At Llecumo, 8 miles from Horno, there is a sand-beach at

the mouth of the river Chilco, on which boats can land at low tide. The waters of the lagoon of Chilco are at this point joined by those of another torrent.

Marimeli islets. Marimeli islets stretch for 2½ miles NE. and SW.; they are of moderate height and covered with trees; their SW. extremity divides the channel into two equal parts, about 9 miles to the eastward of the entrance; at the NW. extremity they are separated from the north coast by a channel filled with rocks, which is only from 220 to 330 yards wide. It is, therefore, best to pass to the southward of the islands, more especially as the current in the north channel is always very strong.

Between the NE. point of Marimeli and point Sotomo, is the Estero d'Arrayan, then the river of the same name, which is the outlet of a lake. The hills which border this coast are about 4,100 feet high.

Sotomo bay. Point Sotomo is 2¾ miles ENE. of the NE. point of Marimeli; it is composed of rocky islets covered with trees; the western one has been called Observatory cay; its latitude is 41° 39′ 36″. To the NW. of this point the bay of Sotomo opens, the first real anchorage which is found on the north coast; it forms a right angle, of which the short side extends ⅓ mile in a NW. direction, the longer one 1 mile NE. A small rock, called Toro cay, is near the latter side. Vessels anchor in 33 fathoms, muddy bottom, with point Sotomo S. by E., Toro Cay N. 40° W. This is less than 400 yards from the land, and sheltered against winds from NW. and SW.

In the southern part of the bay, between high cliffs, and at the head of a gully about 660 feet long, are hot sulphur springs, which rise from the ground 6½ feet above the sea. They have a temperature of 62° and 108°.

Tides. It is high water, full and change, at Sotomo, at 1ʰ 8ᵐ. The average rise is 18 feet, but at equinoctial springs it is 23 feet.

La Factoria. The north point of the bay of Sotomo is formed by an island, separated from the main-land by a narrow channel. From there the shore runs in a straight line to the N. by E. for 5 miles, to Factoria. This is a small settlement built by wood-cutters. There are 55 fathoms of water alongside of the shore.

El Canutillar.

The principal settlement of the wood-cutters is El Canutillar, 2½ miles to the northward of Factoria, where there
is a small beach on the border of a torrent. There are
about twenty houses, inhabited by the men who cut the
alerzal in the mountains about 2 miles distant, and not
far from lake Chapo, whose waters run into the gulf of
Reloncavi.

Vessels cannot anchor off El Canutillar, but must go
opposite, in the bay of Cochamo.

About 1 mile from El Canutillar, between point Pocoi-
huen on the west and point Relonhue on the east shore, the
estuary contracts between the mountains to 1,050 yards; it
then extends 5 miles N. by E., with its banks steep-to on
both sides, to the entrance of the bay of Ralun.

The semicircular bay of Ralun is partially obstructed by
the deposits of the rivers Petrohue, Reloncavi, and that of
Este.

Bay of Ralun

On first entering the bay, vessels must steer NW. for ½
mile, after which they must head for the small cay called
Nahuelhuapi, which is on the northern part of the east
shore; it can be passed on either side, but as the southern
side has only a channel of 183 yards width with 12 fathoms
of water, it is best to pass to the northward, but very close
to the island. After this is passed, steer for Nahuelhuapi
cove, which opens to the southward, where the anchorage
is in 17 fathoms, with the cay bearing ENE.; this position
is within 400 yards of the shore. There are not less than
24½ fathoms in the channel leading to the anchorage.

The bay of Ralun is protected against all winds, and is
the best anchorage in the estuary; but its distance from the
entrance will prevent its further development.

All the southern part of Nahuelhuapi cove is full of mud.
At the end of the immense mud beach are hot sulphur
springs, with a temperature of 90°, having their source near
the low-water mark.

On the north shore the beach uncovers for more than 1
mile. In comparing the map of 1795 with that of 1871, it
will be seen that the bay of Ralun has silted up for more
than 1 mile. It is now perfectly dry at low water where the
Spanish maps show from 30 to 47 fathoms. It is thought
the eruption of the volcano Osorno, in 1790, changed the

8 c

bed of the river Petrohue, and that the eruption of 1835 augmented the effects of the former.

Tides.   It is high water, full and change, at Ralun, at $1^h 16^m$; mean rise, 18 feet ; springs, 19.6 feet.

Basins and mountain chains surrounding the bay of Ralun.   Between 41° and 41° 45′ south latitude, there are many lakes in the depressions of the Andes, which surround the estuary of Reloncavi. On the shores of the most distant lake, Nahuelhuapi, the Spaniards established a mission. The last Chilian expedition has thrown a new light on the geography of this region.

Two principal water-courses empty into the bay of Ralun, the river Petrohue in the NW. and the river Reloncavi in the N. The former has a course of about 16 miles, and is obstructed by many rocks and sand-banks; it is the outlet of lake Todos los Santos. The river Reloncavi is only a torrent 4 miles long, separated from lake Cayatue by a ridge 1,580 feet high, called Cabeza de la Vaca, in whose gorges the river has its sources. The lake just mentioned is 2 miles long and 1 wide, and is connected by a torrent with lake Todos los Santos. This whole country is wooded. The famous Bariloche road, which led to the mission of Nahuelhuapi, passed through the valley of the river Concha to the eastward of lake Cayatue. The level of this lake is 780 feet above the level of the sea, while that of lake Todos Santos is only 370 feet. Lake Llanquihue, to the N. of Puerto Montt, and lake Todos los Santos, are separated by a swampy plain 980 feet above the sea, 6 miles in length. A short distance from it to the northward is the volcano Osorno, and to the south Calbuco mountain; both appear to be isolated. To the N. of lake Todos los Santos is the peak Bonechemo, which is covered with perpetual snow. To the east, in the vicinity of lake Nahuelhuapi, 1,811 feet above the sea, is Tronador mountain, 9,757 feet high. ([1])

([1]) The geographical positions of the different peaks are—

	Lat. S.	Long. W.
Volcano Osorno	41° 08′ 30″	72° 33′
Mount Tronador	41° 10′ 45″	71° 51′
Volcano Calbuco	41° 21′ 02″	72° 39′
Mount Castello	41° 42′ 14″	72° 17′
Lake Todos los Santos, S. extremity	41° 14′ 04″	72° 19′
Lake Cayatue, S. extremity	41° 17′ 13″	72° 19′

To the southward of point Relonhue, opposite El Canutillar, is Cochamo bay, 2 miles wide and 1 long. Vessels can anchor here temporarily, ¼ mile from the mouth of the Cochamo river, in 49 fathoms, muddy bottom. A less depth of water will be found closer to the shore. This cove is very unsafe with SW. winds and an ebb-tide, the eddies and swell being sufficiently strong to capsize boats. *Cochamo bay.*

From Cochamo bay the coast is high and wooded, and turns S. by W. for about 7 miles, to the mouth of the Puelo river, the largest water-course emptying into the estuary. It has its source in the eastern slopes of the Andes, traverses several lakes, and seems to be the most natural highway to Patagonia; it was used for that purpose by Captain Gormaz in 1872. It is navigable for boats, but the current is very strong. *Puelo river.*

About 1½ miles from the river Puelo the river Blanco empties, whose valley separates Castillo and Yate mountains. It has its source in the immense glacier to the eastward of mount Yate. This beautiful mountain, 6,970 feet high, covered by perpetual snow, is the western end of the branch which unites with the Andes in the ESE. It is more elevated than the principal chains in the same latitude; it has three peaks. On the S. and SE. sides the snow reached down to 328 feet above the sea. The Yate can be seen from Ancud. On the side of the estuary it is covered with large alerzales. *Mount Yate.*

From the river Blanco the coast trends WSW. for 10 miles, to point Chaparano, which projects to the NW. and separates two coves. In the western one, called Martin bay, there are 39 fathoms about 400 yards SSE. of a small island; after which the coast runs W. 4 miles to Morro Chico, a small point which forms the southern entrance of the estuary of Reloncavi. *Morro Chico.*

The part of the coast between the entrance of the estero de Reloncavi and the islet Nao is called the coast of Contao. *Coast of Contao.*

Manihueico is the name of a small point and brook situated 2½ miles SSW. of Morro Chico. From it the coast runs S. 19° W., and at a distance of 2¾ miles forms a large cove, into which the sea enters at full tide, and in which the river Contao empties. It is frequented by the wood-cutters. When the tide recedes, it leaves a dry sand-bar of 550 yards extent.

The coast, low and wooded, continues inclining to SSW., and forms the estuaries of Poza de Quidalco and of Mui. These can only be used by large boats. The former is a narrow channel about 666 yards long, ending in a pond of about 500 yards in diameter. The ebb-tide leaves the channel dry but not the pond, which abounds with fish.

Aulen island. Farther to the southward is the Aulen channel, formed by Aulen island and the continent. It is nearly dry during the ebb of the equinoctial springs; in some places it has a width of 300 yards, but it contracts greatly at its southern entrance. The island has nearly 2 miles circumference, is low and wooded, and has at its northern extremity a bank of 550 yards extent.

To the southward of Aulen is Caramuñen cove, which can be used by boats. The point, which is the southern limit of the cove, is extended by a bank about 650 yards.

Point Trentelhue or Aulen. Aulen point, which is 2 miles SW. of Aulen island, is the W. extremity of the continent. It ends in a narrow tongue of sand, which uncovers at low water, and is joined to an islet called Nao. Point Aulen is 11 miles SW. of the entrance of the estuary of Reloncavi.

Nao islet. It is dangerous to approach Nao islet to the northward; its length is 430 yards from E. to W., and its breadth 275 yards from N. to S. In connection with point Aulen it forms an extended anchorage, which is sheltered from SW. winds.

The channel between Nao and Queullin is clear. It is about $\frac{3}{4}$ mile wide, and vessels of any size can pass through it.

The current of the ebb in this channel sets to the south, and has a velocity of 3 to 4 knots. On leaving it, it is divided into two distinct branches; one follows the coast of the continent and the other runs toward the Cháuquis islands. The force of the latter branch diminishes after leaving the island Queullin. The flood-tide is less strong; it never has a velocity over 2 miles, and often after heavy rains it is entirely counteracted by the currents of the rivers.

Queullin or Cullen island. Queullin island is to the westward of Nao; the western part of this island and Puluqui form a clear channel 2 miles wide, which leads to the entrance of the gulf of Reloncavi. Queullin is a little less than 1 mile in diameter, is slightly elevated, and has pointed cliffs on its southern, western, and part of its northern side. The east coast is low, with fine, cultivated hills; the ebb leaves the beach dry for 220 yards, but there are no shell-fish, which obliges its 140 inhabitants to cut wood on the continent and fish on the neighboring banks. There is but one cove on this island, in its eastern part; it is called Martin, and lies to the S. of the point of that name, the NE. extremity of Queullin. This cove is open to all winds excepting those from the west; it can only shelter small vessels; larger ones can only remain temporarily during fine weather. Ridges of rock run out for about 540 yards from points Martin and Chollehuaihue, and from point Huin; on the western part of the island a sand-bank makes out about the same distance.

The bank of San José is 1¾ miles to the N. of Queullin, and is only dry at low-water springs. From a boat anchored on it in 1 fathom, the following bearings were taken: Point Perue, (east point of Puluqui,) S. 74° W.; point Redonda, (east point of Huar,) N. 9° E.; point Martin, S. 25° E. In some places this bank is covered by cholgas to a thickness of 3 feet. The channel between it and Queullin is clear and deep; there is no bottom with 40 fathoms; the channel on the Puluqui side is also clean, and 3 miles wide, but it is best to keep close to the island. *San José, or Sepuhuapi bank.*

The Hualayhue coast extends between point Treutelhue and the estuary of Comau; it inclines rapidly to the SE., and, like that of Contao, is low and wooded. One and a half miles from Treutelhue is Rolecha cove, and, farther to the southward, Queten cove. The latter, which is sheltered from N. and S. winds, offers a good temporary anchorage for all vessels. *Coast of Hualayhue.*

About 3 miles farther to the SE. is point Chauchil, from which a bank projects 530 yards; it abounds in shell-fish, and is the resort of fishermen. From Chauchil the coast inclines a little to the E., forming the coves of Lleguiman and Cheñua, which are completely open to the S.; in the former the tide recedes nearly ⅓ of a mile; point Paehuapi is rocky.

Santo Domingo bank lies off Lleguiman cove about 1½ miles from the coast; it only uncovers at low-water springs; the part which is then dry has an extent of about 330 yards; it is formed of sand and round stones. From it point Chau- *Santo Domingo bank.*

chil bears N. 34° W.; point Pachuapi N. 74° E.; point Ascension N. 21° W. The channel between the bank and the continent has a depth of about 24 fathoms in the middle. This bank, as that of San José, is the terror of the fishermen, who believe them to be haunted.

Cove Hualayhue.   Hualayhue cove is at the entrance of the estuary of Comau and to the eastward of point Hualayhue. From point Hualayhue the coast runs N., forming this spacious cove, which is sheltered on the east by the islands Yanchid, Senhuao, Macun, &c. To the NW., at the bottom of the bay, is the small island los Manzanos; the channel which separates it from the continent is dry at low-water springs; at that time the beach is uncovered for more than 1 mile Behind the island of los Manzanos is a chain of mountains, from 1,600 to 3,200 feet high, which ends at the entrance to the estuary of Reloucavi; between this chain and the coast is a low, extended plain, on which the alerce is cut. The small river of Hualayhue empties into the center of the bay; it is navigable for a short distance by boats and balsas, which tow the logs of wood.

Comau or Leleu estuary.   At point Hualayhue, the northern point of the entrance of the estuary of Comau, and 11 miles from Trentelhue, a group of small islands commence, through which it is said there is a passage, as also some interior ports; but the proper passage is between the island Llancahé and Comau head, the southern point of the estuary, and either to the N. or the S. of the small island Lilihuapi. This island is clean, but a little over ½ a mile to the eastward of it is a cay, called la Ballena, which is above high water, and is dangerous for those who take the N., or Marillmo channel; to vessels taking the southern, or Comau channel, it is less dangerous.

After passing la Ballena the estuary Cahuelmo is seen to the NE., and to the SE. the estuary of Comau. The latter is 18 miles in extent. From Lilihuapi it extends SE. for 15 miles and ends in a funnel-shaped bay 3 miles in diameter; its depth is 22 fathoms. In the NE. corner, at the head of this bay, vessels can anchor in 15 fathoms, off a sand-beach, to the southward of which the river Bodudahue empties.

Vessels can also anchor in Leptepu creek to the southward of the estuary and about 3 miles west of the entrance to the mouth of the Bodudahue. The river Leptepu, which can be navigated by boats for 4 miles, empties into this creek. The estuary is bounded by high mountains covered by alerzales. The flood has a velocity of from $\frac{2}{3}$ to 1 mile, and the ebb from 1 to 2 miles.

The river Bodudahue is about 220 yards wide at its mouth; River Bodud abue. its entire length is 56 miles. At low water it has from $1\frac{1}{2}$ to $2\frac{1}{4}$ fathoms at the entrance, and is navigable without obstacle as far as Malpaso, where are the first rapids and the end of tide-water; navigation then ceases and the river is but 60 yards wide and runs through a thick growth of alerza-les. It has its source in the western part of a reservoir or circular lagoon, 220 yards in diameter, in latitude 42° 23′ S.; longitude 71° 47′ W. To the southward is a cascade 390 feet high, one of three which form an échelon; the other two are 560 and 650 feet above the former. The last serves as a weir at the union of two torrents, one of which is in-terrupted by a cataract of 650 feet. This series of cascades seen from a mountain opposite is a beautiful spectacle.

The Andes appear to fall here and give the appearance of an easy road to Patagonia.

From this estuary the description of the coast is taken Coast to south-ward of the estu-ary of Comau. from Moraleda, and therefore very doubtful. The high coast runs about WSW. for 8 miles from Comau hill to point Chulao, then to the S. with several bends. It is indented by the estuary Reñihue, which extends into the coast about 6 or 8 miles. Thirty-five miles from Chulao point is point Lieuleu, which lies opposite the E. point of Talcan island and is separated from it by a channel hardly 2 miles wide with a depth of 85 fathoms; bottom coral and broken shells.

About $2\frac{1}{2}$ miles south of point Lieuleu is Vilcun mount- Mount Vilcun. ain, in the shape of a sugar-loaf. It can be seen over the SE. extremity of Talcan; is thickly wooded to the summit and rises directly at the edge of the sea. To the southward of this mountain is a deep inlet with a small island before its entrance.

At 13 miles N. 88° E. of mount Vilcun is the volcano of Chayapiren, 8,000 feet high; 26 miles SSW. of the latter is the mountain Corcovado, 7,500 feet high; and 18 miles south of this is Yanteles, 6,600 feet above the sea; these mountains are from 6 to 12 miles from the coast. From

abreast of them the coast trends nearly south for 21 miles
to cape Corcovado, where there is a temporary anchorage.

Tictoc Bay.   Tictoc bay is 30 miles farther south; according to Moraleda's charts, it is formed between a small archipelago and a
small indentation in the continent. To reach it when coming from the N. the coast of the continent must be followed,
leaving all the small islands to starboard; care must be
taken to avoid some rocks in the channel, $\frac{1}{4}$ mile from the
continent, and to pass close to the islet which is opposite
the foul point of. Tictoc. The anchorage is in 22 fathoms,
soft mud, under the northern part of the continent, Mira-
hulay mountain bearing N. by W.

Tides.   It is high water, full and change, at $1^h 45^m$; rise 10 feet.

Port Piti Pa-   Point Huala separates the bay of Tictoc from the estuary
lena.
of Piti Palena. This tortuous inlet is about 5 miles long;
but the perfectly sheltered part is only 3 miles long and $\frac{1}{4}$
mile wide. All its NE. shore is bordered by high mountains, sandy at the summits and covered with snow. Its
SW. shore is an immense swamp, partially covered during
high water. The anchorage is beyond the first turn, about
3 miles from the entrance, in 16 to 20 fathoms, sandy bottom. Everywhere that it was examined, the SW. coast was
found to be shoal 400 yards out, whereas the NE. shore is
steep-to.

Tides.   It is high water, full and change, at $12^h 23^m$; rise 10 feet.

Port Santo Do-   From Piti Palena the coast is foul to 4 miles from the
mingo.
shore; it trends SSW. for 20 miles, where a channel, which
is formed by the large island Refugio and the continent,
commences; in it are two groups of islands. About 3 miles
from the entrance is port Santo Domingo, where vessels anchor in 25 fathoms, 400 yards from the shore, behind a
swampy point, which is often inundated during the rainy
season or during N. winds. Vessels can moor head and
stern a little farther in, between an island called Long island
and the continent, under Calvario peak, in $9\frac{3}{4}$ fathoms. The
coast of the island is clear. The channel is only 200 yards
wide. There is a shoal with $3\frac{1}{4}$ fathoms water over it.

Tides.   It is high water, full and change, at 2^h; rise 7 feet.

The description of this part of the coast ends here. From
the estero of Comau they are taken from the surveys of
Moraleda in 1794, and are incomplete.

The passage of the frigate Numancia has proved that the route inside of Chiloé is perfectly practicable. It should not, however, be attempted except with the object of avoiding violent N. winds, as there is almost incessant rain; the barometer and weather are often affected by local causes; and often bad weather is experienced in the passage when it is comparatively fine 60 miles out.

A vessel taking this passage should pass outside of Huamblin island and then steer for Huafo, which should be closed to within 5 miles before keeping for the entrance of the channel; as otherwise she would be liable to be set to leeward and toward the dangerous shores of the Guaitecas by the strong tides, which are said to attain a velocity of from 3 to 4 knots between Huafo and the Guaitecas; keep as close as prudent, 6 or 7 miles, to the coast of Chiloé, and haul close round for San Pedro, giving the Canoitad rocks a berth of 2 miles. From San Pedro steer ESE. for 8 or 9 miles, before keeping to the northward, to avoid the bank of Laytec island. Point Centinela should not be approached nearer than 3 miles, and a lookout must be kept for Numancia bank. After passing these, steer for the passage between Quehuy and Chaulinec, keeping well over to this island. From this, pass 1 good mile to the W. of Quenac and steer between N. and N. 5° E., leaving point Tenoun one point to starboard; when within 2 miles of it the vessel must be brought on the line drawn through the N. point or summit of Meullin and the western point of the SW. island of Chauqui; this course will clear the channel between point Quicavi and the reef off the NW. Cháuqui island, as also Pulmun reef.

A good lookout is necessary between Linna, Meullin, the Cháuquis, and the continent of Chiloé. If it should be desirable to anchor at Quenche or in Oscuro cove, the Caucahue channel should be followed; however, in case of necessity, vessels can anchor temporarily to the SE. of point Tres Cruces. The passage described, called the western, has the advantage that the land is in sight, and the violence of the N. winds, as well as the short and heavy seas from the gulf of Ancud, are avoided; it is also comparatively clear.

The eastern passage is wider; pass between the Deser-

tores and Chaulinec, looking out for Solitaria islet, and
then to the E. of the Cháuquis; but the fogs and northerly
winds, which obscure the land, render this passage inferior
to the first.

Fishery in the
archipelago of
Chiloé.

The inhabitants of Chiloé and the islands in its vicinity
have but two occupations: wood-cutting and fishing. Wood-
cutting is the usual occupation of the men, and fishing that
of the women and children.

As the tide rises and falls some 10 feet, it leaves large
tracts of beach and banks uncovered, which abound with
masses of shell-fish; placing these in reservoirs washed by
the tide, they are kept for winter provision. The *curanto*,
a dish peculiar to this island, is made of them.

The quantity of shell-fish is diminishing around the inhabited islands.

Tariff of pilots in
the gulf of Ancud
and the channels
around Chiloé in
1871.

The price of pilotage does not vary with the distance.
For a vessel drawing 6 feet or less, $40; from 6 feet to 10
feet, $50; from 10 feet to 15 feet, $70; from 15 feet to 20
feet, $90; from 20 feet to 23 feet, $110; 23 feet and over,
$130.

CHAPTER V

Variation from 19° 03′ to 18° 31′, in 1876,'increasing annually about 1′ 30″.

Continuing along the coast to the northward of Chiloé, Islets Sebastiana and Carelmapu. it may be well to mention again that the islets Sebastiana and Carelmapu should be avoided, as the tides sometimes produce violent eddies near them, and when the swell from seaward meets the ebb-tide it produces an ugly short sea to the NW. of these islands.

Vessels should keep close to point Coroua, and avoid Sebastiana. By avoiding a sand-bank ½ mile SE. of the eastern point of Sebastiana, there is plenty of water between it and cape Chocoy; vessels can also pass to the eastward of the Carelmapu islets, but the westerly swell is so heavy and the tide so strong that the passage should not be attempted without a good local pilot, a good working breeze, and a favorable tide. These islands should not be approached to the westward nearer than 3 miles, and it is more prudent to give them a berth of 4. About 6 miles to Estuary of the Maullin river. the eastward of the westernmost of the Carelmapu islands, is the entrance of the river Maullin, opening between Amortajado head and point Quennir.

To the southward of the entrance to Maullin river is the Peninsula of Amortajado. peninsula of Amortajado, of a yellowish color and peculiar appearance; the land, which from cape Chocoy is covered with small hills, is low at the commencement of this peninsula, after which it rises rapidly for the first third of its length, when, after falling again, it rises and ends in a pyramidal promontory, vertical toward the sea, bare of vegetation, and stony at its extremity. The top of the peninsula is covered with bushes. A chain of rocks extends from it about 300 yards, the last of which are under water. The extremity of Amortajado is 8 miles from cape Chocoy.

Doubling Amortajado to the eastward, the bay of Puelma Pu.'ma bay. is to the south; it has from 3¾ to 7 fathoms in its entrance, and shoals rapidly toward the land. A small vessel can anchor temporarily in 4 fathoms, sandy bottom, to the eastward of, and as close as possible to, the northern point of the peninsula. To the SE. are some coal-mines, near which flows the river San Pedro Nolasco.

Point Pangal. On the other side of this river is a low sand-beach 2½ miles long, which ends in point Pangal. This beach, behind which are masses of sand and plains which extend to the foot of the Cordillera of the Andes, is entirely exposed to the swell of the ocean. The breakers commence 400 yards, and often farther, from the shore, the water being shoal to a distance of ½ mile.

Point Changüe. Point Changüe, NNW. of point Pangal, limits the interior entrance to the river Maullin, whose navigable breadth is contracted between this point and point Pangal by a chain of rocks above water. The two southern and principal rocks are called the Dos Amigos; between these and point Pangal are some submerged rocks which obstruct the channel, which is here but ¼ mile wide, and 2 fathoms at low water.

The river Quenuir empties into the estuary of the river Maullin, between points Changüe and Quenuir; its volume of water influences considerably the current at the entrance of Maullin river.

Point Quenuir or False Godoy. N. by E. 1¼ miles from Amortajado and about the same distance from point Pangal is point Godoy, thickly covered with trees, the eastern limit of the high land which starts from point Godoy; it is rendered more remarkable by a small islet, 400 yards to the westward, of a yellow-gray color, called Javier Igor.

Point Quenuir is clean on the south and southwest, but from the SE. a chain of rocks, the largest of which are submerged, makes out in that direction about 500 yards.

In 1859 these rocks were marked by a buoy, which has been washed away and replaced several times since; it was about 23 feet S. 14° W. of the outer rock. Javier Igor just open of the S. extremity of Quenuir point bore N. 83° W. from it, and the N. extremity of Amortajado S. 31° W.; the depth in which it was anchored was 2½ fathoms at low water. If the buoy is not in place the extremity of the rocks will be known by the bearings.

To enter the river Maullin, a vessel should pass 1½ miles from point Godoy and then head for the island Javier Igor; when 1½ miles from this island keep a little to starboard, passing it, and Quenuir point within 400 yards; continuing this course, which should be E. 2° S., the buoy will bear to the NE. and will be passed 100 yards to the southward; then steer for the Dos Amigos, which can be seen at all times from the buoy, from which they are about 2 miles. On this course there is not less than 3½ fathoms.

Off the mouth of Quenuir river the current is generally very strong and divides into two branches; one follows the principal current, and the other, that of Quenuir river, is modified by the ebb or flood. A good lookout must be kept to see that the currents of the flood do not drift the vessel on the banks off the mouth of this river, and that those of the ebb do not set her on the bank, which terminates on the sunken rocks marked by the buoy, situated between this river and point Quenuir. On the other hand, Pangal bank, which is the greatest danger at the entrance, must not be lost sight of; the sea breaks on it constantly and the currents set across it; it appears to be constantly changing, sometimes enlarging and at others contracting the channel. When abreast of the mouth of the Quenuir river a vessel should be kept to starboard gradually until the center of the channel between the Dos Amigos and point Pangal is well opened; this latter point must be passed within 200 and the SE. of the Amigos within 400 yards; there are not less than 2 fathoms in the passage at low water. When point Pangal bears SSW. all danger is passed by keeping about 400 yards from the south shore, and leaving Chaba Cay to port, the anchorage off the town of San Javier de Maullin, in 3¾ to 4½ fathoms, bottom muddy sand, can be taken. When the buoy is not in place, point Quenuir must be passed as before stated, and then make E. 2° S., allowing for the current; when the center of the Dos Amigos bears N. 58° E. the vessel must be headed for them, and proceed as before directed.

Besides the dangers at the entrance of the Maullin river, in the southern part, a short swell rises suddenly, having the appearance of heavy breakers; it is caused by the currents and prevailing winds. Although it is dangerous, more especially at the change of wind or tide, the land must be approached as directed, keeping over to the northern shore.

It is desirable always to take a pilot for this entrance at Ancud, as it is dangerous. The balandras of the natives are constantly lost here.

Tides.

It is high water, full and change, at San Javier de Maullin at 12ʰ 30ᵐ; rise, 8 feet.

From this point the river Maullin is navigable during high water for vessels drawing 13 to 14 feet.

River Maulliu.

The Maullin river is the only outlet of lake Llanquihue; it is 40 miles long in a straight line, but its numerous bends increase this greatly. In the first part of its course from the lake it is a torrent which ends in a semicircular cascade from 3 to 6 feet high, after which it flows among low hills and its bed is obstructed by *tepuales*,* through which only canoes can pass; finally, for the last 27 miles, it is wider and deeper, receives many affluents, and can carry small vessels and steamers drawing 10 feet 18 miles from its mouth. Tide-water extends 30 miles. The ebb often attains in summer a velocity of 6 miles.

Port Godoy.

Port Godoy is NW. of point Quenuir; it is a large semicircular bay, badly sheltered from the prevailing winds and open to the SW.

To the westward of the bay is a bluff called Varillasmo, from which a small bank, almost level with the water, runs out to the eastward. One mile from the bluff vessels can anchor in from 4 to 4¾ fathoms; bottom fine brown sand. The sand which the NW. wind brings from the dunes of the bay is filling this anchorage rapidly and will soon render it useless.

During favorable weather boats can land to the northward of the bluff. Some fresh provisions, water, wood, and fish can be obtained at the houses in the vicinity. The landing-place is in latitude 41° 31′ 23″ S. and longitude 73° 54′ 00″ W. During NW. and SW. gales the sea renders this anchorage untenable.

Point Quillahua.

Quillahua is a low point to the westward of the port and

---

* Thickets of the tepu, a small tree with many roots, which extend for a long distance. The wood is very hard.

17½ miles N. 40° E. from point Huechucucuy. In 1588 the Spaniards gave the name of the gulf of los Coronados to the space between these two points. Point Quillahua is rocky at its base, somewhat steep and covered with bushes; to the north are some sand-dunes. To the SE., ½ mile from the coast, is a large isolated rock, called la Solitaria; to the south of the point the breakers extend ⅔ of a mile to seaward.

Two other isolated reefs run out 1½ miles WNW. from the coast; these can only be seen with a SW. swell. This coast should not therefore be approached nearer than two miles.

From point Quillahua the coast runs NNW., with frequent bends, for 8½ miles to point Estaquillas; the beaches are sandy and the points alternately rocky and steep. No landing can be made on this coast.

Parga cove.

N. 23° E. of point Huechucucuy and 14 miles north of the large islet of Carelmapu is a small cove named Parga, known for the excellent coal-pits in its vicinity. The coal, however, has been found by experiment to be inferior to that of Lota. As a cove Parga is insignificant, being only 200 feet wide and 1,000 feet long; at the most it could only contain four vessels of eighty tons each at one time. The pilots say that it is sheltered from the prevailing winds. Its shores are steep, but boats can land on its eastern extremity. The outside of the cove is foul; to enter it a pilot is indispensable, as it cannot be seen from seaward. Close to the southward of Parga is a small cove with a sand-beach called Playa del Carbon, on which boats can land during moderate SW. winds, but as its entrance is full of rocks it is necessary to have a pilot to reach it.

Point Estaquillas, immediately to the northward of Parga cove, is of moderate height and steep; it is remarkable for the chain of islets and rocks in which it ends; their extremity being 8½ miles N. 23° W. from point Godoy. About 1 mile ENE. from this point is Estaquillas bay; it is full of rocks and dingy-looking islets which rise like columns from the bottom, and a heavy swell sets in produced by the prevailing winds.

Point Estaquillas.

Three miles NE. of point Estaquillas is another small cove, called Llico after the river which empties into it; it is

Cove and river Llico.

without shelter and cannot be used as an anchorage; its depth is from 10 to 12 fathoms; in the center, rocky bottom; some rocks are visible above water in its southern part. The river Llico empties in the middle of a sand-beach in the eastern part of the cove; it is said to be navigable for 20 to 22 miles; it is full of fish and bordered by woods; the bay is impracticable even for boats; near its source the river is called Rio Frio.

Point Capitanes.   To the northward of point Estaquillas the coast is of moderate height, very irregular, and inaccessible; it is backed by high mountains covered with the alerce; here and there are some small rivulets.

Fourteen and a half miles farther to the north, the direction which the coast takes, is a small island, from which the coast trends N. 29° W. for 5 miles, to cape Capitanes. This part is very irregular and foul, with occasional sand-beaches; 1 mile from the coast there are from 14 to 24 fathoms. Point Capitanes is very undulating; at its extremity is a high island, having a peak of a yellowish color; it has the appearance of being attached to the main-land until close to it and bearing north and south. After passing cape Quedal, when going to the southward, this is the point most readily recognized. Point Capitanes is backed by mountains 2,400 feet high.

Cape Quedal.   Between point Capitanes and cape Quedal the coast runs N. 5° E. for 10 miles, with a slight curve; it is rugged and without coves; it is inaccessible, but there are no outlying dangers.

The seal hunters report that there is a shelter for boats, called San Luis, about 5 miles from cape Quedal, but it can only be reached with a pilot taken from these hunters.

Cape Quedal, the western point of this part of the continent, is clean, steep, and of moderate height; about ¼ mile from the shore there are from 18 to 20 fathoms. The cape extends 1⅛ miles N. 19° W. A little more than 1 mile NE. of cape Quedal is a cove, whose eastern beach is of sand, but it is bad, even with the winds from S. to W. Two rivulets from the Cordillera empty into it.

Point San Pedro.   Point San Pedro, 3 miles N. 54° E. of cape Quedal, is rugged and of moderate height; to the northward a chain of sunken rocks runs out 1½ miles.

Somewhat less than 1 mile to the westward of San Pedro Anchorage off the islet of San Pedro. point is a triangular group of islets, called San Pedro rocks. The longest side of the triangle opposite the coast runs N. 19° E., and extends 1 mile; they form, with the continent, a clear channel without a current. The anchorage to the eastward of the islands is sheltered from SW. winds by the land of cape Quedal; that of cape San Antonio gives shelter from NW. winds. The fishermen of that region say, however, that the sea from the NW. is never very heavy, but that the winds passing over the islands reach the vessels at the anchorage; however, the anchorage to the east of these islets is the only one which can furnish shelter from the prevailing winds on the 140 miles of coast which separates Valdivia from the bay of Ancud. The best anchoring place is in 7 to 8 fathoms of water, bottom sand, off the middle of the islands. In the middle of the channel there are from $8\frac{3}{4}$ to $9\frac{3}{4}$ fathoms, the depth of water diminishing gradually to the rocky coast.

The NE. extremity of the islands is S. 8° E. from San Antonio, and the southern extremity N. 44° E. from cape Quedal. Vessels at this anchorage can get under way easily as long as the wind is not between W. and NW.

The large bay. San Pedro, was discovered by Pastène, a Bay of San Pedro. lieutenant of Pedro de Valdivia, in 1544. It opens to the northward of the islands, is 4 miles wide from N. to S., and 2 long from east to west, and is limited to the northward by cape San Antonio; it has a moderate depth and sandy bottom, and is exposed to the heavy swell from seaward. Opening from it are four creeks, the most important of which are Huayusca and El Manzano.

El Manzano creek is limited on the W. by point San El Manzano creek. Pedro; it is narrow, less sheltered, and much smaller than Huayusca cove; it has, however, an even bottom, favorable for coasters; there is a good landing-place in the center of the beach of pebbles, at the mouth of a rivulet which empties into it. The shores at the sides are studded with submerged rocks covered with sea-weed.

Huayusca cove, the most central, at the southern portion Huayusca cove. of the bay, is in latitude 40° 56′ 21″ S., longitude 73° 56′ W. From the description by Pastène, this was probably where he anchored. The entrance is obstruced by a rock and a small

9 c

bank, on which the sea breaks at times; the depth is moderate and the holding-ground good; landing is easy, and at
half-tide boats can enter the river Huayusca, which is navigable for 500 yards. The seal hunters of Chiloé anchor
their sloops in this river to shelter them from the NW. swell,
the only swell which enters the cove; they visit this coast
without fear, in their frail boats, and they say that vessels
can ride out the most severe summer tempests here; also
that cape San Antonio is a protection from the winds and
sea. They anchor their sloops with a *sacho*,* and ride out
the heaviest gales without injury.

The cove to the eastward of Huayusca is not accessible;
heavy breakers roll constantly on the shore.

River Lliuco.   The NE. angle of San Pedro bay has a deep entrance,
with a fine sand-beach; at first it would seem to offer shelter against the sea and NW. winds, but it is bad at all
times, and the beach cannot be approached. The river
Lliuco empties here; its mouth is well marked by a high
sugarloaf-shaped rock, which is to the eastward of it, and
close to the coast.

From the NW. point of San Pedro bay a series of rocks,
which are awash, and small islets with sunken rocks at their
extremities, extend ½ mile to the SW.; between these islands
and cape San Antonio the coast bends in a little, but is in-
approachable on account of sunken rocks, some of which
are 400 yards from the beach.

Remark.   There is a great confusion in the names of all the points
and rivers of this coast, commencing at the river Maullin.
For instance, the river Lliuco is often confounded with the
Huayusca. The names given here are those generally used
by the coasters, which would seem the preferable.

Cape San An-   Cape San Antonio is 7 miles N. 16° E. from cape Quedal,
tonio.   and projects but a short distance; it is high, and covered
with vegetation; it is rugged at the base, of a grayish color,
and overlooked by mountains from 1,150 to 2,200 feet high.

Point and cove   Point Condor is 3 miles N. 14° E. from cape San Antonio;
Condor.   it is high and steep; about ¼ mile from it there are from 15
to 17 fathoms of water. The coast to the NE. of this point

---

* The *sacho* is a wooden anchor with 4 arms, whose weight is increased
by stones. It is used by the coasters of Chiloé; they use a very strong
rope of *quilineja* instead of chain.

forms a cove of the same name, running ESE.; there is a house on the western beach of this cove. Its position is, latitude 40° 46' 6" S., and longitude 73° 55' 42" W. The north and south shores are rocky and abrupt, their upper part being covered with trees; to the north of the sand-beach is a good landing, near a round rock.

At the entrance of the cove there is from 20 to 22 fathoms, in the center 10, and close to the shore $3\frac{3}{4}$; it is only sheltered against the winds and sea from S. to W. The best anchorage for steamers is in 9 or 10 fathoms; sailing-vessels should anchor a little farther out, so as to be able to put to sea on the first sign of bad weather. Although the holding-ground is good, vessels must not attempt to ride out the heavy NW. gales during the winter, as a tremendous swell rolls in; all who have tried it have been lost.

To the eastward of the beach of Condor cove is an isth- River Chalhuaco. mus of coarse sand, 230 feet wide, which separates a lagoon emptying into the river Chalhuaco from the sea. This river probably emptied formerly in the cove of Condor; the heavy swells from the NW. must have stopped the alluvial deposits and created the isthmus. The mouth is 1 mile to the northward of Condor, in an unimportant cove, without shelter, and full of rocks.

The lagoon is shoal, but can be used by boats, as can also the channel through which the lagoon empties, from half-flood to half-ebb. The small breadth of the isthmus allows boats to be carried across it into the lagoon. The river can be navigated for $4\frac{1}{2}$ miles, by boats drawing 2 feet; higher the water is too shallow, especially during the summer; in winter rises of 16 to 22 feet occur without much current. The banks and neighboring plains are covered with the alerce of a superior quality, which can be easily transported.

Three and one-quarter miles NNW. from cape Condor is Cape Compass. cape Compass, the most remarkable point on the coast between cape Quedal and point Galera. It is elevated and clean, rugged at its foot, and covered with trees at the top $\frac{1}{8}$ mile from it there are from 12 to 17 fathoms of water, sandy bottom. It is probable that this is the same point as that named Huililil by the unfortunate expedition under Juan Ladrilleros, but the position given by him differs too widely to admit of preserving the name.

Immediately to the east of cape Compass is a deep bend in the coast, called Ranu road, in the middle of which are three small rocks above water; the road has an opening of 3 miles from N. to S., and is 1 mile deep. The rocks, or, rather, reefs, in the center leave a passage between them and the coast, but it is not practicable. In the eastern center of the road is a high point, steep to the north and south, with a beach of yellow sand. The river Hueyelhué empties in the southern part of the north beach, and the river Ranu, which empties in the center of the southern beach, gives its name to the roadstead.

The coast comprised between Hueyelhue and cape Compass is very dangerous; the breakers extend over ⅛ mile from the road, but at its southern extremity vessels can find some shelter, during the season of SW. winds, by anchoring in 10 to 11 fathoms, sandy bottom, with point Compass bearing S. 81° 30′ W., and the rocks in the center N. 21° 30′ W.; there is no landing in the cove. The land back of the coast is of moderate elevation and thickly wooded; the Hueyelhue runs in a deep valley; its mouth, closed in by breakers, is 164 feet wide; inside it widens to 490 feet.

N. 38° E., 9 miles from cape Compass, is point Muilcopue; it is rugged, and surrounded by rocks awash. This point shelters the creek of the same name, which is open to the NW. from SE. winds, but these cause a very heavy swell. The anchorage is clean, with moderate depth and good holding-ground; the cove is surrounded by wooded mountains of moderate height; the shores are rocky, with detached rocks at short distances from them.

The southern beach, also called Muilcopue, is flat, and is alone accessible; the landing is on its western part, in latitude 40° 35′ 52″ S., and longitude 73° 47′ 45″ W. A small rivulet empties on the east side.

From Muilcopue the coast, rocky and irregular, trends about N. 25° E.; off it are three islets, named Los Lobos, which form two insignificant coves; the one perpendicular to the islets is called Pulameuu.

To the north of these is the SW. extremity of Manzano creek; the point is rugged, surrounded by rocks, and only

---

* Not to be confounded with Manzano creek, p. 120.

extends out a short distance. The cove is to the NE.; it is badly sheltered, has a bad landing place, and can only be used in calm weather, which occurs very rarely. The depth of the anchorage varies between 7 and 14 fathoms, sandy bottom. This cove forms the southern part of the open roadstead of Manzano; the eastern beach of the roadstead is of yellow sand; it is broken in its center by an abrupt, high point, from which some rocks project $\frac{1}{8}$ mile. Two small brooks, which cannot be approached on account of the breakers, empty, the one on the southern and the other on the northern beach.

The coasts of this roadstead, as well as that more to the northward, are dangerous. The mountains which lie back of it have a mean height of 1,550 feet, and are thickly wooded.

The coast continues NNW., with exception of the small point Pulome, to point Pucatrihue, 7$\frac{1}{2}$ miles N. 21° W. of the roads of Manzano; 2$\frac{1}{2}$ miles S. 16° E. from the latter point, and 1$\frac{1}{2}$ miles S. 86° W. of point Pulome, is a submerged rock, called Covadonga, $\frac{1}{3}$ mile to the southwest, and also to the north of this danger are from 27 to 28 fathoms of water, with rocky bottom. The sea breaks on it with a heavy SW. swell. *Covadonga rock.*

One mile and a half SSE. of point Pucatrihue is a small cove, into which a small river empties. In the western part of the cove is a narrow sand beach, which would appear to offer some shelter to boats; from seaward some breakers are seen near it.

Pucatrihue point is high and rugged; some rocks lie off it. *Point Pucatrihue.*

Banderas bay is bounded to the WSW. by Pucatrihue point, and is open from that, bearing to N. The SW. swell renders this road bad, even in fine weather; it shoals gradually from 18 fathoms in the center to 7 near the shore. The bottom is fine sand, with large stones in the south part of the bay, which is the best sheltered. Boats can only land alongside of the rocks in the southern part, and the surf makes landing at all very difficult; on the eastern side of the bay is a beach of yellow sand, but as it is always exposed to breakers, landing can only be attempted during favorable weather. *Banderas or Choroichalhuen bay.*

The south coast is rough and studded with large stones;

some rocks run out from it into the bay on the side of point
Pucatrihue. The neighboring hills are covered with vegetation, and rise as they advance to the east. The roadstead
is of no importance. The Indians who come here in summer for shell-fish have built some huts.

Milagro cove.   A little more than 6 miles N. 3° E. of Banderas bay is a
cove, in whose southern part is Milagro cove. In the interval the coast preserves the same aspect—wooded cliffs
backed by elevated land; between Corral and Rio Maullin
it is but one extended alerzale forest.

Milagro cove is remarkable for an isolated island, in the
shape of a sugar-loaf, which lies off it about ⅔ mile from the
coast; although it is smaller than Lamehuapi cove, it is a
little more sheltered from the SW.; the landing is worse,
however, as boats ground a short distance from the beach,
and lie exposed to the surf. The river Zehuilanquen empties on the eastern part of the sand-beach, in front of the
rocky coast; its entrance, 165 feet wide, is inaccessible; it
appears to be of some importance. The islet forms, with
the coast, a deep channel, safe for moderate-sized vessels;

Directions.   it should not be taken except in cases of necessity. When
standing in for this cove from the NW., it can be recognized
by a wooden house, which is in the middle of the yellow
sand-beach at its head; the lead-color of this house contrasts strongly with the green of the vegetation; the islet
is the best landmark, however; it cannot be mistaken.

The best anchorage is in 8¾ fathoms, bottom sand, in the
southern part of the cove, with the south point bearing S.
18° W., and the eastern part of the sand-beach S. 47° E.
The landing-place is on the western part of the southern
sand-beach; its position is in latitude 40° 26′ 10″ S., and
longitude 73° 46′ 30″ W.

In 1860 a brig was surprised in this cove by a NW. gale
and lost. This anchorage is passable during SW. winds,
but entirely worthless during NW. gales, especially in the
winter.

About 3 miles S. 6° W. from the island, following the
coast, there is a small creek with a sandy beach, into which
a brook empties; it is supposed that boats can enter it during good weather. Behind a group of low islets, in the
NE. part of Milagro cove, is another cove sheltered from

SW. winds. The river Trahuilco empties into it. A landing can be effected without trouble on the western beach of
this small cove. It is better for taking in wood and water
than that of Milagro.

The southern point of the Rio Bueno is 5½ miles N. 6° W. Rio Bueno.
from the northern part of Milagro cove; the bay is entirely
open to the prevailing winds. The bar of the river is in
latitude 40° 15′ 38″ S., and longitude 73° 46′ W.; outside
of it are from 4 to 8 fathoms of water. To the southward
are some hills from 2,600 to 2,950 feet high. The northern
point of the mouth of the river is rocky, and some isolated
rocks lie off it; one about 230 feet from the shore is dangerous for the small vessels which frequent the river. The
bar runs N. and S., and is 550 yards from the mouth of the
river, completely barring it with a line of breakers. The best
channel for crossing the bar is west of the entrance in from
1½ to 3 fathoms. To the north and south of it the water
shoals and the breakers are heavier.

The mean breadth of the Rio Bueno varies between 490
and 740 feet; at the mouth it is 590 feet wide. The current
of the river is variable, but with the ebb-tide it reaches,
near the mouth, a velocity of 3 knots. During good weather
the bar can be crossed safely by steamers drawing 8 feet,
but is always dangerous to sailing-vessels.

The southern coast is elevated and keeps off the wind; a
small vessel crossing the bar may therefore be left at the
mercy of the current, which would carry her on the stone at
the mouth. So many have been lost that the navigation of
the river was interrupted until the commerce was so far
developed as to warrant the use of steamers.

The river Bueno has its source in lake Ranco, 390 feet
above the sea. Some of its tributaries have their source
near lake Llanquihue. In the first third of its course there
are several rapids, but in the remainder navigation is easy.
It is said that steamers drawing not over 8 feet can go up a
distance of 46 miles, to Trumao; the tide is felt 43 miles
from the mouth.

From Rio Bueno to Lamehuapi point the rugged coast Lamehuapi
point and cove.
bends to the westward. From Lamehuapi point, point Ga-
lera bears N. 4° E., and point Hueicolla N. 38° E. A short
mile outside the point are from 16¾ to 20 fathoms, bottom

fine black sand. The coast then trends N. for 2½ miles to Lamehuapi cove; it is steep, clean, and backed by wooded heights.

Lamehuapi cove is a good shelter against SW. winds, but is entirely exposed to those from the north, which bring in a heavy sea. It is of good size, with a moderate and regular depth, with good holding-ground, and is capable of containing a large number of vessels in from 6 to 8 fathoms; bottom black sand. In a little creek, to the eastward of a sugar-loaf-shaped hill, there is a good landing, during SW. winds, on a small sand-beach between the rocky coast and the large eastern beach, in latitude 40° 11′ 47″ S. and longitude 73° 45′ 51″ W.

Directions. This cove can be recognized by an unpainted wooden house, which is just above the SE. beach; its leaden color, due to the weather, makes it very prominent; a small brook empties into the sea to the eastward of this house; the bay is bordered by wooded hills of moderate height. Sailing-vessels can beat in without difficulty, but must abandon it on the first indications of a northwester.

Hueicolla or Gyüicolla point and cove. Point Hueicolla is 8 miles N. 19° E. from Lamehuapi cove; it gives the cove of the same name a little shelter against SW. winds. About 550 yards N. of the point is an isolated submerged rock, on which the sea breaks constantly; there are 6 fathoms between it and the land, but the channel should only be used by boats. There is no good landing, but boats can run in on top of the surf and beach, if the surf is not too heavy.

Hidden by a stony point to the eastward of the anchorage of the cove, the river Hueicolla empties; it has its source in the mountains about 9 miles from the shore, and is inaccessible.

About 2 miles to the northward of the cove is a small rocky point, which is clear, and projects only a short distance. About ⅛ mile from the coast there are from 16 to 18 fathoms.

The river Colun empties to the NE. of the point; it has its source in the Cordillera of the coast, and, like the river Hueicolla, is deep near the sea, but the rest of its course is a torrent; it is not accessible from the sea, and the point does not protect its entrance.

Hueicolla cove is part of the large indentation from which the coast runs N. 8° W. for 6 miles to point Galera. This, a nearly straight piece of coast, is clear and bordered by a beach of dark sand.

In clear weather point Galera is visible 15 or 20 miles; it was discovered by Pastène, September 21, 1544; it is salient, mountainous, and surrounded by two rocky points, which renders its approach dangerous during foggy weather; otherwise its vicinity is safe; to the ENE. it is overlooked by the mountains of Valdivia, 1,680 feet high, which are a good landmark. As far as point Falsa Galera, the coast is undulating, clean, and rocky. It is proposed by the Chilian government to establish a light of the 2d order on this point. It is to be fixed and flashing, and visible 25 miles. *Point Galera.* *Light proposed.*

Falsa Galera point is 3 miles N. 41° E. from point Galera; it is level, and forms the western prolongation of the mountains of Valdivia. About 1 mile outside of it are some detached rocks above water; these rocks, the point, and the mountains are in a straight line; otherwise the vicinity of the point is clear. *Point Falsa Galera.*

A little more than 4 miles N. 52° E. from Falsa Galera is point Chaihuin, a salient, rocky, and foul point; it is backed by wooded heights of moderate elevation. *Point Chaihuin.*

The river Chaihuin is immediately to southward of the point. During calm weather its bar can be crossed by boats; the channel follows the point; it must not be attempted without a pilot. The N. winds blow into the cove and river, and those from SW. send in a heavy sea and make it inaccessible. *River Chaihuin.*

After crossing the bar, boats can ascend the river, if the tide is with them, for nearly 15 miles; its breadth for the first two miles is 550 yards at high water; at low tide it is but 328 feet. Farther up it gets narrower, and its banks are wooded to its source in the Cordillera. The right bank of the lower part of the river is strewn with sand-banks and the left with mud-banks, which are covered during the rainy season.

The river Chaihuin runs in the gorge of Chaihuin. When seen from the NW., a sugar-loaf-shaped hill will be noticed at its end. The entrance has some resemblance to that of Valdivia.

The coast between point Chaihuin and Gonzalo hill makes a regular curve, broken by four small rocky points. The middle one, Palo Muerto, is the best defined ; a reef runs out from it 800 yards.

Gonzalo head.   Gonzalo hill or head is about 10 miles N. 51° E. from point Chaihuin, and 15½ miles from point Galera; it is of a yellowish color, very rugged, and covered with vegetation to its summit. It is 550 feet high. About 800 yards N. 63° E. from this hill is the isolated and elevated rock Peña Sola. It is clear to seaward but connected with the coast by a chain of submerged rocks; 1,400 yards from the hill is Palo Muerto, low and rocky, the breakers extending from it ⅓ mile; at their outer edge are 2 fathoms. Between this point and the hill are three small coves, in which boats can land during good weather; they are called Cabeza de la Ballena, Maliuo, and La Loberia; all three are rocky, narrow, and foul; their entrances are full of submerged rocks, which run out 328 yards. They are overlooked by high, peaked cliffs.

Light proposed.   It is proposed by the Chilian government to build a lighthouse on cape Gonzalo ; the light to be fixed, white, and visible 24 miles.

Point and fort San Carlos.   The small peninsula of San Carlos, crowned by a fort of the same name, is S. 70° E. from cape Gonzalo. The peninsula is very rocky, and has a diameter of 1,300 feet, and is 42 feet high. The low and stony isthmus has a small creek on its eastern side, in which boats can land during good weather.

The fort, which dates from the time of the Spaniards, has the shape of a regular demi hexagon, three sides of which face the entrance to the bay of Valdivia.

Aguada del Ingles.   The coast between points San Carlos and Palo Muerto forms a cove with dangerous rocky shores ; the depth varies between 2 and 3 fathoms. About ⅛ mile S. 64 W. from San Carlos it incloses a small beach, called Aguada del Ingles ; this cannot be used with NW. winds. It was the landing-place of admiral Cochrane, in 1820, when he took fort San Carlos, with its garrison of 1,500 men, with 150 sailors.

Point Juan Latorre.   Point Juan Latorre is 3 short miles N. 62° E. from Gonzalo head. It is steep and rocky, with a level surface ; a chain of rocks makes out from it ⅛ mile.

Gonzalo head and point Juan Latorre, properly speaking, form the entrance to the bay of Valdivia, which runs in SSE

Two-thirds of a mile SSE. from point Juan Latorre is point Molino; it is broad, steep, and surrounded by submerged rocks, which extend out 300 yards. The northern portion is called Numpulli, and the southern portion is, properly speaking, Molino. When the sea does not break, boats can land at the edge of the sand-beach among the rocks. S. 3° E. 1½ miles from point del Molino is point Niebla; between them is a spacious cove, with a sand-beach divided in its center by two small rocky points. *Point del Molino or del Ancla.*

Point Niebla.

Point Niebla is 114 feet high, and level at the top; its base is surrounded by a bed of large stones, which is uncovered. At the edge of the bank are 2 fathoms, with sandy bottom. Seen from seaward, this point is remarkable for its bold, steep cliffs, the white light-house, the battery facing the river, and the barracks above. To the eastward of the point is a small creek full of rocks, in which boats can land; it is called Huairona. This is the only point from which the fort can be reached.

The fort is constructed in the hard rock which forms the point; it is backed by the cliffs, and has but one entrance.

On the western part of point Niebla is a light-house. The illuminating apparatus is in a tower on the top of a small house 121 feet above the level of the sea; it is *fixed white,* and can be seen during clear weather from 6 to 8 miles; in rainy or foggy weather, however, it can only be seen 1 mile. The tower is 32 feet high, square, and of wood. There is a signal-mast alongside the light-house. *Light-house: Lat. 39° 52′ 10″; long. 73° 24′ 50″.*

DESCRIBING THE WESTERN SHORE OF THE BAY.

Point Amargos is 1 mile SSE. of point San Carlos; the intermediate coast is rocky and foul; near San Carlos is point Barro, and near Amargos is point Pastigo; these prevent Amargos from being seen from San Carlos. *Point and fort Amargos.*

Point Amargos is low and rocky; at its extremity is a large flat rock called El Conde, which can be approached without danger; at its foot there are from 6 to 8 fathoms of water, sandy bottom. On the point is a dilapidated battery 19 feet above the sea, called Amargos.

Point Avanzada or Chorocomayo.   Point Avanzada makes out ⅔ mile S. 17° E. beyond fort Amargos; it is steep and rocky, and on its upper part, which is level, is a battery, called Chorocomayo.

Amargos cove.   The coast forms a cove between these two points, called Amargos; it is not deep, and ends in a sand-beach, on which a brook of excellent water empties. Vessels can anchor in this cove, but are exposed to NW. winds.

Atreal shoal.   Fitz-Roy indicates a rock 100 yards to the E. of point Avanzada; it does not exist. It is probable that it was confounded with Atreal shoal, which is on the line which joins point Avanzada with the rock del Conde de Amargos, and bears E. 18° S. from the sand-beach in the cove of that name. The bank is of rock, the central part being formed of pebbles, and is covered with sea-weed; during spring-tides it uncovers from 3 to 7 inches. It is not dangerous for vessels going in or out of the port of Valdivia; the ebb tide of the river sets toward it, but stops short 100 yards from the rock.

Port Valdivia or el Corral.   Following the rocky and clean coast 600 yards to the southward from point Avanzada is point Laurel, or Calvario. From it the coast first trends S. 18° W., forming a small indentation, which is closed to the southward by the point and Castle el Corral; this is Corral, or the port of Valdivia, affording excellent shelter in all kinds of weather.

On entering, the bay of Valdivia appears to be very spacious, but the anchorage is contracted, and the part entirely sheltered is capable of holding but about 40 vessels moored with one anchor to N. and a stern-fast to the southward, on account of the strong tides. The anchor to the northward should be dropped as close as possible to the N. coast in 5½ to 6½ fathoms; bottom black mud.

Description.   El Corral is one of the greatest ports of the republic, and is of importance from the great progress which the province of Valdivia has made in European immigration; large assorted cargoes are consigned to this port from Europe and an excellent quality of hides are exported. Steamers stop here four times monthly, and communication with Valdivia is carried on by river steamers.

Fresh and salt provisions can be obtained at a moderate price. The water is excellent, and abundant in the western part of the port. Wood for construction is plentiful, and it can be obtained for repairs and spars; but they are not usually on hand, and must be sent for. Coal is scarce, and can only be obtained in small quantities.

The following are the articles of exportation: Wood of all kinds, hides, beer, cider, charqui, or dried beef, salt meat, glue, and many other products of the German colony. There are a few naval stores; carpenters and blacksmiths are found at Valdivia.

The fortifications were the same in 1871 as those marked on the charts, but they are not armed.

To the southward of the port is a low beach, on which a brook empties; a little farther is the rocky point Corral, surmounted by the castle of the same name, back of which is the small town of the port, having 500 inhabitants.

There are one government and two private wharves at Corral. During bad weather ballast is thrown overboard near the shore in Pantheon cove; in summer and in good weather in Amargos cove, stone ballast must be thrown on the beach between the government mole and port Corral. There is a life-boat station, and there are two ship-yards for repairing.

Bound to Corral, the landfall to be made depends on the season, and especially on the prevailing wind when nearing the coast. During the winds from the NW. quadrant the shores should be approached on the parallel of 39° 40′ S., and during the SW. winds point Galera or the 40th parallel should be made. *Directions for entering the bay of Valdivia.*

It must be remembered that on nearing the coast, the cove of the river Chaihuin, which is 10 miles S. 51° W. from Gonzalo head, resembles so much the entrance to the bay of Valdivia that many have mistaken them. During clear weather it is easy to make the distinction, when it is remembered that the sugar-loaf-shaped hill which is seen in the center of the entrance to Chaihuin is more elevated and pointed than the one seen in the center of Corral. Gonzalo head is also of a more characteristic shape than the coast S. of Chaihuin.

With SW. and SE. winds the weather is clear, and point Galera can be seen 15 miles; as soon as it is recoguized, the land can be approached within 2 miles, or steer for Gonzalo head; when it is made it must be rounded closely, and then the coast can be kept aboard without fear. If the winds should be contrary, and the tide ebb when off San Carlos, the anchor should be dropped as soon as possible. During the flood vessels can beat in and extend their tacks to within 400 yards of the shore, until the narrowest part of the channel, between Amargos and Niebla, is reached. Vessels cannot beat farther on account of the narrow channel; they must then anchor and await a tow or a change of wind. If the ebb should set in while beating, the anchor must be dropped immediately.

The landfall is much more inconvenient with NW. winds, as they are always accompanied by fog. When the wind is between W. and NW., the land can seldom be seen farther than 3 or 4 miles; then it is most prudent for sailing-vessels to stand on the off-shore tack until the weather clears. The fog or mist seldom lasts longer than 48 hours.

With the winds from NW. to W., which are a little less foggy, it is best to stand in for Bonifacio head, which, though projecting but little, is more pointed and higher than the land around Valdivia. When this hill is made, the bay can be entered with a fair wind.

Point San Carlos can be passed within 400 yards, and Amargos as close as desirable, but not inside of the line joining the points which form the bay.

It must be remembered that the ebb-tide is always stronger and of longer duration than the flood, and during the season of freshets, that is during the winter, the only effect the flood has is to diminish the strength of the ebb. It is not well to attempt to enter after dark, especially in thick or foggy weather, as the light cannot be depended upon.

Tides.   It is high water, full and change, at Corral, at $10^h 35^m$; rise $5\frac{1}{2}$ feet.

Tres Hermanas or Mancera bank.   Mancera bank is situated almost in the center of the bay of Valdivia; the water shoals gradually toward it, excepting to the S. and SW., where it shoals suddenly; the western edge of the bank is tangent to the line which joins port Corral with Huairona cove, to the eastward of point Niebla; it is $\frac{1}{8}$ mile from Corral, and the same distance from Laurel point; the SW. part of the bank is the shoalest, it having

on it but from 1 to 3 feet of water, bottom fine sand mixed
with small shells.

During quiet weather the bank can be distinguished by
the color of the water, and when it is rough by the breakers.
The NE. and NW. winds cause considerable sea, and when
they occur with the rising tide it is best not to try to cross
the bay.

A buoy has been moored in 3 fathoms of water on the
western edge of the bank; it lies S. 71° W. from point
Piojo, and N. 49° W. from the SW. extremity of Mancera
island. When the wind is light it is best not to go too close
to this buoy, as the tide sets over the bank, and at times
with great force; it must never be passed to the eastward.
This bank is continually increasing; where there were from
18 to 20 feet in 1788, there were only 4 feet in 1835, and in
the same place at present there is only 1.4 feet. The depth
on the remainder of the bank varies between 3 and 7 feet at
low water. Although the deposits of the rivers Valdivia
and Torna Galeones which form it, are irregular, there is no
doubt but the bank will soon be an island. In September,
1862, a very dry year, the top of the bank appeared at low
tide. It has, therefore, the tendency to diminish the already
small harbor of Corral.

In 1835 the bank was 2,800 feet from point Calvario, and
from the left bastion of castle Corral. In 1868, the first
was reduced to 2,265 feet and the latter 2,296; the bank
has, therefore, annually increased 13.4 feet; under these
conditions the anchorage will disappear in a century and a
half; but as the deposits are only fine sand without stones
or pebbles, the dredge can be used with great advantage.

Observations prove that the entrance to the bay of Valdivia has also changed in depth. Since 1835 the water has,
however, deepened on the Niebla and Amargos side, while
it has shoaled near San Carlos.

The island Mancera is situated in the SSE. part of the Mancera island
bay of Valdivia about ½ mile from Tres Hermanas bank;
it is somewhat more than ⅓ mile long from north to south
and 2,000 feet wide from east to west. The center is taken
up by a hill 300 feet high, which runs in the longitudinal
direction of the island. The slopes are surrounded by
plains, on which are some cultivated spots and a few houses.

At the northern extremity, in a gloomy valley, is a little hamlet with its single street. The N. and W. shores of the island consist of cliffs; the S. and E. shores terminate in sand-beaches, on which boats can land, excepting with a swell from NW. The best place for landing, however, is a rocky point, called la Cal, at the NE. extremity of the island; it is a natural mole and near the most thickly inhabited part of the island. On the NE. point is a battery, called Mancera, which commands the channel leading to the river Valdivia.

The N. coast is rocky but clean; vessels drawing 10 feet can approach within 119 yards. From the S. coast a reef runs out to half the width of the channel separating the island from point de la Rama or Trinidad. This reef is dry at low water; beyond it is a large rock, called Los Lobos. A low point, called Castillito, makes out from the SE. of the island; to the southward of it is another reef, which is covered at high water.

This island, la Guiguaçabin of the Indians, was discovered by Pastène in 1544. It derives its present name from the Marquis of Mancera, viceroy of Perú in the 18th century.

From the island there are three openings: That to the NNE. is the mouth of the river Valdivia, which leads to the city of that name; that to the eastward is Torna Galeones, leading to the river Futa, to l'Angachilla, and also to the river Valdivia; the third is the inlet of San Juan.

San Juan Inlet.   Following the high, rocky coast to the SE. is point de la Rama or Trinidad, 1½ miles from fort Corral; from thence the coast runs S. ½ E. and forms the inlet of San Juan, which is about 2 miles long and 1 wide; it is of little importance to navigation; its shores are steep and rocky, with exception of the southern, which is sandy and marshy, but can only be approached during high tide. Its depth at the entrance is 2½ fathoms, 1 mile to the southward 1 fathom, and within ½ mile from the beach 3 feet. Three small rivers, the San Juan, the Catrileufeu, and de los Llanos, empty into it.

The mouth of the San Juan is 1 mile to the southward of point Trinidad, in the middle of a small sand-beach; it is limited to the southward by a low rocky promontory. This river is shoal and very tortuous; canoes can go up with the tide for about 1 mile. Some few huts and cultivated land are on its banks.

The rivers Catrileufeu and los Llanos empty at the extremity of the sand-beach at the head of the cove, but are navigable for small boats during high water. Los Llanos, the eastern, is the most important.

Point San Julien is the eastern end of the inlet. The shore to the southward is foul and encumbered with reefs.

The mouth of the Torna Galeones is ⅔ mile wide be- Mouth of Torna Galeones. tween points Carboneros to the northward and Fronton to the southward, and ⅔ mile from the island Mancera. Carboneros is a low, steep promontory, bordered to the westward by a bank of large stones. The southern extremity of the promontory is called Puerto Clara. A mud-bank, called Simon Reyes, the top of which can be seen at low-water springs, makes out from Carboneros; this bank ends near Mancera, but is separated from it by a channel with from 2¾ to 3¾ fathoms.

Fronton point is low and covered with vegetation; 200 yards from it, NE. by N., are some rocks, which can be seen at low water. The depth between this and the preceding points varies from 1¼ to 2½ fathoms. Vessels drawing 13 feet can go up the Torna Galeones.

The shore between Fronton and point San Julien is rugged and thickly wooded; in the center there is a sand-beach, from which a bank of stones makes out for 100 yards; the channel between it and Mancera is shoal; a bank of rocks in the eastern half of the channel uncovers at half-tide; the channel can only be used by vessels drawing 6 feet.

One-half mile SSE. from point Niebla is point Piojo; it is Mouth of the river Valdivia. low and rugged, and a bank of rocks extends from its southern part. The tide runs at the rate of 2 knots an hour along the edge of the bank. The mouth of the river Valdivia is between this point and Carboneros; it trends about ENE., and is contracted to the width of 875 yards between two rocky points 1 mile NE. of Mancera. The bank Simon Reyes encroaches on this channel. The maxi-

10 c

mum depth is first on the NW. shore, but soon shifts to the southern, or side of point Alcones.

Currents.   The currents in the estuary of Valdivia are very irregular, and vary according to the season, the tides, the rains, and the prevailing winds. The flood has a velocity of 1 knot at the entrance, in the center of the channel, and reaches 2 knots near point Niebla and the Peña del Conde. It comes from the SW. along the outside coast; on entering the gulf where the river Valdivia empties, it runs to the SE., directly toward the river; it is divided by the bank Tres Hermanas; the larger branch running toward the Rio Valdivia and the smaller to the cove of San Juan. They join again on the other side of Mancera island, off the Torna Galeones.

The ebb comes down the rivers, unites its waters in the bay, and runs to seaward in the center of the channel. From the entrance of the estuary it turns to the SW., and, when off point Galera, it runs S. At that point it is almost imperceptible, excepting during the winter.

Both the ebb and flood are influenced by different causes; with SW. winds it runs out to the NW. with very little force; otherwise it follows the coast from Gonzalo head to point Galera.

During the rainy season, which is also that of the NW. winds, the flood has very little strength; the ebb hardly stops running; when the latter is aided by the waters of the swollen rivers it has sometimes a velocity of from 3 to 4 knots; during that time it is found nearer to the S. coast, and its velocity increases toward Gonzalo head, attaining from 5 to 6 knots; from there it runs toward point Galera and passes it with considerable strength.

After N. winds have prevailed for some time the ebb is increased. The current, which runs parallel to the coast from point Niebla to Bonifacio head, has always a tendency to separate.

Vessels leaving Valdivia must keep in the center of the channel, as the currents first set toward points Laurel and Avanzada, and afterward to the Peña del Conde.

River Valdivia.   The river Valdivia, the Ainilevo of the Indians, was discovered in 1544 by the Genoese Pastène, lieutenant of Val-

divia. The latter gave it his name up to the town of Valdivia, 12½ miles from the mouth.

The river Valdivia is the estuary of the river Calla Calla,
and is the largest and, at the same time, the most intricate
part of that river. From the bay of Valdivia it runs ENE.
for 5 miles, then E. for 2 miles, and, finally, N. to the city.
The navigation, during the fine season, is very good; its
shores are bordered by thickets and flowers, in the midst of
which are some houses; the water is brackish in the first
part, but becomes fresh as soon as the turn to the eastward
is reached, but it does not attain its full transparency until
after its confluence with the river Cruces.

Two small passenger steamers run between the bay and
the city, and many balandras and schooners are continually
transporting wood to El Corral. Sometimes sea-going vessels go up to the town, and they could ascend even much
higher. The depth of the river varies according to the rises
and winds; but no vessel drawing over 13 feet can ascend.

All vessels drawing from 4.5 to 9.8 feet must leave El Directions for going to Valdivia.
Corral 2 hours before high water, being careful to leave the
buoy of Tres Hermanas bank to starboard, and steer for the
light-house of Niebla. At ½ mile from it they must change
the course so as to head for the northern point of Mancera
island, or a little to starboard of it, according to the strength
of the current, so as to avoid the bank which makes out
from point Piojo; at ⅔ flood there are not less than 2½ fathoms.

The island Mancera must be approached within 660 yards,
and the vessel must then be headed N. 41° E., or for the
middle of the river Valdivia, but keeping gradually to Carboneros point, which must be passed within 110 yards;
after that the left or SE. shore must be followed at a distance of 45 to 110 yards; not less than 2¾ fathoms will be
found.

At 1½ miles from Carboneros is a small, steep cliff, called
point Alcones; the channel runs from it obliquely to point
Agua del Obispo, on the opposite bank. This is the worst
place in the river; two large banks, which often do not give
1½ fathoms at high tide, obstruct this passage; the bottom
is very irregular, of fine sand; masses of muddy sand, with
only a depth of 1¼ fathoms, are often found; vessels are

therefore frequently grounded, and the constant changes in the channel render a pilot absolutely necessary. Just before arriving at point Alcones the vessel must be kept a little to port, to head for point Agua del Obispo, but never to the eastward of the line joining the two points; when near enough the latter point, the coast is kept about 84 yards distant, avoiding the ground at the confluence of river Cutipaï, which is low and marshy. The right bank must be followed until point Palo del Diablo bears S. 52° E., when keep to starboard and head for that point; thus the channel will be followed, clearing Cancahual bank and that off the estuary of Estancilla. The minimum depth at high water will be from 2 to $2\frac{1}{4}$ fathoms.

When off point Palo del Diablo, which will be recognized by a red cliff to its right, the coast must be followed at a distance of about 82 yards; after passing the mouth of the Cantera, that is, the high coast to starboard, the vessel must stand for the southern part of Mota island, which must be passed within 110 yards, then gradually steer from the right bank to the center of the river.

After passing the mouth of the river Guacamayo, the center of El Islote channel must be kept; the town of Valdivia will then be in sight, but it cannot be reached on a direct course, on account of a bank which obstructs the passage between El Islote and the confluence of the river Cruces.

The center of El Islote channel must be followed; to the northward of it the principal branch will be again entered; a good lookout must be kept for the banks which make into the channel from the N. and S. part of the island; it is best to keep from the main bank a distance of about $\frac{1}{3}$ the width of the channel. Vessels can anchor anywhere to the northward of El Islote, where there are no banks or dangers.

If going into the river Cruces, after passing the Guacamayo, keep to port, to the entrance of the river; if not drawing over 9 feet, the center of the river can be kept.

Tides   It is high water, full and change, at Cancaguel, 32 feet above the sea, at 11^{h} 15^{m}; rise and fall, 4.3 feet; at Valdivia, 60 feet above the sea, 11^{h} 40^{m}, rise and fall 3.9 feet.

Valdivia   Valdivia was founded in 1552 by Don Pedro de Valdivia; in 1599 it was destroyed by the Araucanians; it was rebuilt in 1644, but did not flourish until 1850; it experienced a conflagration and terrible earthquakes in 1737 and 1837. In 1850 was the first arrival of the German colonists, who, with the addition of subsequent emigration, have raised it to its present state of prosperity. It has become a center of export, and would grow without obstacles if the Germans agreed with the natives.

The town has a picturesque aspect, built as it is on undulating ground 36 feet above the river, which preserves it from inundations. The houses are good; the population about 7,000.

From Valdivia the river takes the name Calla-Calla, and runs nearly east and west. Navigation is possible for vessels drawing 8 feet as far as Arique, 23⅔ miles from Corral; it is navigable for boats to a distance of nearly 45 miles ; its breadth is variable, the shores are high, and the water deep; the current is weak, and the tides are felt as far as Arique; there are some banks, but with a pilot and a good chart the navigation is not dangerous. At present the commerce does not extend higher than Valdivia, but it is to be supposed that the emigration will develop the riches of the entire territory. River Calla. Calla

The flood counteracts the current of the river entirely as far as Arique, and partially to Chincuin, a little more than 1 mile beyond. The ordinary force of the current near Valdivia is 1 knot, but in some places it reaches a velocity of 2½; the ebb is always somewhat stronger. In the winter the flood is only felt in the river for 2 or 3 miles; during the freshets there is only a rise of the water; in some cases as much as 16 feet, and the water, which is generally clear as crystal, becomes muddy. The current of the river runs about 6 knots above Arique.

The Calla-Calla springs from the deep lake Reñihue, in latitude 39° 45′ N., and longitude 72° 21′ W. The latter is situated at the foot of the volcano of the same name, 432 feet above the sea. The length of the river is nearly 73 miles ; its only tributary worthy of mention is the river Quinchilca.

It is high water, full and change, at Arique at $1^h 40^m$, at Chincuin $1^h 50^m$; rise 1.9 feet at Arique, and 1.3 at Chincuin. Tides

Teja island, which forms a delta at the confluence of the rivers Calla-Calla and Cruces, is limited to the north by the Island of Teja or Valenzuela.

river Caucau; it is large, fertile, inhabited, and cultivated. Situated opposite Valdivia, it is the site of the factories whose products are exported from the town.

River Cruces. The river Cruces empties into the river Valdivia about 1 mile to the southward of the town; the confluence is called the Palillo.

This river has its source on the northern slopes of the volcano Villarica; it is only navigable from a point called Panul to port Cruces, and then only by crafts drawing 5 feet. From there a vessel may draw from 6 to 7 feet. The schooner la Janequeo, drawing 12 feet, ascended to within 13 miles of Palillo, the junction of the rivers Cruces and Pelchuquin. At that point the Cruces is barred by a bank with 7 feet of water on it.

The currents in this river are not strong; the flood-tide runs about $\frac{1}{2}$ knot and the ebb 1. At springs the rise is 3 feet, and during neaps 2.7 feet; it is felt as far as the hamlet or port Cruces, 17 miles from Palillo, in latitude 39° 36' 31" S., and 73° 08' W. longitude. The maximum rise at this place is 2.9 feet; the freshets deepen the water temporarily to 8 and 9 feet.

The banks of the Cruces are generally less elevated than those of the Calla-Calla, and it has numerous tributaries. At a point called Tres Bocas de Cruces the river splits in two branches and forms the island of Rialejo; it also receives two water-courses. On the upper Cruces, which has been abandoned by the Araucanians, is the town San José de Mariquina, founded in 1850.

From port Cruces the river runs N. 19° E. nearly parallel to the coast, from which it is separated by a chain of mountains.

River Caucau. The river Caucau joins the Cruces with the river Valdivia, leaving the island of Teja to the westward; the current follows the tide, running alternately from one river to the other. Its length, from NW. to SE., is 1⅔ miles; its breadth varies between 130 and 297 feet, and it has from 2½ to 3½ fathoms of water over muddy bottom. The banks are low and overgrown with trees; the current has generally a velocity of 1 knot. Near the most elevated banks of the island, at a place called Coihues, under the largest trees, is a sunken rock, but it is so close to the shore that it is not dangerous.

Vessels wishing to pass from the river Valdivia to the

river Cruces pass through this channel instead of going to Palillo; it has been passed by vessels drawing 12 feet.

The Cutipaï, a tributary of the river Valdivia, 2½ miles to the northward of its mouth, is a small estuary filled and emptied by the tide. Boats can run up for 3½ miles to the village of Cutipaï, to which it owes its only importance. It is high water, full and change, at the village at 11ʰ 15ᵐ; rise 2.6 feet.

About 1½ miles to the southward of the confluence of the river Valdivia with the river Cruces is the channel of Guacamayo, and 1¼ miles to the westward of it the channel of Cantera. These channels lead to the confluence, called the Tres Bocas de Futa, at the same time forming the island of Guacamayo. At their junction they receive the waters of the Futa, and together these three water-courses take the name Torna Galeones, which is in reality but the mouth of the river Valdivia.

The many arms of these rivers render the Tres Bocas de Futa very picturesque. On the east side is the small island Valverde, which is always inundated by the rise in the winter. To the southward, between it and the continent, the channel is only 82 feet wide and the depth 4.2 feet; this island must be kept to the eastward. The passage to the westward has no difficulty for boats, and can even be used by vessels drawing 12 feet, when they have pilots.

It is high water, full and change, at Tres Bocas at 11ʰ 30ᵐ; rise varies between 2 and 4 feet, according to the age of the moon.

The river Guacamayo, which unites the river Valdivia with Torna Galeones, is a little over 3 miles long; its breadth is more than 430 feet, excepting at an elbow, where it is only 250.

In the middle of its course, where it runs nearly N. and S., it receives the river Angachilla, and then splits, forming the island of los Venados, which is 1¼ miles long and 820 feet wide. This island also runs N. and S., and is very low; an estuary cuts through its southern portion, and banks covered by aquatic plants run from the N. point nearly to the river Angachilla.

At the point of junction of this river with the Guacamayo there is a mud-bank with 6 feet of water on it, which runs from the northern extremity of the confluence nearly to the island of Guacamayo, which is on the western bank of the river of the same name. The passage between them is very narrow, and has only 1½ fathoms at low water, on the side of the island. This is the shoalest part in the branch of the river Valdivia, called Torna Galeones. A vessel drawing from 11 to 13 feet can pass this spot only during high tide. The depth of the river Guacamayo and its western branch is often more than 8 fathoms. The rise of the tide averages about 4.2 feet, and its strength between 1 and 2 knots.

The eastern channel of the Guacamayo, to the eastward of the island of Los Venados, cannot be used by vessels drawing more than 8.2 feet.

River Anga-chilla.   The Angachilla river, 7 miles long, as well as its tributaries, is subject to the tides. On its banks are wood-sheds and some villages. For half of its course it is 570 feet wide. Vessels drawing 8 feet can use it. The current is weak, and its tributaries are navigable.

River Cantera.   The river Cantera, at the Guacamayo, carries the waters of the Rio Valdivia, during the ebb, to the junction of the Tres Bocas with the Rio Torna Galeones. The strength of the current varies between 1 and 2 knots. The flood carries the waters through the same channel in an opposite direction, but with less rapidity.

The river Cantera is at the least 8 feet deep at low water, and in many places there are 6 fathoms. Its length is almost 4 miles. The banks are high. In its northern part is Valdez island. The rise of the tide is 4.2 feet.

D'el Rey and Guacamayo islands.   The Cantera river with the Guacamayo forms the island Guacamayo, and d'el Rey with it and the Rio Torna Galeones; some small rivulets from these islands empty into the Cantera; there is considerable wood here, and some timber-sheds. D'el Rey is covered with high hills; its few cultivated fields do not raise sufficient nourishment for the inhabitants.

River Futa.   After the rivers Cruces and Quinchilca, the river Futa is the most important tributary of the Valdivia; it has many tributaries of its own.

Vessels drawing 5 feet can ascend the river during high water to within ½ mile of Futa, which is about 13 miles from the mouth, in latitude 40° 00′ 20″ S., and longitude 73° 10′ 06″ W.

The tide runs up to Futa during the summer; vessels drawing 6.5 feet can only go as far as Pichi, 6½ miles from the mouth.

The current of the flood is not very strong in the lower river; and that of the ebb varies between ½ and 1 mile, excepting during the winter, when the currents are stronger, and the water often rises from 13 to 26 feet.

The river Futa empties into the Torna Galeones at the Tres Bocas. All this country is covered with wood, which is cultivated; also with coal, which is not now worked, but is often found on the surface.

It is high water, full and change, at Futa at 12ʰ, rise 0.2 to 0.3 feet. *Tides.*

The river Torna Galeones is 10¼ miles long from the Tres Bocas to the sea; but its length, united with that of the Guacamayo, is greater than that of· the Valdivia; there is more water in the latter and the navigation easier. *River Torna Galeones.*

From Tres Bocas the river runs S. for 3 miles; here its breadth varies between 820 and 1,360 feet, but the depth which averages 2½ fathoms at low tide increases to 15. The course of the river lies generally between high rocky hills or thick woods. In the interior the country is varied by high mountains. This river is seldom used by vessels. The bottom is tolerable even in this part of the river, being of hard stone and mud. There are no indications of dangerous rocks.

The flood and ebb tides run with a variable velocity from 1 to 2 knots. The rise varies with the age of the moon, from 4.2 to 4.5 feet.

From the narrow part mentioned the river runs W. for 1,640 yards, then WSW. for 2,150 yards, to the estuary de la Romasa; in the latter reach it receives the river Naguilan. The junction is called Poco Comer. From this junc‐ tion to the estuary the middle of the channel, in 2 fathoms, must be kept, as the banks are not steep-to.

From the estuary the river runs NNW. for 2¾ miles, and empties into the large bay of Corral. This latter part is the most difficult; it is full of rocks and banks, which neces‐ sitate a pilot for all vessels drawing more than 13 feet.

From the estuary de la Romasa the breadth varies between 1,640 and 3,280 feet. The banks are alternately rocky and muddy, and covered with reeds. The depth varies with the width, but the nature of the bottom remains the same.

The mouth of the Torna Galeones has been described. In its center is a bank with 8 feet over it at low tide; on either side are from 11 to 14 feet of water. The two shores are foul; the northern is flat with bottom of a muddy sand; the southern is bordered by a chain of rocks extending out 110 yards, some parts of which are uncovered at low water.

About ½ mile ESE. of Puerto Claro, on the N. side, is El Huapi island; it is oblong and wooded.

At 1½ miles ESE. from Puerto Claro is the island Liguiña, rocky and inaccessible except on its NE. end; it is flat and without trees, other than a few apple-trees. The space between it and Huapi island is full of rocks and banks. Its SW. coast.is foul to a distance of 164 feet.

The navigable channel runs to the south of Liguiña; it is 656 feet wide and bounded to the southward by a hard stone cay which uncovers at half-tide and is united with the bank that fills the cove back of the cay; all this part is thickly wooded. From here to the estuary de la Romasa the middle of the channel must be followed.

From this description it will be seen that after the passage of Liguiña the navigation of the Torna Galeones is easy, with exception of the Tres Bocas de Futa, where a pilot is indispensable. Vessels of more than 12 feet draught cannot attempt the passage.

River Naguilan. The river Naguilan is the largest tributary of the Torna Galeones; there are a few huts of wood-cutters on its banks. It is navigable for 3¾ miles for vessels drawing 5 feet; they go up to the small village of Naguilan to bring down the alerce. In the winter vessels of 8 feet draught can ascend; the water often rises 13 feet at that time. The rise of the tide is 9 feet.

As the attention of commerce will probably be drawn to this territory on account of its rich forests, it has been deemed desirable to give this short description of it.

FROM VALDIVIA TO CONCEPCION.

Variation from 19° 03' to 17° 15' easterly in 1876. Increasing annually 1' 30".

The 6 miles of coast between point Juan Latorre and Bonifacio head recedes a little, throwing out three small points, which are separated by sand-beaches. The first of these, which is immediately to the N. of Juan Latorre, is called Loncollen; it is steep like the former and surrounded by rocks which extend out 800 yards; to the northward there is a bad landing-place for boats, having the same name. Point Loncollen.

Following is Mission point, rugged and surmounted by a small hill; it is surrounded by rocks which extend out 800 yards. To the northward of it is another landing-place, but it is worse than that at Loncollen. A mission was founded here in 1777. Mission point

From Mission point to Bonifacio head the coast runs in a little to the eastward, forming a semicircular sandy cove, with some outlying rocks. The principal one of these is point Calfuco; on it are the ruins of a mission for the conversion of the Indians. From this a horse-path leads, in two hours, to the mouth of the river Cutipaï. Point Calfuco.

Bonifacio head is a steep promontory, 2 miles long N. and S.; its approaches are without danger. It lies 8 miles N. 16° E. from Gonzalo head; ¼ mile from it there are from 11 to 13 fathoms of water; bottom of large stones; at 2 miles the depth increases to 22 and 26 fathoms with sandy bottom. The upper part of the hill is thickly wooded. To the eastward there are mountains from 2,000 to 2,300 feet high. The coast between Morro Bonifacio and Mayquillahue bay is only accessible by boats. Bonifacio head.*

There is a small cove immediately to the northward of Bonifacio head, called Ronca, which is much frequented by fishermen, who report that the landing is excellent. Ronca cove.

---

* The explorations on the coast of Araucania were made by Captains J. E. Lopez and F. Vidal Gormaz, Chilian navy.

Lican cove is half-way between Bonifacio and Mayquillahue; the intermediate coast appears to be steep and clear.

The low, small point 11 miles from Bonifacio is called Chanchan point; it shelters a creek of the same name from southerly winds, which is in latitude 39° 34′ 20″ S. and longitude
73° 19′ 09″ W. According to the coasters it offers good
shelter to small vessels and a good landing for boats.

The southern extremity of point Mayquillahue, until recently called Chanchan, is a little more than 4 miles to the
northward of Chanchan point; it is 1 mile long N. and S.;
some rocks extend from it on the N., S., and W. sides. The
islet and reefs to the northward extend at least ⅔ of a mile,
in a NNW. direction; they must be looked out for when
running into the anchorage. The northern of these islets is
6 miles N. 5° E. from point Chanchan.

Point Mayquillahue gives its name to a large bay which
opens to the eastward; it is 4 miles wide and 2 deep.

In the southern part of this bay is a cove, affording shelter from S. to W. winds ⅓ mile E. of the northern islet, in
from 3 to 6 fathoms of water, bottom sand and shell strewn
with large stones. The cove is entirely open to N. winds,
which render it untenable. A shipwreck which took place
in 1838, during one of these winds, probably led to the report that the holding-ground was not good; the islands enlarge the anchorage, and render it better than that of
Queule.

The river Mehuin or Lingue empties in the eastern part
of this bay; it is navigable by boats for 15 or 16 miles;
its bed is full of large fallen trees and the current runs from
3 to 5 knots an hour. The banks are fertile and inhabited
by the native Indians, who will sometimes furnish provisions; but their acquisition often leads to trouble with these
unsubmissive people.

The mouth of the Mehuin is crossed by a chain of reefs
which start from the S. point and make its entrance dangerous. The depth on the bar is 2 feet at low and 7 at high
water.

At the northern extremity of the bay is a small creek with
a sand-beach, called El Mariscadero, formed by the projection to the westward of point Ronca. Reports to the contrary, it would seem that this cove was even more danger-

ous with N. wind than the cove of Mayquillahue and Queule, when considering the gyration of the winds on this part of the continent, and that there is no obstacle to the entrance of the sea.

Vessels anchored at Mayquillahue should take refuge at Corral on the first indications of a norther.

Point Ronca, which separates the bay of Mayquillahue from that of Queule, appears like an island, but is connected with the continent by a chain of hills. The southern part is called point Loberia and the northern point Choros. *Point Ronca or Queule.*

The bay of Queule is formed to the northward by point Nigue and to the southward by point Ronca; it is 4½ miles wide and 1 deep. *Queule bay and cove.*

The cove of Queule, in the northern part of the bay, is named after the river which empties into it; the mouth of the river is in latitude 39° 25′ 20″ S. and longitude 73° 16′ 06″ W.

The depth of the bay is moderate; the bottom is of fine, hard sand; in the cove there are from 3 to 5 fathoms; the best anchorage is in 4¾ fathoms on the following bearings: The barren ravine of the southern mountain, S. 21° W.; the center of the barracks at the mouth of the river, S. 6° E.; the NW. extremity of point Ronca, S. 70° W. Small vessels can anchor a little nearer to the shore, in 4 fathoms. The anchorage is sufficiently sheltered from the SW. swell, but it must be left on the first signs of an approaching norther. It is imprudent to attempt to ride out gales from the northward and westward, excepting during the summer. During fresh winds the sea breaks in the entire bay in 6 to 7 fathoms of water. It is best to go to port Valdivia until the bad weather is over.

In going in or out of this bay it is necessary to give Choros a berth of 328 yards, as a small, rocky bank, called Martinez, with 2 fathoms of water over it, lies 240 yards to the northward of the point; the depth of water over it is irregular; there is a passage between it and the point with from 6 to 7 fathoms. *Martinez rock.*

The remainder of the rocky coast is steep and without danger; this is not the case on the shores of the river, which are very low and covered with alluvial deposits.

The cove of Queule has become the key of operations against the Araucanians; it owes its importance solely to *Description.*

this fact, as it is no better than any of the other inlets along the coasts.

On its shores are store-houses for merchandise, barracks for the troops, and boats belonging to the government and private parties; fresh and other provisions can be obtained, brought here by the coasters. The water is excellent, from a ravine to the southward of the mouth of the river. Ballast is thrown overboard about 1 mile to the northward of the anchorage.

Directions.

During clear weather the bay of Queule can be easily recognized by point Ronca, which seems to be an island; by point Nigue, which is larger and more extended, and by the two naked islands to the northward of point Mayquillahue. The entrance is easy, even during foggy weather, which is very frequent during all seasons, as the depth of water is very regular to a distance of 2 miles from the coast.

Tides.

It is high water, full and change, at Queule at $10^h 28^m$, rise 5 feet.

Queule river.

Queule river has become of some importance since the occupation of the coast of Araucania; its bar is only impassable with winds from the N. to W.; it is, however, generally smooth; at low tide there is but 1 foot of water on it; during high water vessels drawing from 4 to 5 feet can cross, the depth being then 6 feet. They are guided by the stakes placed at the ends of the banks between which the river empties.

The bar is formed by two sand-banks, one of which starts from the southern bank and the other from the small point to the northward; the latter ends in a line of rocks which runs into the end of the creek, and, approaching the S. bank, only leaves a channel of from 130 to 165 feet. After the bar is crossed the depth increases from 6 to 9 feet at the sandy point which forms the mouth of the river to the northward; an isolated rock which is in the center of the river must be left to starboard; from it the river enlarges, but in its center there is a large sand-bank, leaving a channel for boats on either side; the northern channel, which is alongside the bank, is the longest, but the deeper; both can be forded at low tide. From this bank a small, rocky point will be seen, which, in line with the isolated rock, indicates the southern channel. This point is passed close aboard, and then the middle of the river followed to the estuary of Catrehue, which is on the left bank, a little more than 1 mile from the cove of Queule, near which is the mission of Queule. To land, the river must be ascended a little farther, leaving a small island, which is in the center of the channel, to starboard. Boats land just before reaching this, at the mouth of the river Piren, which empties into the Queule on its left bank.

From this point for 3½ miles, to the settlement of Cayulpu, the channel, with a minimum depth of 2 feet at low water, is very tortuous and approaches the sea. The settlement is only 656 yards from the sea, separated from it by sand-dunes, which are constantly encroaching on the river-bed. From Cayulpu the depth is between 1½ and 3 fathoms, and the current is hardly perceptible at the port or fort de los Boldos, 17½ miles from the cove of Queule and 2¾ miles from the settlement of Tolten, which can be reached by a road.

The river Queule then runs parallel with the Tolten, and continues to be navigable for more than 9 miles; its principal tributary is the river Voroa, which is equally navigable. The current of the river Queule is the same during flood and ebb, and varies between 1½ and 3 miles.

From Queule creek to the point called Nigue by the natives, the coast is bordered by sand-dunes. The western part of the point is of medium height and thickly wooded; its shores are rocky and bordered by some hillocks; it is formed by a mountain of the same name, which extends a little more than 1 mile to the eastward and is separated from the mountains of the coast by the river Queule. *Point Nigue or Tolten.*

From this point to the river Tolten the coast is a low semicircular beach, backed by sand-dunes.

The mouth of the Tolten river, situated in 39° 16′ 50″ S. latitude, and 73° 16′ W. longitude, was mistaken for the river Cura by Fitz-Roy; it is 4½ miles N. 3° E. from point Nigue; the entrance is hardly perceptible at a distance of 1 mile. During the dry season, from January to April, it is from 260 to 300 feet wide; the surrounding country is low, formed of uniform sand-dunes, with the exception of one, which is ⅓ mile N. 74° E. from the mouth. The coast *River Tolten.*

is without shelter, washed by the waves of the ocean. The river deposits considerable sand-banks at its mouth, which are continually shifted by the sea, making a bad bar and shifting channels, with a constant line of breakers. There are generally, however, two passes; the first opens S. 56° W., and the second N. 50° W.; their depth varies between 1 and 1¾ fathoms. Steamers drawing 6 feet have entered. A pilot is indispensable. At the entrance for 1½ hours the current of the ebb attains a velocity of 7 knots; it never ceases, and the flood only tends to weaken the velocity of the descending waters. The current inside averages about 2 knots. After crossing the bar, and before arriving at the narrowest part of the mouth, there is a small bank of pebbles with 6 feet of water over it, in the center of the channel; the depth then increases suddenly to 2½ and afterward to 5½ fathoms. From this bank the middle of the channel is followed to a small, rocky, and remarkable point on the left bank, which must be borrowed on to avoid some sandbanks on the opposite shore, until the entrance of the river is shut out by a small point on the right bank. There are here some sand-banks, probably shifting, which dry at low water, and occupy nearly half of the river. The river then takes a turn, after which the middle of the channel is kept, and then the right bank.

About 1 mile from the first elbow there is an extensive sand-bank on the left bank, marked by trees; after passing it, and when the N. part of the south bank bears ENE., it must be followed to the entrance of the river Catrileuvu and to the village of Tolten, off which vessels can anchor 110 yards from the shore, with 3 to 3¾ fathoms at low tide and good holding-ground.

Farther up there is still a good depth, but a pilot is necessary. The reconnaissance of this part of the river is incomplete.

This river has terrible freshets during the winter months. The water rises over 9 feet and sweeps everything with it. In the description the depth of water has been given at its lowest point. The river Tolten has its source in the western part of a lagoon called by the Indians Larquen; it has a circumference of 40 miles, and is situated at the foot of the volcano of Villarica.

It is high water, full and change, at 10ʰ 28ᵐ ; rise at the Tides.
mouth, 5 feet.

The mission of Tolten is separated from the village by Village and Mission of Tolten.
the river Catrileuvu, which can be ascended by vessels
drawing 6 feet. The village, located between the Tolten and
a large lagoon, is in the middle of fertile lands, overgrown
by excellent wood of all kinds. It was founded in 1867.

A road 45 miles in length, following the shore, unites the
mission of Tolten with that of Imperial; another road unites
Tolten with Queule by the Cerro de Nigue; it has a bad
pass, but the distance is accomplished on horseback in three
hours; the road by Los Boldos and the river Queule requires
5 hours.

Off the coast in clear weather the summits of the Andes
can be seen; it is said that the volcano of Villarica, 60 miles
to the eastward of Tolten, is visible that distance to seaward.

The coast between the river Tolten and the river Imperial Coast.
is low, a sandy shore backed by abrupt and pointed cliffs.
The Cordillera of the coast is from 5 to 10 miles distant, excepting at a point called Puancho, where it sends to the sea
a spur from 150 to 300 feet high. Several rivers empty on
this part of the coast.

The mouth of Yenellenchico river is 6½ miles N. 16° W. River Yenellen-chico.
from the river Tolten, in latitude 39° 10' 27" S., and longitude 73° 18' 5" W. For 3 miles this coast is covered with
small sand-dunes, then it becomes stony.

This river has its source in some lagoons about 3 miles
from the coast; its banks at the entrance are remarkable; it
is a rapid torrent in the winter, and nearly dry in the
summer.

The Ruca Cura river is 2½ miles N. 10° W. from Yenel- Ruca Cura river.
lenchico river, in latitude 39° 07' 50" N., and longitude 73°
18' 30" W. The intermediate coast keeps its rocky aspect;
on the beach at the south side of the mouth is a small hill.
This small river is without importance to navigation; it
has its source in the Cordillera of the coast in the middle
of a marsh 7 miles in the interior.

There are some small hills about 3 or 4 miles back from
this coast; at their foot is an Indian village.

The mouth of the river Chille is in latitude 39° 00' 09" S., River Chille.
and longitude 73° 19' 30" W., 16½ miles from Tolten and 8

11 c

miles N. 4° W. from the river Ruca Cura. The two rivers are separated by almost uninterrupted high cliffs; but 1 mile on either side of the river Chille the coast is low and covered with small sand-hills. During the summer the mouth of the river can be forded, but at other times boats are necessary to pass it. It is navigable in the interior, and its banks are fertile and populated. The Cordillera of the coast, in which it has its source, is 10 miles from, and runs parallel to, the coast.

Point El Barco de Puancho. Point El Barco de Puancho is in latitude 38° 54′ 00″ S. and longitude 73° 22′ 10″ W., 6½ miles N. 18° W. of the river Chille. It does not project into the sea, and is remarkable for its cliffs, 260 feet high. Puancho rock, called El Barco by the natives, steep on all sides, is 656 feet to the westward of the point. This is, properly speaking, the point, but the name is given to a stretch of coast 4 miles long, which descends gradually to the southward and prolongs its cliffs to the river Chille.

River Budi. The river Budi is 5 miles N. 22° W. from El Barco; it is only the mouth of a salt-water creek, situated between the Cordillera of the coast and the sea. Its entrance is in latitude 38° 49′ 26″ S., longitude 73° 24′ 30″ W.; after passing for about 2 miles between hills of moderate height, it opens into the lagoon of Budi or Colem.

Its breadth at the entrance is about 200 feet; it is deep, and the tides penetrate to the lagoon; during the dry season it can be forded at the mouth, there being but 2½ to 3 feet of water, but the ford is dangerous, as it is reached by the swell from the ocean. The Budi river cannot be entered from the sea; to the northward of its mouth is the wreck of a small steamer.

The SW. swell washes the sand to the entrance of the river Budi and closes it completely from February to April, and the waters of the lagoon bank up and inundate the fertile valleys of the neighborhood. A channel is then cut in the sand which arrests the water. This is the occasion of a great festival, with an abundant supply of fish for the Indians. The lagoon is 6 miles long N. and S. and 4 miles wide from E to W.; it is full of islands, and deep, like the river.

Cholgi head. Six miles N. 22° W. from El Barco is Cholgi head, the end of the mountain chain which separates the rivers Budi and Imperial, whose mouth is about 1½ miles from the former.

Cholgi and Trugue head, which is 1 mile to the NNE., are 390 feet high, and are excellent landmarks for the entrance to the river Imperial.

The river Imperial was discovered by Pastène in 1544; Valdivia founded a city, called Imperial, on its banks, 20 miles from its mouth, but it was destroyed by the Araucanians. The river takes the name of Cautin above the ruins of the city. The mouth, which is 433 yards wide, in latitude 38° 47' 45" S. and longitude 73° 25' W., opens to the SW. It is difficult to make out from seaward; at first there appears to be but a dangerous coast beaten by the breakers, which generally extend out $\frac{1}{4}$ of a mile from the shore, and, during bad weather, for 1$\frac{1}{2}$ miles. On looking more attentively, two places will be seen where the sea does not break, and where it is less agitated during bad weather; these are the channels; they are formed by sand-points, which shift from one tide to the other. The Indians assert that the mouth has moved to the southward; in fact, the river runs parallel to the coast for 3 miles, before emptying into the sea, from which it is only separated by sand-hills less than 1 mile wide. The mouth of the river has the shape of a cornucopia whose outer portion terminates at Cholgi hill, on the left shore; the inner part is terminated by a sand-point on the right bank, the end of the cornucopia being turned to the northward. The two points do not project outside of the coastline.

River Imperial or Cautin.

The waters of the river coming from the N. are arrested by Cholgi hill, and, turning to the SW., break through the sand, which, coming from the south, forms a bar with two channels. Inside of the bar the deepest water is found in a small indentation under the Cholgi named El Caleton; the bar runs N. and S., the channels opening at either end, about 500 feet apart. The first, examined in 1845, 1866, and 1869, seems to undergo little change; it has 12 feet at low water and 18 at high, with a breadth of not less than 500 feet; it opens nearly WSW. from the foot of Cholgi head. The second has 6 feet at low water and 12 at high; its direction is about W. by S. from Trugue head; it was examined in 1869. These channels unite at El Caleton, along Chogli head, which must be passed as close as possible, as a sharp turn has to be made here to enter the river.

Vessels should always have a river pilot when crossing the bar. After the bar is passed, the river, 433 yards wide, runs nearly N. and S. for $5\frac{1}{2}$ miles, to the confluence of the estuary Mocho, with a depth varying from 3 to $5\frac{1}{2}$ fathoms. The left shore is a swampy plain, bordered by banks which contract the channel; the right shore, which is covered with sand-dunes, is clear. A vessel should anchor before arriving at the confluence, in 4 fathoms, with her head to the promontory which separates the two water-courses. The estuary Mocho, 218 yards wide, is navigable for at least 4 miles; its depth at low water is 4 feet; the channel is tortuous; there are some houses along its banks.

From the confluence, the Imperial is clear of banks and 380 yards wide; it runs east for 5 miles, with a minimum depth of 3 fathoms, to the island Doña Ines, which must be left to the northward. After which the river flows between ranges of hills, and is clear and deep for 7 miles to Maule bank, which takes up two-thirds of the river; a small channel on the S. side can be passed at high tide by vessels drawing 12 feet. There is no danger for the next 3 miles, at the end of which is the ruins of the city Imperial; here, behind an island, which was the last place held by the Spaniards, are $2\frac{1}{4}$ fathoms of water.

In 1869 the Chilians had not passed this point, as the Araucanians of this locality were more hostile than those of the coast.

The freshets have little influence on the currents of this river, which vary between 2 and 4 knots. As the reconnoissance of which the results are here given was made during the dry season, the depths given can be considered as the minimum; therefore, a vessel drawing 12 feet can go up to the ruins of Imperial. A vessel drawing 15 feet has to stop at Maule bank.

A mission founded in 1852, and tolerated by the Araucanians, is located about 2 miles N. 48° E. of point Trugue. The road from Tolten ends here. The Fathers occupy themselves in cultivating the land. The fiercest and most indomitable of the Araucanians are found between the rivers Tolten and Imperial. They are numerous, and, aided by a fine climate, they cultivate this rich land.

It is high water, full and change, at about 11^h; the rise at the mouth is 5 feet, and opposite the ruins from 2 to 3 feet.

Cautin head is a barren and rugged promontory 328 feet high, and clear to seaward; it is 9 miles N. 22° W. from Cholgi point. The intermediate coast is low and backed by sand-dunes, the highest of which is half-way between the points; the beach is sandy and slightly undulating.

The coast continues N. 4° W. for 11 miles, clean and rug- ged, to point Manuel, which is also 328 feet high; it is over-looked by a high mountain chain called the Cordillera de los Pinales, in which the celebrated Araucaria or piñon tree is found.

Cape Tirua is a little more than 7 miles from point Man- nel; it projects well to the westward, where it ends in a small island; it is the nearest point of the coast to Mocha island, with which the Indians, from the river Tirua to the northward of the cape, communicate by rafts, balsas, and canoes.

The road de los Riscos, 20 miles long, and the shortest connecting Tirua with Imperial, passes by cape Tirua over precipices and through deep gorges.

The other road connecting these points. called Pinales. traverses that part of the Cordillera where the trees of this name abound; it runs up the valley of the Tirua for 9 miles; it is longer than the former but less difficult.

The channel between the island Mocha and the continent is 18 miles wide; it is clean, with a depth of from 9$\frac{3}{4}$ to 20 fathoms over a sandy bottom; the flood runs through it, setting to the northward at the rate of 1 knot; the ebb sets SSW. with less strength.

Mocha island, 1,200 feet high, is an important landmark for navigators; it is about 7 miles long and 3 wide; it should not be approached too close, as there are dangerous rocks off its S. and W. coasts; the most distant are 3 miles S. of the island. These rocks are particularly dangerous during the flood, as it sets over them coming from the SW., and then bends to the northward in the channel; sometimes the ebb is not felt for days, then the flood is a continuous northerly current. The depth increases rapidly 2 miles from the west coast, which must be considered when approaching during foggy weather. The fog often lasts several days on this coast. On the east side of the island the rocks do not run out farther than ¾ mile. The anchorages of Mocha are of no importance. Vessels can anchor on the NE. coast or near the SE. extremity, off a point called Anegadiza by the Spaniards. This anchorage, which is good during N. winds, is opposite to the first small hills in from 5 to 7 fathoms. In anchoring, a lookout must be kept for a bank near the coast, on which the sea does not always break, but it is well marked by large bunches of sea-weed. The other anchorage, which is tolerable during S. winds, is off English Creek, about 1¼ miles from the coast, and has from 12 to 20 fathoms, sandy bottom. Nearer the land the bottom is rock. Landing is difficult, but can be effected when necessary. Wood can be obtained, but no provisions; the water, which is excellent, can only be taken in with great difficulty. There is much good land on the island which produces large quantities of vegetables and serves as pasture for cattle.

Before the 18th century the island was inhabited by the Araucanians, who were driven from it by the Spaniards. During the time of Fitz-Roy it was only occupied by some few stray animals. At present it is visited by hunters and fishermen.

Tirua bay   Cape Tirua, to the ENE. of which the bay of that name opens, protects the latter from southerly winds but not from the sea; the anchorage is always exposed to the swell, and the beach is so bad that even the best boats should not attempt to land. The river Tirua empties here. The Indians cross its bar in their canoes.

Bay of Quidico,   The bay of Quidico, or Quirico, is in latitude 38° 14′ S.,
or Quirico.   longitude 73° 27′ W. After passing cape Tirua the coast runs in to the NNE., afterward projecting to the NW. Eight miles to the northward of cape Tirua is a high point, 1 mile long, called Neña, which trends about N., and is prolonged ¾ mile by a bank making out to the northward. Generally the sea breaks over it every 8 or 10 minutes. Rounding this point is the entrance to the bay of Quidico, across which there is always a heavy swell; a short distance inside the water is always smooth. It is about 1¼ miles wide at the entrance, and is 1 mile deep. After passing the high

auds of the point is the western mouth of the river Quidico, which forms a delta, in front of which the sea always breaks heavily; then comes a sand-beach, on which is the best landing. About 656 yards to the westward of the eastern point of the river, at the foot of a mountain which seems to divide the valley into two parts, the river divides into two arms, one branch forming the western mouth and the other the eastern, which runs along the beach for 3 miles before emptying.

The anchorage of steamers is E. of point Neña or Quidico, in 5 fathoms, bottom sand and shell; it is sheltered on the south side but open to northward.

The bay is deep to the beach, where there are 3 fathoms. Sailing-vessels must anchor to the northward of the point to be in position to get under way with northerly winds. The plain of the valley, on which there are some Indian huts, rises in a gradual but constant slope, and is free from inundations from the rivers which traverse it. The surrounding hills, covered with trees, contain coal-mines, and the land is well adapted to cultivation. These hills are followed by high wooded mountains, which are a branch of the Cordillera de los Pinales. As the river Quidico is only fordable at the mouth, all the roads following the coast of Araucania meet at the head of the bay. This circumstance, and the fact that it is the only passable harbor between Lebu and Queule, gives it some importance as a military point.

The Lleuleu is a small river whose waters join those of a **River Lleuleu.** lagoon emptying about 9 miles to the northward of Quidico; it is nearly dry during the summer. The bar is inaccessible.

The river Paycavi, which has its source to the N. of **River Paycavi.** Tucapel, runs N. and S., and receives the waters of several lagoons or estuaries; it is not wide but tolerably deep, having but few fording-places; it can be navigated by canoes drawing from 2 to 3 feet. The Spaniards made it a line of defense against the Indians, and built a number of forts, the ruins of which still exist; they had a flotilla which crossed the bar, but that is now impassable. In the summer it is dry at low tide, and during high tide it is closed by an uninterrupted line of breakers.

The coast between Quidico and Morguilla forms a semi-circle, with a sand-beach ; it is free from outlying dangers, the rocks indicated by Fitz-Roy not having been found; it is, nevertheless, bad, as the sea breaks for more than 1 mile from the beach. There are some other small estuaries besides the rivers Lleuleu and Paycavi.

Morguilla island. Morguilla island is about 26 miles from point Quidico; it is nearly round, with a diameter of 1 mile, and is about 32 feet high; its soil is good; it is almost connected with the continent by a sand-bank of recent formation. To the north and south of it are two islands frequented by sea-wolf. The northern island forms Curaco cove, which is only accessible during very fine weather. All dangers are visible, and vessels can approach the island within ½ mile. H. B. M. S. Challenger was lost on this island in 1835. This loss was attributed to currents setting toward the land, a fact which was confirmed in 1868 by the observations made between Valdivia and the bay of Arauco by captain Mayne, in H. B. M. S. Nassau.

Point Chimpel and Lorcura. Between Morguilla and Bocarripé the coast curves a little to the eastward, placing points Chimpel and Lorcura, or Tucapel, in full view; ¾ mile to the seaward of the former is a rock ; the latter is low, rocky, flat, and of a dark color.

In this vicinity the interior of the country is very fertile, and is strewn with hills and valleys covered with wood and pasturage well watered.

Bocarripé head. Bocarripé head is 2½ miles south of Lebu or Tucapel head, and 12 miles from Morguilla. From it point Lorcura bears S. 17° 30' E.; the islet near that point S. 9° 30' E., and Morguilla S. 8° E.

Three small water-courses, Lorcura, Chimpel, and Curaco, empty between Morguilla and Bocarripé. From the latter to Lebu hill the coast, which is rough but clean, trends Lebu or Tucapel head. about north. Lebu head is 625 feet high, steep to the westward, and descends gradually toward the south. The pirate Benavides took refuge in a cave near this point during the war of independence. In 1835 the crew of H. B. M. S. Challenger camped on these heights until they were taken away by H. B. M. S. Blonde. The latter anchored in 27 fathoms N. 27° W. from the head.

Port of Lebu. The cove between point Millongue to the north and Lebu head to the south takes its name from the river Lebu, which empties into it. It is 1 mile wide and 2 long, in latitude 37° 35′ 26″ S. and longitude 73° 40′ 04″ W. The depth of water is 7 fathoms over a sandy bottom, which shoals gradually to the beach. Lebu head and a plateau of rocks, which extends out from it for ½ mile to the northward, protects it from the SW. swell. The best anchorage for a small steamer is in 3½ fathoms, about 200 yards E. of the highest rock of the plateau. A sailing-vessel should anchor about 1 mile to the northward of this, to enable her to get under way readily in event of a north wind, to which the bay is open. Under these circumstances a steamer can find a good anchorage 13 miles to the northward, in Yanès cove, for making which steer for Carnero hill, to clear Maule bank.

The communication with the shore is only interrupted by a strong N. wind; boats can always enter the river or land in a small creek close to the mouth. Ballast is thrown on the sand-beach between the mouth of the river and the Morro Cueva, the extreme point of Millongue; stone ballast is thrown on the banks at the entrance; there are four stranding-places and ship-yards and one tug.

Lebu was founded in 1863; its population is 1,000. The coal-mines have been the cause of the rapid development of the city.

It is high water, full and change, at Lebu at 10ʰ 30ᵐ; rise, Tides. 5.5 feet. The tide runs up the river to Salto de Gorgolen, about 10 miles from its mouth. The mouth of the Lebu river is to the NW. of the anchorage : the left bank, which runs along Lebu hill, and is strewn with rocks, can be seen from the anchorage ; the right bank is entirely of sand. The mouth of the river is about 100 feet wide, and the depth of the channel at low tide is never less than 5 feet. The tides have a velocity of from 4 to 6 knots : as soon as the mouth is passed this velocity diminishes : the river enlarges and is filled with banks. The depth at the ford, which is less than 1 mile from the mouth, is never less than 4 feet ; from it the depth increases. These results were obtained during the dry season ; during the rainy season the river rises about 4 feet, but the strength of the current is not increased.

This very tortuous river has its source in the Cordillera de Nahuelhuta, about 15 miles from the sea. At low water there are many banks and fords; but it can always be used by vessels drawing not more than 8 feet and not longer than 100 feet, for a distance of 3 miles; beyond, not more than 3 feet can be taken.

The old settlement of Tucapel, or Lebu Viejo, at the foot of the plain to the southward of the mouth, has been rebuilt.

On the south bank of the river is a mole, from which the coal of the surrounding mines is shipped.

Point Millongue, which separates Carnero bay from Lebu cove, is surrounded by reefs, which form a small port, in which vessels can anchor in 5 fathoms; sandy bottom. At the head of the port is a small landing-place, well sheltered from the southwest swell, called fort Viel; it is open to the N., and would have no importance but for the coal in its vicinity.

Carnero bay is formed by Millongue point to the southward and Carnero hill to the northward; it is a large bay, 11 miles in width from N. to S., and 4 miles deep, and embraces the coves of Ranquil and Yanès.

Port Ranquil is situated between point Millongue and point Huenteguapi; it takes its name from a small estuary which empties here; the water is shoal and the bottom strewn with rocks. The north beach is bad, but boats can always land among the numerous rocks which border the south shore and extend out for ½ mile. It also owes its importance to the coal-mines in the vicinity.

From point Huenteguapi to point Liles the coast is studded with rocks to about 200 yards to seaward, and cannot be approached by boats; two points make out from it, Batro and Locobe. Three miles west of the latter, and 4 miles to the southward of point Carnero, is a chain of sunken rocks, over which the sea breaks, called Maule bank; there is a passage between it and Locobe point; to clear it, keep outside of the line joining Lebu head and the foot of Carnero hill, until at least 2 miles south of the islets of Yanès.

Point Lacobe is to the northward of point Liles; on rounding it, in the north of Carnero bay, is Yanès cove, closed to the westward by Carnero head and two islands called Pichiguapi and Uchaguapi, situated SE. of the hill; the northernmost of these is joined to the coast by a recently formed sand-spit. Yanès cove is an excellent anchorage, sheltered

(marginal notes:)
Point Millongue.
Carnero bay.
Port Ranquil.
Maule bank.
Yanès cove.

from winds from west to east by the north; there are 13
feet of water close to the beach, bottom sand. The islets
are rugged and clean, and although the cove is open to the
westward, the sea is so broken by them that the anchorage
can be kept when the wind is not too strong. Boats can
always land in the corner of the cove, where the sand-tongue
joins the northern island, as near as possible to the latter;
this corner is large enough to shelter a vessel of moderate
size. Steamers which enter this cove should anchor on an
E. and W. line with the northern island. Good water is
abundant, and can easily be taken in at the estuary of
Tralicura.

From Yanès the coast projects to the SW. as far as the
cliff of Carnero, the southern extremity of a chain following
the coast. *Carnero head.*

To the northward of Carnero, between it and the island
Piures, are two small coves with no hidden dangers, but they
offer no shelter against wind or sea. To the N. of Piures
island, which is a little nearer to Carnero than to cape Rumena, is a small cove in which boats can find excellent shelter after crossing numerous reefs; a pilot is, however, indispensable. *Piures cove.*

The coast between Carnero and cape Rumena trends N.,
and is of moderate height; extending from it ½ mile to seaward are rocks and breakers.

Cape Rumena, 8½ miles N. 3° E. from Carnero hill, is
steep and surrounded by rocks, as follows: *Cape Rumena.*

About 2 miles S. 17° 30' W. from cape Rumena is a bank
which is 1 mile from the shore, and which only shows above
water at low-water springs. *Rock awash.*

Captain Hall, of the P. S. N. Co.'s steamer Cloda, in 1859,
saw the sea break over a sunken rock situated N. 27° W.
from the outermost rock off the extremity of cape Rumena,
and 2½ miles from the land; this is dangerous as being in
the route of vessels crossing the channel between Santa
Maria and the continent. It shows at low water, and with
a heavy swell the sea appears to break about every 15
minutes. The Chilian reports confirm this vaguely; they
place Hall rock 3 miles W. 17° N. from cape Rumena, and
say that it only breaks during heavy gales. *Hall rock and Four Fathom bank.*

Mr. Petch, master R. N. of H. B. M. S. Shearwater,

searched unsuccessfully for this rock in 1864, both in the vessel and with boats. He discovered about ⅔ mile from the coast a bank of rocks in 4 fathoms, on which the sea breaks heavily during moderate weather. Cape Rumena bore S. 17° 30' W. 1½ miles distant, and point Lavapié N. 57° E. Mr. Petch concludes that Hall rock may exist farther off shore, as several coasters and captains of mail steamers assured him that they had seen the sea break over it in bad weather.

Raimenco cove. From cape Rumena the coast runs nearly NE. for about 6 miles, to point Lavapié; it is formed of high, steep cliffs, behind which there is elevated and thickly wooded land.

About 4 miles from point Lavapié is a point, to the north-ward of which is the beach of Raimenco cove; boats can here land easily. The anchorage is about 1 mile in extent, and is sheltered against the sea and winds from the south, but is open to the north; to the SW. are two small banks. The beach is small and surrounded by high mountains containing coal.

Four fathom bank. The charts show another 4-fathom bank 1½ miles SW. from point Lavapié and ½ mile from the land. It is recommended not to approach the land between Carnero and Lavapié closer than 4 miles; the depth of water is uncertain and the currents set to the eastward at a rate of from 1 to 2 miles an hour.

Point Lavapié. Point Lavapié is low and surrounded by reefs extending to the W. and NE. for ¾ of a mile; there is a rock surrounded by others ½ mile N. 17° E. from the point; the sea breaks constantly over it. The point must not be approached nearer than 2 miles.

Channel be- tween the island Santa Maria and point Lavapié. The channel between point Cochinos or Lobos, the south-ern extremity of Santa Maria, 5 miles wide, is subject to strong tides. When these currents combine with the wind from seaward, the ebb, for instance, with SW. and W. winds, the sea breaks more or less in the entire channel, especially to the southward of the Delicada reef, where there are but 5½ fathoms in mid-channel; many rocks have been reported whose positions have not been verified. It is, therefore, more prudent not to use this channel during the night. Coming from the southward, being 4 or 5 miles from the land, steer for the middle of the channel and pass it steering east.

When coming from the bay of Arauco, after the middle of the channel is passed, the vessel heading west, no northing or southing should be made until at least 4 miles from cape Lavapié.

Hector rock has been reported 1½ miles N. 17° E. from the eastern part of point Lavapié ; it would, therefore, be nearly in mid-channel and requires a careful lookout. This rock was unsuccessfully searched for in 1860 by the Chilian commander Rebolledo, and in 1864 by H. B. M. S. Alert and Shearwater; the latter vessel anchored as near as possible to the position in 13 fathoms, bottom fine sand, and searched for the rock with vessel and boats. They only found a clump of rocks in 2¾ fathoms N. 6° E. ¼ mile from the eastern rocky point of the bay of Luco, and a rock awash, on which the sea breaks violently, 1,200 yards N. 27° W. from Lavapié point.

Hector rock.

The English charts erased this rock, and it was again reported in 1870 by the commander of the Chilian war-steamer Ancud, in latitude 37° 06′ 35″ S., longitude 73° 32′ 40″ W., 4 miles N. 57° E. of point Lavapié and 2½ miles from the position where the Shearwater sounded. During the latter part of 1871 Lieutenant Riches, of H. B. M. S. Scylla, searched, without success, sounding constantly with two boats. Twice it was almost calm with a heavy swell, and the sea broke heavily on point Lavapié and the reefs, but nothing was discovered. These contradictory reports necessitate a good lookout in this passage.

The P. S. N. Co.'s steamer Araucania touched during the night of May 18, 1871, off point Lavapié, it bearing S. 17° W., 2 miles distant. Lieutenant Riches could find no obstructions here.

Araucania rock.

This rock, whose existence is doubted, was discovered in 1849 by Master Rundle, R. N., commanding H. B. M. schooner Cockatrice, when searching for the rock on which the John Renwick was lost, which proved to be the Dormido rock, off the north end of Santa Maria island.

Cockatrice rock.

Cockatrice rock is S. 29° W. from Cadenas point, the western extremity of Santa Maria island, and more than 3 miles S. 85° W. from point Lobos or Cochinos. This rock is dangerous though not in mid-channel between Lavapié and Santa Maria.

Santa Maria island. Santa Maria island is comparatively low, and dangerous on account of the many reefs surrounding it. The coast is bordered by cliffs, excepting to the eastward; the currents around it are very irregular. It is 7 miles long in the direction of the meridian and from 1 to 4 miles wide.

Proposed light. It is proposed to establish here a light of the fourth order, showing a fixed white light, varied by flashes, visible 16 miles.

Point Lobos, or Cochinos. Point Cochinos, the southern extremity of Santa Maria, is surrounded to the S. and SW. by rocks which extend out 1 mile; the sea does not break over all of them. One of them, over which the sea does break at intervals, is about 36 feet square, $3\frac{1}{2}$ miles N. 40° E. of point Lavapié and S. 13° E. from point Cadenas, the western point of Santa Maria; there are 2 fathoms over it and from 5 to 7 fathoms near it; it was reported by the Chilian commander W. Rebolledo. From this the reef runs NW. to Dormido rock, and extends about 2 miles from the shore of the island.

When going through the channel between Lavapié and Santa Maria, Cochinos point should not be approached within 1 mile until it bears N. 45° W.

Roadstead of Santa Maria. There is a passable roadstead off the SE. part of the island, with anchorage in from 4 to $8\frac{1}{4}$ fathoms and good bottom, but the only sheltered spot is under the eastern part of point Cochinos. Formerly there was a good anchorage between this point and the sand-spit Delicada, the SE. point of Santa Maria, but the earthquake of 1835 raised the bottom, leaving but $1\frac{1}{2}$ fathoms water. The depth diminished gradually in the cove, and vessels can choose their berths according to their draught. It is necessary to pass 1 mile SE. of Cochinos, as a bank makes out for half that distance; doubling this bank there is a small cove close to the point where landing is possible, and to the SE. of which a vessel may anchor in 5 fathoms. To the north of this cove the coast of the island is strewn with outlying banks as far as a gorge, where there are a few houses. This gorge is about 2 miles from the anchorage, and boats can generally land opposite it, but there is sometimes a heavy surf which renders it difficult.

When doubling round inside and to the eastward of Santa Maria, it will not be prudent to haul to the northward of a line drawn from it E. $\frac{1}{2}$ S., until at least 3 miles from the point, to clear Delicada spit, which runs out about SE., and the extensive bank with 3 to 4 fathoms to the southward of it; the spit commences at the gorge already mentioned; its perimeter has increased more than 1,100 yards during the past 35 years; it is at present about 8 miles.

There are many rocks off the northern extremity of Santa Maria island; the principal ones being Dormido, 3 miles N. 33° W. of that point; and the Vogelberg, two rocks, about 4 miles N. of the same point. They are not always marked by breakers, and it is not prudent to pass inside of them. Dormido a n d Vogelberg rocks.

The N. and W. coasts of Santa Maria should never be approached nearer than 3 miles.

Arauco bay is comprised between the island Santa Maria, point Lavapié, and the coast to the eastward and to the northward as far as Coronel. In the southern portion of the bay there is good anchorage protected against southerly winds, but there is no shelter from northerly winds except in the harbors of Laco, Lota, and Coronel. The depth of water in the bay is very irregular, bottom sand; there are 2$\frac{1}{4}$ fathoms up to the beach, which is so beaten by the surf that boats can only land in certain places and at favorable times. Bay of Arauco.

All the dangers between point Lavapié and the bay of Luco are indicated by breakers; a vessel must be cautious when wishing to anchor off the small cove of Trauco and Trana, situated to the eastward of the eastern part of point Lavapié. Boats can easily land in these coves, which are somewhat sheltered from N. winds by the island Santa Maria. The anchorage is better and more sheltered against S. winds than that of Luco. The land is high and probably contains coal. Trauco and Trana coves.

Luco bay is to the eastward of Trauco and Trana coves; it is a fair anchorage, but not entirely sheltered to the NNW., and with sharp southerly winds it is liable to violent gusts which sweep down the mountains that overlook cape Rumena. There are 6 fathoms and good holding-ground near the land. Landing is easy. Bay of Luce.

Tubul river.   Tubul river is 7 miles from the anchorage of Luco. The
coast for 3 or 4 miles on either side of the river is formed
of peaked cliffs, with high hills resembling dunes. Good-
sized vessels were formerly able to enter, but the earthquake
of 1835 so raised the bar that at present vessels of 30 tons
only can cross it. It was supposed that this bar would wash
out. The mouth of the river sheltered from the swell from
the W. is generally smooth and affords an excellent landing;
the surrounding country is very fine and fertile. The river
is frequented by small schooners, or by balandras, which
carry the produce of that section to Coronel and Talcahuano.

El Frayle rock.   El Frayle rock or islet, on which the sea breaks, excepting during very calm weather, is about 1 mile off shore and
to the northward of the outer point of the large cliffs to the
westward of the river Tubul. A bank of 4 fathoms has
been reported 2 miles NNE. from El Frayle.

Arauco.   About 4 miles to the eastward of the entrance of the river
Tubul, on the Laraquete beach, and near the mouth of the
Rio Carampangue, is the town of Arauco; it was formerly
a place of importance, but now nothing remains but a fort
or rather earthwork of about 600 feet square, surrounded by
a few houses. It is probable that great geological changes
have taken place in this section since the Spanish conquest.

Laraquete   Laraquete, a sandy beach in front of groves of trees,
beach.   extends about 10 miles ENE. from the cliffs of Tubul. At
2 miles from the beach there are 10 fathoms of water, sandy
bottom. The river Carampangue, which empties near
Arauco, is not navigable at the mouth, which is full of
sand-banks, with varying depths of from 2 to 6 feet; outside
of it there are from 3 to 6 fathoms, but N. winds prevent
anchoring in this locality; two miles in the interior the
river is wide and deep.

Tides.   It is high water, full and change, at Arauco at $10^h 15^m$;
rise at springs, 6 feet.

Coast.   The river Laraquete, the northern boundary of Laraquete
beach, is without importance. From it to point Coronel
the coast, high and rugged, trends north; there are no outlying dangers, and there are several coves. The first is 2
miles from the river Laraquete, at the mouth of the small

river Chivilingo; this cove offers shelter to very small vessels, excepting during SW. gales.

The hill and point Villagran separate the coves of Chivilingo and Colcura; the latter has an opening of 1 mile and is ¾ mile deep; it is bounded to the northward by the Siles islets and Lobos rock, or Piedra Blanca, behind which are the mountains of Fuerte Viejo.

Cove of Colcura or Fuerte Viejo.

Colcura cove is exposed to both SW. and NW. winds; it is less sheltered and not so good as that of Lota; the anchorage is ½ mile south of the Siles islets, in 5 fathoms, muddy bottom. A river empties in its NE. part, into which vessels not drawing more than 6 feet can enter on high water. An important flour-mill has been established at the entrance of the river; the village of Colcura is a little higher up. Ballast is thrown on point Villagran.

Lota bay has become very important on account of the coal-mines which surround it; it opens to the NW. of Fuerte Viejo; according to Fitz-Roy, it is better than those of Colcura or Coronel.

Lota bay.

Lota can be recognized by the white houses of Lota Alta and by those of Don Luis Causiño, on the hill which overlooks the bay, as also by a large iron jetty, having a crane-wheel and drop at its extremity for shipping coal. Vessels of medium draught anchor in 5 fathoms about ½ mile to the south of the extremity of the jetty; the bottom is good holding-ground, but the sea is so heavy as often to interrupt communication with the shore. A large vessel must anchor ½ mile S. by W. from the jetty, in 7 to 7½ fathoms, on a line from the Morro Lutrin to the white rocks of Siles. The N. point of Santa Maria bears N. 66° W. from the anchorage of Lota.

Lota is the port for shipping the coal of several coal-mines, the three principal of which are at Chambique, Lotilla, and San Carlos; a little to the northward of Lotilla a railroad brings the coal from these three points to the edge of the iron jetty, alongside which vessels make fast. Two hundred and fifty tons of coal can be loaded daily by the crane, under which there are 3¼ fathoms at low-water springs. War vessels generally take their coal from lighters; this is easily done in good weather, but the strong southerly winds are a serious interruption. There are copper-smelting works at

Description.

12 c

Lota which utilize the small coal. The foundery has two wharves, one for unloading the copper-ore and the other for shipping pigs of metal.

All the coal-mines, the smelting-furnaces and three jetties belonged to a Chilian, Don Luis Causiño, who sold them to a Chilian company, having an English gentleman, Mr. Munro, as director. The mines are worked by English subjects, miners from Newcastle, who have made a contract for getting out the coal. This coal, which was first worked in 1841, can be easily mined, and its veins cover a large area. At first there were many accidents, as it contained pyrites in large quantities, but since getting deeper into the veins the quality has become better; it is a quick-consuming coal, and should be mixed with slow-burning coal. The high price of the English coal increases the demand for the Chilian; the mail steamers take it now; the price of this coal, however, has also increased as the quality has become better. In 1870 it could be bought at $4 to $4.60 at the dock, and at $5.60 to $6 along the coast; since that time the prices have increased greatly. In 1870 from 80,000 to 90,000 tons were exported; the mines are now well worked. The company of Lota alone employs 1,500 workmen, and Lota and Colcura have about 6,000 inhabitants. Lota is divided into two parts: Lota Alta on the heights near the mines; Lota Baja on a sand-beach at the head of the bay. At the foot of the cliffs of Lota are store-houses and work-shops. Fire-brick are also manufactured at Lota. Ballast is thrown overboard near Lobos rock.

Tides.   It is high water, full and change, at Lota at 10^h; rise 4.6 feet.

Chambique cove.   Lutrin head, the SW. extremity of Lota cove, separates it from the cove of Chambique, which is open to WSW.; it can hold three vessels anchored in from 4 to 5 fathoms. One of the mines of Lota is on the shores of this cove.

Lotilla cove.   A short mile from point Piquete, the northern extremity of Chambique cove, is Lotilla cove, protected by islets and rocks; it is quite safe but extremely small, and has only from 3 to 3¾ fathoms of water. It is near the mines of Lotilla and San Carlos.

Islet del Cuervo.   A little to northward of the entrance to Lotilla lies an islet, called del Cuervo, which is connected with the continent by a chain of rocks. All the coast between Villagran and del Cuervo is clean a short distance from the land.

When coming from the northward this island is easily recognized by its reddish-brown pointed cliffs, forming the angle of a bastion. They are surmounted by a hillock of a brick-red, on which there are trees.

From the same position is seen the small castle with tower and the English garden of Causiño on the promontory which separates the two bays; behind it the steeple of the church of the workmen's town and the smoke from the brass-foundery are seen.

The bay of Coronel opens to the northward of the island Cuervo; its entrance between this islet and point Puchoco is a little over 3 miles wide; the bay is about 1½ miles long. Port Coronel is in its NE. part.   Bay of Coronel.

The coast of the bay is in the shape of a semi-ellipse, whose longer axis passes through the island Cuervo and point Puchoco; the SE. part is a sand-beach, called Playa Blanca, which is bounded to the northward by a hill that extends to the sea; the mine of Playa Negra is at this place. The coast under this hill is dangerous; a bank of rocks, which begins off the houses, extends ⅓ mile to the westward. There are several isolated rocks in the vicinity of this bank, the farthest being ¼ mile from the shore and ¼ mile to the northward of the bank; alongside of it there are 4½ fathoms.

Off the Playa Blanca, where the water deepens rapidly, is a small bank of rocks in 2 fathoms, full of sharp points and steep-to; it lies about ½ mile from the land. In 1862 a buoy was placed in 7 fathoms of water, 30 feet to the westward of it, under the following bearings: Coronel bank.

Western part of point Puchoco................ N. 39° W.
East extremity of de Cuervo.................. S. 37° W.

These bearings will give the position of the bank if the buoy should be swept away. Two houses on this hill overlooking Playa Blanca are a good mark for the bank: it is on the line passing through the northern angle of the lower house and the middle of the upper one.

To the northward of the heights of the mine of Playa Negra is a bank of black sand 1 mile long, called Playa Negra; it is limited to the northward by the ravine of Corcovado, on the other side of which is the town of Coronel.

The bay here forms a semicircle 1½ miles wide; port Coronel is on its eastern limit, sheltered from N. and NW. winds by point Puchoco. The surrounding land is of medium height.

The anchorage is in 7½ to 9¾ fathoms, muddy bottom, ½ mile S. 35° E. from the mole of Puchoco, and a little to the southward of the point of that name; many vessels, however, anchor within half that distance, there being from 7 to 7½ fathoms 300 yards from the shore. This anchorage is preferable to that of Lota. This bight is divided into two distinct parts; one toward the town of Coronel, where there is a mole in a dilapidated condition, and a clean shore; the other on the side of Puchoco, which is the center of the mines, where there are some rocks a short distance from the beach of Puchoco, and a bank with some 2½ to 3 fathoms over it near the mole. The N. point of Santa Maria bears N. 77° W. from the anchorage of Coronel.

The vicinity of Coronel, like that of Lota, is full of mines, to which it owes its importance; the principal ones are those of Coronel, Rojas, and Puchoco; the latter, which are on the exterior of the NW. point of the bay, are worked out under the sea 875 yards from the coast. The coal is brought by railway, which runs through a tunnel under the point, to the end of the mole. To the northward are the mines of Rojas, one of which belongs to the company of Lota, and its coal is shipped from the end of the wharf off the center of the town. The mines of Puchoco belong to a company called *compañia de carbon de Puchoco;* they employ 1,500 persons; the shafts descend 130 feet; the breadth of the veins is between 1 and 5 feet, and they run in the direction of Lota. There are six steam-engines of from 12 to 60 horse-power in use.

The mines furnish 80,000 tons annually, of which 30,000 are for steamboat companies, 35,000 for the foundries in the north of Chile, and 10,000 for gas companies and domestic purposes. Since 1867 a factory for fire-brick has been established in connection with the mines.

The coals of Coronel give rise to more maritime commerce than that of Lota, to which they are inferior, but they cost $1 less; they are said to be more liable to spontaneous combustion than the latter. The steamers from Liverpool to Valparaiso coal at Coronel.

Ballast is thrown overboard between Puchoco and the point of the same name, or on the rocks off the settlement of Playa Negra, or between point de los Mirquenes and the chain of reefs to the eastward.

Chile produces only tertiary lignites, the superior quality of which is recognized. The district containing it is a narrow, broken, littoral belt between Talcahuano and Quirico. Farther south, in the provinces of Valdivia, Llanquihue, Ancud, and Chonos, there are only insignificant veins of the lignite formation, not large enough to warrant their being worked.

The annual production is 240,000 tons. The price varies between $6 and $8 per ton, according to the value of English coal. The Chilian coal can be mined easily; the beds are regular, pure, and from 3 to 5 feet wide: they are not subject to inundations, and lie under soft, not liable to crumble, and slightly aquiferous rock.

At Lebu there are about 1,000 acres of coal-land, containing about 5,000,000 tons. It is thought that, at the present rate of working, the mines of Arauco can only furnish coal for 8 years, and that the total working of the coal-mines can last but from 20 to 30 years.

Point Coronel is 1¾ miles NNW. of point Puchoco; some rocks extend from it S. by W. 600 yards. The coast between the two points is clear and of moderate height. *Point Coronel.*

From point Coronel the coast runs E. for about 2 miles and N. 5 miles, with about the same elevation; it is then low for 4½ miles, and is bordered by an extensive sandbeach to the estuary of the river Biobio, whose mouth is about 10 miles from the N. point of Coronel. A sand-bank, forming the bar of the river, runs out from this beach, which is always beaten by the swell. The entrance is between this tongue and the northern part of Pompon head, a peninsula which is connected with the land to the northward by a narrow neck of sand 766 yards long. *Biobio river.*

Pompon head is surrounded, to the southward and eastward, by large rocks from 66 to 88 yards distant; between the point and the two eastern rocks is the only passage; the depth varies between 3¾ and 5 fathoms to the mouth, close alongside of the head; inside it shoals rapidly to 6 and then to 3 feet. A vessel wishing to enter must bring the southern island on a N. and S. line, and steer N. for this narrow passage, but a vessel can go no farther; only boats can enter. As the water does not increase in depth inside, and as the current runs 3 knots, it can be said that the river Biobio is inaccessible. The swell always sets into its mouth.

The Biobio is the largest water-course of Chile, being about 160 miles long. It leaves the lagoon of Huchueltui in the midst of the Cordillera Grande, in latitude 38° 1′ S. and longitude 70° 36′ W.; after a course in which it is said to attain a velocity of 12 knots, it crosses an immense fertile plain, called the Central valley, and clears the Cordillera of the coast before emptying into the sea. It is full of sand-banks, and can only be navigated by flat-boats, of which there are about 100, some of them with steam-power. Its principal tributaries are the Vergara, and especially the Laja, which comes from the lagoon of Antuco at the foot of the volcano of the same name, in latitude 37° 10′ S. The town of Los Angelos, near which there is an important German colony, is in the Central valley between the Laja and the Biobio. Concepcion, the third city of Chile, is on the right bank of the Biobio, 6 miles from its mouth.

Tides. It is high water, full and change, at the mouth/of the Biobio at 10ʰ 15ᵐ; rise, 3 feet.

Paps of Biobio. A peninsula 2 miles long, and of the same breadth, separates the estuary of Biobio from the harbor of San Vincente; it is limited to the SW. by point Cujento, and to the eastward by the Biobio, and is surmounted by the two paps or Tetas del Biobio. The SW. hill is 793 and the NE. one 816 feet high; they are excellent landmarks for the mouth of the river, port San Vincente, and the bay of Concepcion. There are no dangers near them other than a few rocks close to the land.

Port San Vincente. Port San Vincente is to the northward of the paps; it is a bad and exposed anchorage, entirely open to NW. winds and the swell from the westward; it is clean in its southern parts, but has some rocks about ½ mile from its N. shore; there are 14 fathoms in the middle; the cove is 2½ miles deep and 2 wide.

Concepcion Bay.

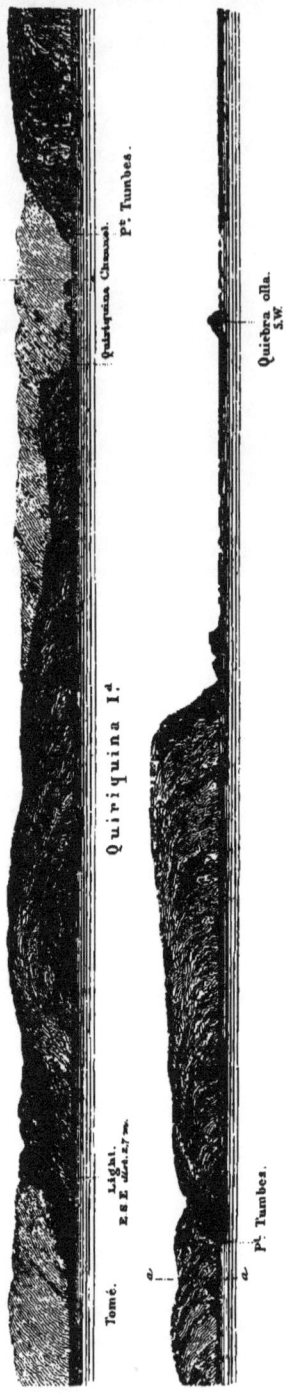

Quiriquina I⁴.

Tomé.

Light.
E.S.E. dist. 2.7 m.

a.
a.

Pᵗ. Tumbes.

Quiriquina Channel.

Pᵗ. Tumbes.

Quiebra olla.
S.W.

Off Maule Riv. entrance.

S.F.S.E.

Entrance to
River Maule.

Topboom

Recommendaciones pile mark

Sandar Bank Bank the same
whilst the church of greyish sand

From point Lobo, the northern boundary of the entrance Peninsula o Tumbes.
to San Vincente, the coast trends N. 18° E. for 6 miles, to point Tumbes, near which are many scattered rocks, some above and some below the water. The Sugar-Loaf rock is 5½ miles from Lobo, and the rocks which surround it are rather less than ½ mile from the coast. Quiebra Olla is a rock 20 feet above water, lying N. 27° W. ¾ mile from point Tumbes. Vessels must not pass between it and the point, as there are several sunken rocks; outside of it there are no dangers.

The name of the peninsula of Tumbes is given to a stretch of land comprised between point Tumbes, to the northward, and point Lobo and the town of Talcahuano, to the southward. The highest elevation, its northern extremity, is 412 feet.

The entrance of the bay of Concepcion is between point Bay of Concepcion.
Tumbes, the NW. extremity of the peninsula, and point Loberia, 6½ miles ENE. of it. The bay is 9 miles deep and 5 wide; the depth of water is good everywhere; the anchorage is extensive and well sheltered from nearly all points; strong N. winds render the bay rough, without endangering vessels that are well anchored. Mount Newke, 1,790 feet high, and about 5 miles from cape Loberia, is the highest land in the vicinity.

Coal of the same quality as that of Coronel and Lota is Coal.
worked in the vicinity of the bay. At Talcahuano it is taken from the NW. part of a hill near the bay. There are two veins, 3 feet thick, separated by a bed of loose stones 14.7 feet thick; the coal is bituminous. There are some other mines at Tierras Coloradas, near the river Andalien, which furnish the best coals. There are two veins on the east coast of the bay, one near Penco and the other in the bay of Coliumo, but they do not clear the expense of working them. The exploration of a mine has been commenced on the island Quiriquina.

The island Quiriquina trends N. 17° E. and S. 17° W., is Quiriquina Island.
3 miles long and 1 wide. It is situated at the entrance of the bay of Concepcion, and shelters it to the northward; its summit is 395 feet high; a short distance from its northern extremity are some rocks, called Pajaros Niños; vessels can anchor off point Arena, its SE. extremity.

Quiriquina, and the reefs which extend from its SW. extremity for about 400 yards, protect the anchorage opposite
the town of Talcahuano against N. winds. The bay can be
entered on either side of the island, but the best passage
for those not familiar with the locality is that to the eastward. The westernmost passage, called Quiriquina channel,
between that island and the peninsula of Tumbes, is 1 mile
wide; the water on the side of the island is deep, but the
Buey rocks, which run out from the NE. point of Tumbes,
reduce the width to ½ mile; the tides in this channel are
irregular. The Great or Eastern channel, between Quiriquina island and point Loberia, is 3 miles wide, with no
danger at a reasonable distance from cape Loberia and Pajaros Niños. There is less tide in this passage.

Light: Lat. 30°
36′ 18″ S.; long.
73° 06′ 05″ W.
A *white light*, varied by *flashes* every 30 *seconds*, the duration of each flash being 9 seconds, with a partial eclipse of
20 seconds, is shown from a round brick tower, painted
white, the balustrade black, and the cupola and ventilator
green. The tower is placed at the NE. angle of the keeper's
dwelling, which is also painted white; the height of the
light is 26 feet from its base, and 211 feet above the level
of the sea, and is visible 15 miles. The light is only visible
to the northward of N. 79° E.; to the southward of this
bearing it is shut in by the high land of Tumbes. With a
northwest wind the sea is often so phosphorescent as to render the light difficult to make from a distance. There is a
signal-mast near the tower.

Belen and Man-
zano banks; Mari-
nao and Viuda
rocks.
The banks and rocks north of the anchorage of Talcahuano, along the Tumbes shore, average about 1½ miles from
the mole. Marinao rock is above water, and marked by an
iron staff 14 feet high, surmounted by a ball painted black;
it is ⅓ mile from the land, and N. 9° E. from Talcahuano
head. Manzano bank is 300 yards north of Marinao rock,
and extends about 200 yards from the shore; Viuda rock,
which is marked by a white staff 15 feet high, uncovers at
low water, and lies 1⅖ mile S. 17° W. from Marinao rocks,
300 yards from the shore, and 492 feet N. of the government
mole. These rocks are part of a bank which terminates,
with a depth of from 2¾ to 3¼ fathoms, about ½ mile from
the shore; between its extremity and the west edge of Belen
bank, there is a passage ¼ mile wide, with 8¾ fathoms. The

U. S. S. Portsmouth discovered a shoal spot, with $3\frac{1}{4}$ to 4 fathoms over it at low water, which would necessitate vessels taking this channel borrowing toward Belen bank; it lies S. 14° E. from Marinao perch, a short $\frac{1}{2}$ mile distant, Talcahuano head bearing S. 18° 35' W., and Viuda rock S. 29° 18' W.

Belen bank is marked by a black buoy, anchored on the north edge of the bank, in $15\frac{1}{2}$ feet of water, at low tide, bottom mud; the bank extends 328 yards in a NW. and SE. direction, bottom mud and shells, and has from 14 to 21 feet of water over it; no vessel drawing more than 13 feet should approach the buoy to within 400 yards, when it bears from NW. to NE. through north. The buoy is on the following bearings, viz: *Belen bank.*

Chief mole of Talcahuano, (end)............ S. 12° 35' W.
Point Fronton, (S. extremity of Quiriquina
   island) .... N. 27° 10' W.
Tomé church-tower....................... N. 17° 25' E.

If the buoys should not be in place, the bank will be cleared by keeping mount Espinosa open of Talcahuano head until the highest hill to the right of the town, on the southern part of the heights of Tumbes, bears S. 82° W.; as fort Galvez has been destroyed, and Lookout hill cannot be distinguished, there are at present no leading marks to the anchorage; large vessels, when choosing a berth, must bear in mind that a bank extends $\frac{1}{2}$ mile from the town.

Choros bank, which is marked on the charts $\frac{9}{10}$ mile N. 16° W. from Belen bank, was unsuccessfully searched for by the officers of the U. S. S. Portsmouth. *Choros bank.*

When facing the town of Talcahuano a pointed hill is seen on the left, at the foot of which is the entrance of a small canal; this is Talcahuano head, an excellent leading mark for the roadstead. *Talcahuano head.*

The port of Talcahuano is in the SW. angle of the bay of Concepcion, and is the best anchorage in the bay: outside the bank, which extends about 500 yards from the town, there is $3\frac{1}{4}$ fathoms. Vessels usually anchor about $\frac{1}{2}$ mile off the town, in from $4\frac{3}{4}$ to $5\frac{3}{4}$ fathoms of water; there is nothing to prevent landing or receiving cargo at all times. *Port of Talcahuano.*

A vessel can be hove down or careened in this port. There are three good watering-places, and water is brought *Description.*

to vessels, in a tank of 30 tons' capacity, at 30 cents per ton.
Ballast or ashes can be landed on the bank of rocks off the
castle of San Augustin. The government has constructed
a mole. There is a pilot of the port, and there are two
yards for repairing vessels and boats.

Talcahuano, which has from 2,000 to 3,000 inhabitants, is
the seaport of Concepcion, from which it is 6½ miles distant.
Talcahuano, with Concepcion, was destroyed by an earthquake in 1835, and it suffered greatly in 1868 from an inundation by the waters of the bay.

Directions. With a southerly wind the land should be made to the
southward of the entrance. The vicinity of the bay is well
marked by the paps of Biobio, the south boundary of San
Vincente bay, to the northward of which the heights of
Tumbes trend to the north and extend 6 miles, as before
stated; the outlying dangers off Tumbes are the Pan de
Azucar and Quiebra Olla rock, which should be given a
good berth. Taking the channel to the east of Quiriquina,
the most advisable for a stranger, after passing Quiebra
Olla steer to round Pajaros Niños as closely as possible,
and, with a southerly wind, haul to the wind immediately
and beat up for the anchorage of Talcahuano; if the wind
is from the northward, it is still advisable to keep close to
Pajaros Niños, where the water is deep, as thus Concepcion
rock, off Point Loberia, is avoided. After passing the north
end of Quiriquina the anchorage off Tomé comes in sight.
With a leading wind, after passing the north end of Quiriquina, and when the south end of the island bears west,
distant 1 mile, steer directly for Talcahuano head; this
course will bring the vessel well up to the buoy on Belen,
which must not be approached nearer than ¼ mile; round
this buoy, and steer directly for the railroad depot, on the
west side of the town.

The best anchorage is in from 5 to 6 fathoms of water,
with the paps of Biobio to the right of the Catholi cchurch-
tower; but the nearer to Tumbes the smoother the water will be, and the better the protection from northers.
The holding-ground is good throughout the bay. On the
west side of the bay all the dangers are to the westward of
Belen bank. Beating up for the anchorage off Talcahuano
it is not prudent to approach Punto Parra nearer than 1

mile, as Rundel bank is nearly that distance from the point, and give Belen a good berth.

Approaching Los Reyes island, keep the bluff on the north side of San Vincente bay closed in by Talcahuano head, the shoal-water along the island making out a considerable distance.

Concepcion, the capital of the province of that name, is **Concepcion.** a town with about 14,000 inhabitants, situated on the right bank of the Biobio, in the plain of Mocha; it was built here after the destruction of old Concepcion; Penco, by the earthquake of 1730. Concepcion was destroyed by the earthquake of 1835, which changed the depth of the bay considerably. The town is well constructed, with numerous churches and public institutions; it lies in the center of a valley, which produces cereals of all kinds in large quantities; beef and mutton are of fine quality and cheap. Twenty years ago the former cost 4 cents a pound and the latter $5 a head; but everything has increased; pigs and fowls are more expensive; vegetables of all kinds are abundant and of moderate price, as are the fruits of the season.

From Talcahuano head a low beach commences, called **Penco.** la isla de los Reyes, which is terminated by a small hill pertaining to the chain in the interior. On the other side of this is Penco, or old Concepcion. Penco is not a port of entry, and has neither the advantages of Tomé nor the security of Talcahuano.

The depth at the anchorage varies between $5\frac{1}{2}$ and 11 fathoms, bottom of sand and mud. Vessels can anchor here, but the northerly swell is very heavy, especially off the Boca Grande; there is no shelter from that direction; it is, therefore, best in bad weather to go to the anchorage off Talcahuano. The water is shoal off the town of Penco. The small water-course of Andalien empties in the Boca Grande at the east extremity of the island de los Reyes.

Lirquen is $1\frac{3}{4}$ miles from Penco, from which it is separated **Lirquen.** by points Cerillo Verde and Lirquen, from each of which rocks extend out 220 yards, and are visible at low water. Between the points is a coal-mine. The remarks on the anchorage of Penco apply equally to Lirquen.

Beechy rock, $1\frac{1}{2}$ miles S. 85° W. of Lirquen point, was **Beechy rock.**

sought for by the U. S. S. Portsmouth, commander Skerrett, but not found.

Point Parra. Point Parra is wooded, and is 2 miles to the northward of Lirquen. A shoal extends out from Parra point 1,300 yards W. by S., and 1,800 yards SW. by S., having from 2 to 3 fathoms on its outer edge, just within which, and bearing WSW., 1,300 yards from point Parra, is Rundle bank, with from 4 to 6 feet over it at low water. Loberia head, N. ¼° E., will clear the bank 600 yards in about 12 fathoms of water.

Port of Tomé. The anchorage at Tomé is considered safe, but vessels are always exposed to the southerly swell; the small hill called El Morro de Tomé protects it from N. winds, but the heavy sea from this direction is somewhat felt. With south winds this anchorage is attained more easily than that of Talcahuano, which is nearly 9 miles to the SW. Wishing to anchor at Tomé, bring the conspicuous white church-spire in line with the end of the pier, and keep them on until the extreme of the land of Huily head comes in line with the extremity of Loberia head; these marks will give the best anchorage for a large vessel in 10½ fathoms, mud and good holding-ground. Care must be taken in anchoring here, as the water shoals suddenly. A stranger should not attempt to take an anchorage at Tomé at night. The bay of Tomé is easily recognized by the mill of Bella Vista, near point Parra, as also by Tomé hill.

Description. The town of Tomé, which, in 1835, consisted of but one street, has developed rapidly, and contains at present about 6,000 inhabitants; it is the harbor of export of the province of Ñuble. The principal street terminates in a jetty 150 feet long, on which a water-pipe has been laid; a tank takes water to vessels, and there are three convenient watering-places in the coves of Tomé and Collen, (Bella Vista.) There is a cloth-factory at Tomé; provisions are cheaper than at Talcahuano, and there are two yards for repairing vessels and boats. Ballast and ashes are thrown NW. of the anchorage, on a beach between two chains of rock.

Zealous rock. Zealous rock lies 700 yards to the southward of Huily head; from it point Loberia bears N. 27° 30′ W., Huily head N. 29° E., and Quiriquina light N. 81° W.; it has over it 3 feet at low water. Close to the westward of it is another rock with from 4 to 6 feet over it: between these and the land there are other submerged rocks, and the bottom is rocky. Zealous rock is to be marked by a buoy.

To the northwestward of point Huecas, the westernmost extremity of the bay of Tomé, is Huily head, an ill-defined, rocky point. Two miles N. 34° W. from point Huecas is point Loberia, the NE. limit of the bay of Concepcion; it is of a dark color and has some outlying rocks, all of which are near the land excepting Lozzi, or Concepcion rock. *Point Loberia.*

Concepcion rock is conical and sharp, so that the lead will not rest on its summit, with 19.7 feet of water over it at low tide. At the distance of a boat's length around it the depth varies from 13¾ fathoms to 19 fathoms. It was necessary to look for it for three days before it was found, and sextant angles were taken that place its position on the following bearings, viz: *Concepcion rock.*

Farallon islet, off Loberia point. .... N. 30° 50″ E.
Seal rock................................ ..... S. 62° 20″ E.
Huique point (Huily point) .... S. 40° 50″ E.
Quiriquina island light .... S. 48° 55″ W.

The sea only breaks on it in heavy storms, and in ordinary weather there is no ripple or eddy to mark it, nor is there any weed on it.

About 3¼ miles NNE. of cape Loberia is point Talca, where the coast turns suddenly to the eastward to point Cullin, and then to Coliumo head, where, turning again to the southward, it forms Coliumo bay, with an opening of 1 mile between the points Coliumo and Lingueral; it is 1⅛ miles deep and terminated by an inlet about 1 mile long with 1½ to 2¼ fathoms of water. Coasters can anchor with security in this bay, but large vessels are not much sheltered during N. winds. The bottom is of fine black sand, and the depth is from 2½ to 8¾ fathoms. The best anchorage is in Rare cove, immediately behind cape Coliumo, where there is a good landing and watering-place. This bay has always been the center of operations for smugglers. *Coliumo bay.*

It is high water, full and change, in Concepcion bay at 10ʰ 14ᵐ; springs rise 5 feet. *Tides.*

CHAPTER VII

Variation from 17° 15′ to 14° 13′ easterly, in 1876; increasing annually about 1′ 30″.

The coast, high and wooded, continues to trend with a slight inclination to the eastward; 16 miles NNE. from the bay of Coliumo is point Boquita.

Port Buchupu-
reo.

Port Buchupureo, in latitude 36° 05′ S. and longitude 72° 45′ W., is 12 miles from point Boquita; it is at the entrance of an extended valley, watered by a river of the same name, which empties under the shelter of point Maquis, its S. point. This point shelters the anchorage against S. winds. On the approach of a norther, vessels should, and can easily, get under way; the bottom is good holding-ground—sand and mud. On a N. by E. bearing from point Maquis are 14½ fathoms; this depth diminishes gradually to near the land, where there are but 3 fathoms; the best anchorage is in 12 fathoms.

Port Curanipe.

Following to the northward the coast is a continuous line, and offers no shelter as far as the river Maule; it is elevated and partially wooded. About 30 miles from point Boquita is a small indentation, called port Curanipe, where vessels sometimes anchor, but are only sheltered from SE. winds; the bottom is good; it can accommodate about 20 vessels in from 6 to 14 fathoms. In the southern part of this port is a small cove, in which coasters load and take refuge. The river Curanipe empties into it, having a bar of ⅛ mile in length, which increases and extends so far to the northward in the summer that the small vessels cannot cross it. This generally takes place with S. winds, which blow almost constantly in these latitudes without raising much sea, the contrary being the case with N. winds.

Boats can water in the river above tide-water; the beach is always dangerous for boats. There are two mooring-buoys in the roads, and a telegraph station on point Trarado, the southern extremity of the bay.

The isolated hill, Centinela, a little to the northward, is a good landmark for recognizing the bay.

The bay called Zorro, or Fox, which is 10 miles in extent Cape Carranza. and limited to the northward by cape Carranza, can hardly be called a bay: it is an exposed bight. Just to the northward of Curanipe and about 56 miles from Caliumo, this cape is quite a low projection of the coast, where small vessels can find temporary shelter among the rocks which surround it; but it is best to avoid it, as for nearly 10 miles on either side there is an inhospitable shore of sand and stone.

Cape Humos is 17 miles N. 28° E. from cape Carranza; Cape Humos. it is a remarkable promontory running out to the westward, and is more elevated than the other land on this part of the coast; it is steep, and there are no dangers in its vicinity.

Close to the coast, 4 miles N. 39° E. of cape Humos, and Yglesia rock. 1 mile S. 62° W. of the entrance of the river Maule, is Yglesia rock; it is remarkable and serves as a landmark for the mouths of the river.

The entrance of the river Maule cannot be mistaken, as Entrance of the river Maule. the coast to the southward is rocky and the land high, in contrast with the long sandy beach to the northward. Not far from Yglesia rock is a remarkable barren space of gray sand on the side of a hill, whereas the heights between cape Humos and the rock are generally covered with vegetation and partially wooded. The highest mountains in the chain reach an elevation of from 1,000 to 1,300 feet; those between the coast and the river being from 500 to 900 feet.

Vessels can anchor off the bar, during good weather, in Anchorage off the bar. from 10 to 14½ fathoms, sandy bottom. There are no hidden dangers, excepting a large sand-bank to the northward of the river, formed by the alluvial deposits, which extending to seaward prevent the approach from that side; at this anchorage vessels can get under way readily.

Cape Maule, which is surmounted by a flag-staff, forms the southern part of the entrance to the Maule river. It is a steep granite cliff, called Mutun by the Chilians, having a beach on either side; it is terminated by a sandy point, called Entrance point, which runs out N. 5° W. ¼ mile.

At the southern extreme of Mutun, and at the entrance of the port, are two high rocky pyramids, the one called Ventana, from having a hole through it, the other Piedra Lobos, it being frequented by the sea-wolf.

Sometimes boats land on the outside head of cape Maule,

but it is very dangerous, as the sea is always high and uncertain; the beach also is very steep and the sand soft, so
that it is difficult to haul even a whale-boat on shore. There
is a better landing near Yglesia rock, though that is bad; it
is best to procure balsas and moor the boats outside of the
surf.

Bar.   The bar extends a distance of 766 yards to the northward
of cape Maule. From May to October it has on it from $2\frac{1}{2}$
to $3\frac{1}{2}$ fathoms; but from October to March only from $1\frac{1}{4}$ to
$2\frac{1}{2}$ fathoms; it should not be crossed without a pilot, and
vessels have often to wait, even during the dry season, for
a week before they can cross it.

Constitucion.   If it was not for the bar of the river Maule the commerce
of this locality would be very flourishing; notwithstanding
this the small village of Constitucion, on the south bank of
the river, 1 mile from its mouth, would become prosperous,
with the aid of a few small steamers and some engineering
work on the bar.

The anchorage off the town is about 656 yards long and
220 wide; the bottom is mud with a depth of from 3 to
6 fathoms. Vessels must anchor with one anchor in the
current and the other inshore, so as to ride out the freshets
in winter, and in summer the violence of the south winds.

Vessels can be careened and pure water procured from
the river. The government keeps a tow-boat, but it is often
absent. There is a pilot.

As the current is too strong to allow permanent buoys,
the authorities of the harbor place small ones when necessary, removing them when no longer required. Ballast is
thrown on the shore north of the entrance.

Description.   Seen from mount Caracol, the town has a picturesque aspect; behind it is a series of mountains and hills which
terminate at the sea; farther there is a small sand-hill and
then Mutun head.

Constitucion, which is opposite an island in the river, is
in the center of a country rich in agricultural and mineral
products, which reach it by the river. The town is not far
from the Upsallata pass in the Andes, which was discovered
in 1805, and is the only pass practicable for wagons between the isthmus of Darien and Patagonia. Small coasting-steamers run between Constitucion and Valparaiso.

The forests around Constitucion have excellent timbers from which are built large numbers of small vessels and boats; the town has a dock for careening vessels and 12 repairing yards. Bricks are manufactured. The principal products of the country are vegetables, cereals, fruits, and liquors. Constitucion is the harbor of export of the provinces of Talca and Maule; the population is about 10,000.

It is high water, full and change, at the river Maule at 10ʰ; springs rise 5 feet. *Tides.*

The river Maule has its source in the Maule lagoon; it is 160 miles long and is navigable for vessels for 9, and for large boats for 77, miles. It is subject to very rapid currents, the bottom is sand and pebbles, and bad holding-ground. It is navigated by some 300 boats, most of which are constructed at Constitucion. *River Maule*

From the river Maule the coast trends to the NNE. for 12 miles to Peñon point, the northern point of the entrance to the river Mataquito or False Maule, so called from having frequently been mistaken for the entrance of the Maule river. In this stretch of coast there is 5 miles of sandy shore backed by mountains 1,300 feet high. There is no anchorage where a vessel could remain without danger, though the depths are moderate, as she would be exposed to the full force of the sea and prevailing winds. *El Peñon.* *River Mataquito and False Maule.*

To the northward of Mataquito river, about 3 miles N. ½ W. of El Peñon is Hoca point, rugged and surrounded by rocks; close to it is a house in ruins. *Hoca point.*

Llico road, in latitude 34° 46' S., longitude 72° 06' W., affords an anchorage in from 8 to 12 fathoms, bottom fine sand; the holding-ground is good, and the anchors can be weighed easily; vessels should always get under way on the first indications of bad weather. *Llico road.*

At the head of this small indentation, which is full of sand-banks formed by deposits of the river Mataquito, is the entrance of the channel leading to Vichuquen lagoon; it is closed during the summer, or season of southerly winds, and open during the winter. The passage to the lagoon is shoal, from 3 to 6 feet deep; it is 4 miles long, and its entrance is always dangerous for boats as the swell sets in.

The greatest length of the Vichuquen lagoon is 5 miles

13 c

NE. and SW.; it is deep and could be made a fine anchorage by improving the channel to the sea. Near this lagoon
are two others, named Torca and Tilicura, and also the town
of Vichuquen, which is 4 to 6 miles from the mountain of
the same name, on which is the best wood for ship-building
found in Chile. The bar of Llico is dangerous. Vessels
should not, therefore, use their boats, but await the native
boats, which will be sent off when the bar is safe, at a charge
of nine dollars. Fresh meat alone can be obtained, and that
only occasionally. Cape Lora is 14 miles N. of Llico.

Coast.

There are several unimportant coves to the northward of
Llico—Salinas, Palos Blancos, Serena, Cahucil, and Petrel—
which have no shelter, and landing in them is dangerous.
Colcura, Polcura, and Palos Negros are a little more extended than the former, and landing is less dangerous. At
½ mile from the land there is from 10 to 12 fathoms of water
over a bottom of fine sand.

Topolcama
road.

Topolcama road is to the northward of the point of the
same name, in latitude 34° 09′ S., and longitude 72° 01′ W.
The anchorage is rather less sheltered against south winds
than that of Tuman, but it is not much exposed to the
northward; it is not large, and is insufficient for a vessel of
more than 200 tons; vessels can get under way easily. A
Chilian observer states that the N. winds do not cause a
heavy swell, or at least not so much as to endanger a vessel.
The effects of the winds are most sensible on the beach; a
landing can always be effected with south winds, but it is
difficult with those from the northward, except under the
point. The shore of this roadstead is mostly surrounded by
a bank, which can hardly be crossed by boats at half-ebb;
it is formed from October to April by the influence of the
southerly winds and the coast current; the north winds
destroy it, blowing the sand back to the beach. This bank
is very inconvenient, and becomes dangerous as soon as the
sea is the least agitated. The channel between it and the
beach is always subject to a northerly current, with a velocity of 1½ to 2½ miles at spring-tides.

There is fresh water in several places within from 600 to
800 yards of the sea; a little beyond there is an estuary.
There is wood in the vicinity, but of inferior quality.

Topolcama and the ports to the northward and southward

are all subject to the same inconvenience, namely, sand-banks, tremendous surf, and a coast-current. All the coast from Llico is high and steep, with plenty of water.

It is high water, full and change, at Topolcama at 9^h 55^m; springs rise 6 feet, neaps 4 feet. *Tides.*

Santo Domingo point is NNE. 4 miles from Topolcama point; it is 334 feet high, and is the south point of Tuman bay; there are 6 fathoms close to the north and west sides of the point, with no outlying dangers. *Santo Domingo point.*

The road of Tuman, 4 miles N. of Topolcama, lies between Santo Domingo and Barrancas points; it is sheltered to the S. and open to winds from the N.; the bottom is the same as along the rest of the coast, fine sand, with from 5½ to 10 fathoms. The anchorage is under Santo Domingo point, bottom sand and clay. There are no currents as to the northward and southward, but the beaches are so flat that small boats ground at half-ebb 60 yards from the shore. In the winter the sea and wind throw up the sand of the beaches. Under the shelter of the pointed south point is a small indentation, which is not encumbered by rocks, and where there are 3½ fathoms at the foot of the point; the landing is tolerable, and better perhaps than at Curanipe, Llico, and Buchupureo; fresh water is scarce. Vessels must get under way on signs of a northerly wind. *Road of Tuman.*

It is high water, full and change, at Tuman bay at 9^h 55^m; spring rise, 6 feet; neaps, 4 feet. *Tides.*

El Farallon del Infernillo is an island in the form of a pyramid, ½ mile N. of Barrancas point and close to the shore; it is easily recognized when nearing Topocalma point, which is the best point for a vessel to make when bound for Tuman bay. *Farallon del Infernillo.*

Papuya cove, completely open to north, is 13 miles N. 30° E. from point Topolcama; to the southward of its entrance is a group of islands of the same name, the largest of which is 600 yards long; it shelters the cove somewhat from southerly winds; some coasters visit it. Vessels can pass through the channel between the island and the continent, and in an emergency anchor to the north or south. *Papuya cove and islands.*

This cove is worse than those to the south of Topolcama; boats can land under the shelter of a large rock, which lies ½ mile from the S. point.

Matanza cove.   Matanza cove is about 2 miles from Papuya; it is a little
more protected from south winds by some islets which close
it in that direction, but it has only capacity for one small
vessel, and it must be left on the first sign of a northerly
wind. The cove may be recognized by a ravine with a
small inlet, called Matancilla, on the north side of which is
a house with its north side painted white and its roof red;
the hills north of the ravine are high and green, while those
to the south of it are of sand.

Bay of Navidad.   Navidad bay is 4 miles NE. of Papuya; it is a bad and
exposed anchorage; no better than the open sea. The south
winds, which predominate, cause the heaviest swell; the
water is shallow and the breakers commence a long distance
out. North winds are more dangerous, as the anchorage is
near the south coast, and the beach is bordered by sunken
rocks. The landing-place is similar to that of Papuya.

Rapel river.   Point Rapel is on the northern coast of Navidad bay, a
little to the southward of Rapel river, whose mouth is 60
yards wide; the bar, which is bad, is 50 yards in front of
it, and has on it $1\frac{1}{2}$ fathoms at high tide.

About $1\frac{1}{2}$ miles from the land, opposite Cerro de Colenar,
is a sunken rock, over which the sea breaks. All the rest
of the cove from this shoal to cape Bucalemo is clear.

Cape Buca-   About 3 miles NW. of point Rapel is cape Bucalemo, a
lemo.
steep cliff 200 feet high. Two miles N. 74° W. from the
cape is Rapel shoal, often called Tapolcama bank.

Rapel shoal.   Rapel shoal is 1 mile long, and has three rocky points
above water, on which the sea always breaks 2 miles from
the coast. There is plenty of water around the shoal, the
soundings gradually increasing from $9\frac{3}{4}$ to 50 fathoms. Vessels must not approach this bank, as with fresh SW. winds
the swell always sets over it, and the current, which runs
more than 1 knot, sets across the reef.

Toro point and   Toro point is $5\frac{1}{2}$ miles N. by E. from cape Bucalemo. N.
shoal.
20° E. of the most salient part of point Toro, 2,737 yards
distant, is a reef, 820 yards long E. and W., and 347 yards N.
and S. Toro shoal lies 3,062 yards N. 53° E. of point Toro,
and 875 yards from the nearest land; the channel between
this shoal and the reef has 12 fathoms water, sandy bottom,
and between it and Toro point there are 10 to 13 fathoms,
sand. These two dangers, located by captain Luis Pomar,

of the Chilian navy, would appear to be those heretofore reported, but in an uncertain position.

There is less water on this coast than on that farther to the south, there being but 14 fathoms 1 mile from the land. To the eastward of point Toro there is another indentation, in which there is a sand-bank about 1 mile off its shores; there is no description of this cove.

From point Toro the coast trends about N. for 15 miles, to point Roca Blanca, forming a slight bend, in the center of which the Maipo river empties; it is closed by a bar, which continues to the northward about 2 miles, parallel to the coast.

Maipo river.

About 3 miles NNE. of this river is the small cove San Antonio Viejo, which offers tolerable shelter to a small number of coasters, when anchored under the lee of a pointed hill.

Ports of San Antonio Viejo and Nuevo.

About 2 miles to the northward of this hill is the small cove of las Bodegas, or San Antonio Nuevo, which has been frequented for some time by large vessels; the holding-ground is good, with a moderate depth. With good ground-tackle and a long scope a vessel can ride out a norther as well as at Valparaiso. The bay is surrounded by high sand-beaches of different colors; there was no pilot in 1872. This harbor was used by the Chilians during the blockade of Valparaiso by the Spaniards; guns of large caliber were landed and large cargoes of grain shipped.

About 13 miles to the eastward of San Antonio Nuevo are heights of 3,270 feet.

The beach, called Carthagena, to the northward of las Bodegas, is completely open to SW. winds. Point Tres Cruces is low and rocky; 5 miles NW. of it is point Roca Blanca, so called from its white rocks, by which it is easily recognized; these rocks seem black when seen from the SE. A sunken rock has been reported 4 to 5 miles off Roca Blanca and about 24 miles south of Carraumilla point.

The land makes in to the S. eastward from Peña Blanca, forming a small bight, called El Canelillo, having in it from 9 to 14 fathoms of water, and open to the NW. Six hundred yards NNE. of Peña Blanca is an island, called Pajaros Niños, lying less than 200 yards from the shore. About 1,200 yards NNW. ¼ W. from Pajaros Niños is a rocky plateau, called

Algarrobo roads.

los Farallones, which extends 400 yards to the NNW., with from 8 to 2¼ fathoms near its edges, and from 11 to 23 fathoms, with some spots of 7 to 9 fathoms, between it and Pajaros Niños.

From Pajaros Niños the coast trends to the eastward and again to the north, forming Algarrobo roads, protected only to the southward, where temporary anchorage may be had in from 18 to 28 fathoms, opposite the small village of Algarrobo, where wood, water, and fresh provisions can be procured.

Shoal.

A rocky shoal with 9 feet of water over it lies N. 27° E., 1,800 yards from the west of point Pajaros Niños, and N. 84° E., 1,350 yards from the northern rock of Los Farallones.

Quintay cove.

Point Gallo, a peaked cliff, is 7 miles to the northward of point Algarrobo; between them are two sandy indentations divided by a rocky point. A small vessel can find shelter against N. winds in the northern one, named Tunquen, close under point Gallo; there is not room for a large vessel. The coast, which trends to the northward from point Gallo to Quintay cove, is bordered by cliffs; the cove affords no shelter, and has a reef of rocks in its center.; boats can, however, find refuge in one of its corners.

Quintay Fraile rock lies 1,695 yards N. 54° 25′ W. of point Lobos, the southern point of Quintay cove; as yet no soundings have been taken round the rock.*

Carauma head.

From the cove of Quintay the cliffs extend 3 miles, to cape Carauma, a remarkable promontory, which deserves special notice, as it is generally the first land made by vessels coming from the south, or even the west, during the summer, bound for Valparaiso. The cape itself is a high cliff, and above it the land rises rapidly to the two high chains of Carauma, the most elevated of which is 1,830 feet, which is 2 miles inland, and N. 61° E. from the cape. From seaward the high part of the chain of Carauma is usually seen above the cape in clear weather, and in the distance the Campana de Quillota, 6,230 feet high, 30 miles inland. If the Andes be seen, the volcano of Aconcagua, 80 miles inland, and 23,200 feet high, will be recognized.

Point Caraumilla.

Caraumilla point is 4 miles N. 51° W. from the heights which overlook cape Caraumilla; although it is not low, it

---

* Captain Luis Pomar, Chilian navy.

appears so when compared with the surrounding country; it is abrupt and rocky; there are two or three islets close to it. It is proposed to build a light-house here with an apparatus of the 1st order, showing a white flash-light, visible for 40 miles.

Angeles point, the NW. extremity of the coast, which forms the bay of Valparaiso, is 7 miles N. 50° E. from point Caraumilla. Between these points the coast forms a deep and angular bight, bordered to the westward by scattered rocks and to the eastward by steep cliffs; this is the bay of Lagunilla, which may be taken for that of Valparaiso. *Point de los Angeles.*

On point Angeles, the western of the bay, and at the end of a plain called Playa Ancha, from a round white tower 61 feet high from base to vane, and 180 feet above the sea, is shown a *fixed white light*, varied by a *flash every minute*, preceded by a short eclipse. During clear weather it can be seen 16 miles, but ordinarily only from 6 to 7; its position is good. When coming from the southward, the light is eclipsed by the land until it bears to the eastward of N. 44° E.; and the breakers off Caraumilla point are not cleared until it bears to the eastward of N. 52° E. *Light: Lat. 33° 01′ 10″; long. 71° 38′ 15″ W.*

Baja rock is a small rock showing a little above water; it is about 164 feet long from N. to S. and 82 feet broad from E. to W.; its position is 164 yards to the eastward of the eastern part of point de los Angeles; near it there is plenty of water. It can be passed as near as convenient, and then the bay, which opens to the southward, can be entered. *Baja or Laja rock.*

Valparaiso bay is of a semicircular form, with a capacity for more than 300 vessels between point de los Angeles and the cove of El Baron; it is well sheltered except to the N. During the winter, when the wind from that direction prevails, a heavy swell sets in. *Bay of Valparaiso.*

The depths at the anchorge vary from 2 to 47 fathoms, bottom stiff mud. The rapid rise of the bottom prevents a vessel from dragging during north winds, but makes the anchorage a poor one with southerly winds. The great depth makes it difficult to weigh anchor with the heavy swells of the winter season.

The best anchorage is close to port San Antonio or Castillo Viejo, in the SW. corner of the bay, in from 16 to 19 fathoms of water; but this is generally full of lighters and it can hold but a few vessels; it is necessary to choose an anchorage in from 27 to 33 fathoms. In summer the best is near the land, but in winter outside of other vessels and to the southward of the buoy near Baja.

Merchant vessels must anchor in the berths assigned to them by the captain of the port. They generally wait for this order near the buoys off the Laja; this is obligatory to all vessels having powder on board. Men-of-war always anchor outside of the merchant vessels, or in line with the English store-ship.

During the summer heavy squalls sweep down the mountains from the southward; clear weather and high barometer indicate strong winds from that quarter. When the barometer is low and the weather cloudy and distant points, such as the hill of port Papudo and the heights near the bay of Pichidanqui, are distinctly seen, a north wind must be expected.

Northers often occur without causing damage, but their effects are often most disastrous. Some captains prefer anchoring near the shore, for the reason that the sea breaking against the coast produces an under-tow which eases the chains, but in that position the outside vessels are liable to drive down upon them and the sea is still heavy. During a norther the best anchorage is about 200 yards from the point next outside of San Antonio; there is more protection from the W. and NW. swell which accompanies NW. winds than in deeper water. As the water at the edge of the bank in this part of the bay deepens rapidly from 16 to 35 fathoms, vessels wishing to anchor here should drop their anchor on the bank, and not immediately outside of it; in the latter case the anchor would be on the slope of the bottom and the vessel would drag when the chains surged.

Leading light. As all small vessels have to discharge at one place a white, blue, and red revolving leading light with intervals of 16 seconds has been placed on the N. end of the esplanade in front of the exchange, this being the authorized spot. The apparatus is dioptric of the sixth order, placed on an iron column 32 feet high. The light is 39 feet above the sea and can be seen from 4 to 5 miles.

Description. The bay of Valparaiso was first visited in 1536, but the town was not founded until 1544, when it was officially declared the port of Santiago.

Valparaiso presents a magnificent spectacle from the bay. It is built in the shape of an amphitheater on the slope of the hills or cuestas which rise to a height of 1,000 to 1,400 feet; on one of these is a signal-mast, from which approaching vessels are signaled; alongside of it is the house of the guard.

In 1832 the city was composed of one long street of scattered houses, extending along the beach; since that time it has been improved and enlarged, fine buildings have been raised, the streets paved, and improvements are still going on.

The cuestas, on which the city is built, are called Arrayan, Carretas, Cordillera, Bellavista, Panteon, Allegre, and Concepcion. The two latter are covered with gardens and two-story houses. The city is divided into three principal quarters, el Puerto, San Juan de Dios, and el Almendral.

The first, the western part of the city, is occupied by public offices, barracks, and buildings pertaining to shipping and commerce. On the side of fort San Antonio are numerous store-houses; a fine custom-house faces the quay, which has been extended; off it are two floating docks, called the Santiago and the Valparaiso; the latter, which can take vessels of less than 1,200 tons, was launched in 1863, the other was launched in 1865, and is 305 feet in length. These docks were much needed considering the difficulties and large cost of hauling vessels up or heaving them down, and that the latter could only be done during the good season, from September to May. This was done at the arsenal on the west side of the mole. The principal mole, which was rebuilt in 1870, was again destroyed in 1871. The swell caused by N. winds washes it and often interrupts landing and discharging during the winter, and even in the spring.

The second quarter, San Juan de Dios, has only stores and store-houses. The third, El Almendral, is the richest and finest; the fine squares, and the large street of la Victoria, 68 feet wide, serve the public for a promenade, as there is no *paseo*.

During the bombardment of 1866 it was proposed to

guard the city from the repetition of such an insult, and it is now protected by 14 well-armed batteries.

Although the surrounding hills of Valparaiso contain much good water, the city is always in want of it; a company was formed in 1868 to remedy this inconvenience, so that now water is brought to the city, in iron pipes, from reservoirs built on the hills.

Valparaiso has an exchange, a large number of hotels and churches; English, French, and American hospitals, not including the Chilian; a splendid fire department, banks, benevolent institutions, &c.

A railroad, 99 miles long, connects the capital, Santiago, with Valparaiso. The latter has a city tramway, carriage companies, &c. There is a telegraph between the two cities.

The city of Valparaiso had a population of about 97,000 in 1875.

Sand ballast is thrown on the Baja and on the Cabriteria, a small point limiting the Almendral to the NE. Stone ballast is thrown between the place del'Órden and the government mole; the boats must go as close to the land as possible.

Quarantine has to be passed in the bay of Quintero and in the cove to the northward of point Cabriteria, but the latter is a dangerous anchorage with N. winds. Generally speaking, there is no quarantine, and vessels are only kept under observation at Vinés del Mar.

Water-tanks, with pumps, bring good water to vessels. It is the best port for provisioning on the west coast of South America. Beef, vegetables, and other provisions are abundant at comparatively low prices; but these are constantly increasing.

There are numerous mooring-buoys in the bay. The outer ones serve in getting under way. Two Chilian companies possess 7 tow-boats, besides lanchas. The boats are repaired on the beach between the Quebrada de Juan Gomez and the battery Chacabuco; a new mole is being constructed at this place. There are five pilots.

Directions.   Vessels coming from seaward should make the land in latitude 33° 20′ S. during the ten months of southerly winds; the mountains will always be seen before any other part of the coast. Among them is the volcano Aconcagua, whose summit is nearly always covered with snow; the highest, which is its western part, has an irregular form and several peaks; the SE. portion is level and regular. When the summit, which is 90 miles from Valparaiso, bears N. 74° E., it is in line with the light-house.

About 27 miles from Valparaiso is the remarkable mountain la Campana de Quillota, already mentioned; the middle of its rugged summit is especially called the Campana, and when it bears N. 84° E. it is in line with the light-house. As the two mountains are generally visible, they are good leading marks for the light.

On coming from the northward, when point Molles or Quintero, 18 miles N. 5° E. from the light, has been recognized, care must be taken not to get too near the land, on account of Quintero and Concon rocks.

All vessels bound to Valparaiso should try to make point Caraumilla; sailing-vessels must not approach the coast near Rapel bank, as the SW. swell and the current set toward this dangerous part of the coast; from Topolcama the current runs at times toward this shoal with a velocity of more than 1 knot.

Soundings, 14 miles west of cape Bucalemo; on the parallel of Rapel bank, show 110½ fathoms, fine sand; 6 miles to seaward of point Caraumilla, 116 fathoms; and off the west point of the bay the same depth, with muddy bottom. During the winter months there is sometimes a current of 1 mile an hour to the southward. Soundings should be taken when approaching the land in dark and foggy weather.

Although a vessel may have a good wind outside of the bay in the morning and forenoon, she will generally lose it at the entrance. It is therefore best to use the breeze as much as possible to pass close to Baja rock, and then steer directly for the roads, taking in sail if the wind dies out, and running the chance of reaching the anchorage. In the afternoon, and with southerly winds, reefs must be taken; for though the wind may be moderate outside, there are violent gus's which sweep down the mountain in the bay; when the wind at sea requires one reef in the topsails, treble-reefed topsails and foresail will probably be sufficient sail for the bay; and if the wind from that quarter is fresh outside, a vessel cannot carry more than close-reefed

topsails over reefed courses, or perhaps over a reefed foresail only, in the bay. When a vessel finds that the wind is
too strong to reach the anchorage by tacking, it is best to
keep under short sail, close to point los Angeles, until the
wind moderates, which it does generally in a few hours.

When a vessel is approaching with a northerly wind, with
indications of a gale, it is best to keep an offing until it
shifts to the westward of NW., which it always does after
a few hours of strong northerly wind. It appears that the
sea is constantly receding in the bay; in 1845 there were
houses where the sea broke in 1830. Every day buildings
are erected which encroach on the bay, and the heavy
winter rains wash such large quantities of earth into it as
to form banks along the shore. This is mostly taking place
in the parts of the bay least exposed to the N. wind.

Tides.

It is high water, full and change, at Valparaiso, at 9^h
40^m; rise about 5 feet.

Concon rocks.

The NE. shore of the bay of Valparaiso to point Concon
is formed alternately by beaches and rocks. To the NE.
of point Cabriteria is the shore of the siete Hermanas, off
which vessels must not anchor; the road from Valparaiso
to Santiago follows this shore. Behind Concon point is a
small cove in which anchorage may be had during ordinary
weather, in 11 fathoms of water, bottom fine sand. The
rocks of that name are N. 6° W., $3\frac{1}{2}$ miles from the point;
during light winds they should be given a wide berth,
though they are above water, as vessels are drifted toward
them by the south swell and a current to the northward; a
line joining Concon and Quintero points clears them to the
eastward; the breakers around them extend about 330 yards
N. and S., and about 550 yards E. and W.

Bay of Herra-
dura de Quintero.

The land between Concon and the bay of Herradura de
Quintero, 10 miles to the northward, is generally high and
steep, and, as all this coast, presents a barren and inhospitable aspect during bad weather; here and there a few trees
are visible; during the winter and spring there is but
little verdure near the coast. The bay of Quintero opens
to the westward of the hill on point Liles or Quintero, which
can be passed close-to; it is large and sheltered from S.
winds, but entirely open to the NW. In the NE. part of
the bay, however, under point Ventanilla, there is a spot

somewhat sheltered to the northward, containing a watering-place when the season is not too dry. This bay offers a good and spacious anchorage during the summer, and some prefer it even to Valparaiso. The best berth for letting go the anchor is in 13 fathoms, fine sand, ½ mile east of the eastern part of point Liles, 1 mile from the head of the bay. A small vessel can stand farther in.

The Chilian government has decreed the establishment of a city at Quintero.

Malenas bank, N. 71° W. from point Liles, has been examined by the commander of the steamer Ancud, who found the bank 1,530 yards from the shore, extending 220 yards east and west, with soundings varying from 6½ to 9 fathoms. As the bank is rocky it is not improbable that there are some points covered by less water. The sea breaks heavily, with strong winds on the coast or winds from the NW. There are 11 fathoms at its edge, and 30 fathoms 380 yards to the westward of the bank. The channel between it and the coast is navigable and clear of danger; its depth is 15 fathoms, but it shoals to 8 fathoms 220 yards from Liles point.

Captain Luis Romar reports that during his late hydrographic researches off the coast of Quintero, he discovered a new bank of rocks WNW. of Malenas bank. It is situated N. 60° W. from Liles point, and has 11 fathoms of water over it. It extends from SW. to NE. for 110 yards, more or less, and the sea breaks heavily on it during the storms from the north to west. Between this new bank and Malenas bank there is a depth of from 20 to 27 fathoms, with rocky bottom. The sea from the SW. on the banks is heavy and violent, and it would be well to avoid approaching them in bad weather. Many soundings were taken without finding a less depth than those given, but it is not unlikely that there are rocks approaching nearer the surface.

The eastern part of point Liles has three dangerous projections. The middle one is terminated by Tortuga rock, besides which, a shoal of rocks lies about 400 yards from the land, separated from it by a channel with 5 fathoms, and about 656 yards from the junction of the cliff with the sand-beach. This shoal is also called Tortuga; it is never uncovered, and must be looked out for when keeping near the land.

Captain Luis Romar, of the steamer Ancud, states that, according to the inhabitants of Quintero, the true Tortuga rock is not the one marked on the charts, but that it is detached from the point and situated farther to the southward. It is about 260 yards from the coast, and bears S. 55° W. from the Tortuga of the charts, distant from it about 450 yards. It shows above water only at spring-tides, and the depth around it varies from 3½ to 4 fathoms; the beacon on Tortuga rock has been restored; it consists of a bar of iron supporting a small white cylinder.

At the head of the bay is a sand-beach, from which to point Ventanilla the coast, in the shape of an arc of a circle, is called la Herradura de Quintero; it is bordered by low hills, amongst which there are lagoons. Fresh water is found here when the season is not too dry.

Quintero rocks. Quintero rocks are 4 miles N. 5° E. from point Liles, and 1½ miles N. 74° W. from cape Horcon. They are above water, but low, scattered, and dangerous; they are of a dark color, and extend about ½ mile.

Cape Horcon. Cape Horcon is 3½ miles from point Liles; it is formed of cliffs 100 feet high, of a dark color. At the extremity of the most projecting cliff is a remarkable formation; back of it the land is level and higher. In the interior are elevated plateaus, and beyond is the Cordillera of the Andes.

Horcon bay. One mile N. 83° E. from cape Horcon is Horcon bay, where a landing can be easily effected among the projecting rocks. Plenty of excellent fresh water, fish, and firewood, and small quantities of fresh provisions can be obtained. The roadstead is good during the south winds which blow nine months of the year. At ½ mile N. of the landing-place are from 10 to 14 fathoms, bottom fine sand; vessels can anchor here.

Zapallar cove. Zapallar cove* lies 10 miles to the N. of cape Horcon; it extends E. and W. 1,094 yards, with a width of 656 yards N. and S. The anchorage is sheltered from the southwest winds by a verdant islet, 138 feet high, called Litis, or Morro del Potrerillo, which is connected to the main-land by a narrow sandy strip, over which the sea flows during gales of wind.

---

* Lieutenant Luis Uribe, Chilian navy.

The western headland of the cove is called Punta Zapallar.

The shores of the cove are rocky and fringed with break-ers, the eastern part presenting a sandy shore, bold, and unapproachable in any weather; in its northeast extremity is the only watering-place; the water is good but not abun-dant.

The landing-place is at a stony point in a creek which empties itself into the cove in the middle of the southern shore, and leads to the houses of the village. It is easily recognized by a miserable wooden mole, used for the pur-pose of loading and discharging lighters.

Pigs, sheep, and poultry can be obtained at Zapallar; beef and vegetables are scarce.

Making Zapallar from the southward, the best landmarks for the cove are the Cerro of Baldo, 1,017 feet high, which is 875 yards northeast of the anchorage, and Litis island, which lies off its entrance to the SW.

Soundings taken in the center of the cove gave 13½ to 16 fathoms water, sandy bottom.

Position (landing-place): Latitude 32° 33′ 01″ S., longi-tude 71° 28′ 22″ W.

Port Papudo is 13 miles north of Horcon; the coast is Port Papudo. steep and free from dangers. The pointed hill, 1,020 feet high, called El Gobernador or Cerro Verde, is an unmistak-able landmark for this small bay, which is an indentation in the coast between point Zapallar to the southward and point de la Cruz de la Ballena to the northward; but port Papudo proper extends only between point Zapallar, the island Lobos, and another point, without name, to the north-ward, on the same meridian as the islet. The depth at the anchorage varies between 6 and 22 fathoms, bottom sand, mud, and shells. A vessel wishing to remain some time should anchor in 13 fathoms, muddy bottom, with Lobos island bearing N. 19° E., and point Zapallar N. 18° W., 600 yards from the land.

Fitz-Roy says anchor 900 yards N. 5° E. of the landing-place, in the eastern part of the bay.

In the SW. part of the bay is a small mole, used for ship-ping grain, wood, and copper for Valparaiso. A vessel could possibly moor alongside of this mole with one anchor to seaward and a line on shore, but it is only possible during the fine season. Fish are caught with the seine; wood and fresh provisions can be procured in small quantities, but at a high price; there is a fresh-water course near the landing.

Point Zapallar, which forms the western extremity of the bay, is low; it must be passed at the distance of $\frac{1}{2}$ mile to avoid the rocks around the Litis islets, which extend from the point. This bay offers a good anchorage during 9 months of the year, but is dangerous during the other three. The NE. point of the coast, which forms this bay, is called Liten; the island Lobos is N. 25° W. $\frac{1}{2}$ mile from the point.

The passage which separates Lobos island from the mainland is divided into two parts by a group of rocky shoals. The western passage has 11 fathoms water, irregular, rocky bottom, and the eastern one is shoaler, having irregular depths of 4 to 6 fathoms, rocky bottom.

Both channels are impracticable with strong winds.

Tides.   It is high water, full and change, at Papudo and Quintero at $9^h 25^m$; rise at springs, 5 feet.

Bay of Ligua.   Ligua bay is 5 miles north of Papudo and east of point Ligua, which is low, rocky, and foul for 400 yards. A river of the same name, which is not navigable, empties into its SE. corner. The bay has no anchorage except for the smallest vessels, principally on account of the heavy swell which sets into it.

Bogot rock.   Bogotá rock, discovered by Captain Hollaway, of the steamer Bogotá, P. S. N. Company, was found by Lieutenant Luis Uribe, Chilian navy, to lie 1,640 yards S. 47° W. from Pichicui point, (Cruz de la Ballena.) At low-water springs it has 12 feet water over it, with $12\frac{1}{4}$ to 14 fathoms in its vicinity. The channel between it and the mainland is wide and deep, there being 20 to $23\frac{1}{2}$ fathoms in mid-channel.

Point Muelles.   From point Pichicui or Cruz de la Ballena the coast recedes and trends to the northward for 6 miles and then W. for 2 miles to point Muelles, which is low, rocky, of a dark color, and steep-to. The shore round Muelles bay is edged by sand-beaches, with some low, rocky points; it, like the rest of the coast, is backed by high mountains.

From point Muelles to point Salinas, the southern extremity of the bay of Pichidanqui or Herradura, the rugged and rocky coast runs N. for about 8 miles.

IV

Approaches to Piohidauque Bay from the northward.

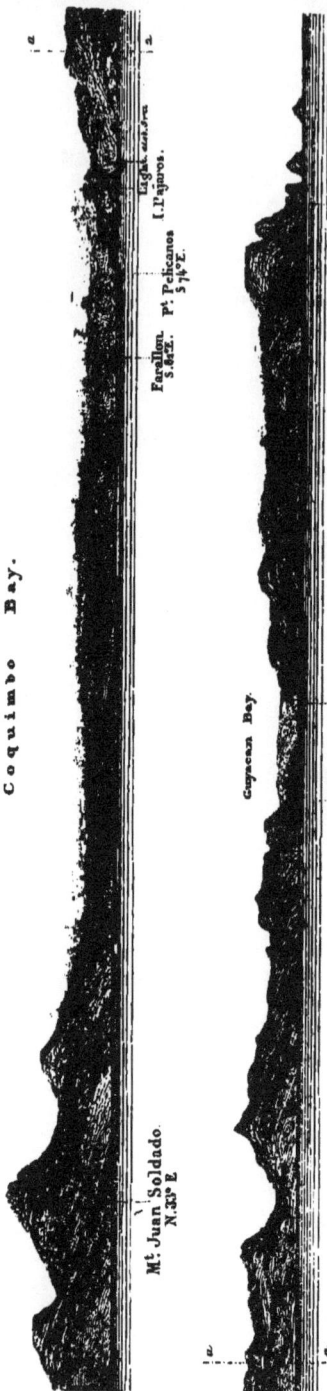

S.W.E. *S.37.E.* *S.11.E.* *S. 4° E.*

Echidauque S. scale

Con...

Coquimbo Bay.

Mt Juan Soldado.
N.35° E

Farellon.   Pt Pelicanos   I.Pajaros.
S.6.E.   S.74°.E.

Light... cabra

a   a

Guyacan Bay.

Pt Mirado.   S.½.E.
S.½.E.

Pt Salienta.   White rock
S.8.E.   S.6°.E.

a   a

Pichidanqui bay would be hard to find if it was not for the excellent landmark La Silla de Santa Inez, a remarkable conical hill in the shape of a saddle; it is 2,000 feet high, and lies 2 miles SE. from the harbor. When this hill bears S. 43° E. the entrance is opened, and will be seen as the land is approached.

In order to facilitate the recognition, a beacon has been erected on the highest part of Locos island. It is a mast 68 feet high, surmounted by an iron cross-piece, with a barrel at each end, and a vane between them. It is all painted white, and can be seen from a distance of 5 or 6 miles.

At 6,000 yards from point Salinas is point Lobos, with the island Locos immediately to the northward; it forms the southern shore of the entrance to the bay, which is ¾ mile wide, and extends to point Quelen to the northward.

Casualidad rock, which is 350 yards N. 57° E. from the northern extremity of Locos island, contracts the entrance of Pichidanqui. It has 6½ feet of water over it at low-water springs, and extends NE. and SW. about 55 yards. This danger requires attention, as there is no indication of it whatever in good weather, though the sea breaks over it in bad weather from the northwestward, and with a heavy swell from the SW. The channel between the island and the rock is clean and deep.

Somewhat less than 2 miles N. 6° W. of Quelen point, the north point of Pichidanqui bay, and about ¾ mile from the shore, is Tapado bank, a bed of sunken rocks, over which the sea breaks heavily; the channel between it and the shore is foul. One hundred and ten yards to the NW. and S. of this bank there are from 42 to 51 fathoms.

The breakers on the Tapado bank can be seen from Pichidanqui bay.

As the northern portion of Locos is steep-to, Casualidad will be avoided by passing close to it. With a sea-breeze, this rock can be easily avoided. The best anchorage is close to the east coast of Locos, in about 5 fathoms, short 200 yards from the shore.

The country produces a large quantity of grain and fruit; there is water, but it cannot be easily obtained; there are plenty of sheep and cattle. Vessels visit this bay for miner-

14 c

als, or for provisions, which can be procured at the village of Quilimari, situated behind the neighboring hills.

Caution. In running along this part of the coast, there are several outlying rocks, which can be seen during the day, near point Salinas, and others which are $\frac{1}{2}$ mile off the coast, 3 miles to the northward of Locos island.

Tides. It is high water, full and change, at Pichidanqui at 9^h 20^m; rise at springs 5 feet.

Point Changos. From Pichidanqui bay the coast trends N. to point Changos, which is 7 miles from point Salinas, low, rocky, and surrounded by breakers. It is in latitude 31° 59' 46'' S.

Cove del Negro. The coast between point Changos and point Lobos, 3 short miles, trends N. 16° E. and forms a spacious roadstead surrounded by rugged cliffs, but without importance to navigators. It is named Negro cove, after a ravine a little to the eastward. A short distance to the southward of this cove and close to the land is a steep rock, to the southward of which is a small creek which can be entered by boats and small coasters.

To the northward of the ravine is the rugged point del Purgatorio; to NE. of which is another small creek or landing-place, which is used by the fishermen of the coast; it is necessary to have a pilot to cross the breakers surrounding its entrance.

Point Lobos. The coast between el Purgatorio and point Lobos is bordered by rocks and breakers which make it especially bad, the more so as it is open to the constant SW. swell; during good weather, however, the fishermen land to eastward of the small hill de la Cachina, 147 feet high, where there is a landing-place of the same name.

Point Lobos is 75 feet high, rugged and of a sombre color. About $\frac{1}{4}$ mile S. 84° W. of the point lies the islet de los Lobos. At $\frac{1}{4}$ mile N. 51° W. from the island is a small bank, on which the sea breaks every half hour during bad weather. This point must not be passed nearer than $\frac{2}{3}$ mile.

S. 18° W. of the same point and $\frac{1}{2}$ mile distant is another island, named Isla Negra; between it and the coast are dangerous breakers, leaving a deep channel close to the island.

Point Vilos. Half-way between points Lobos and Vilos is a curve of the coast, which forms the cove of Quereo; which is inaccessible and encumbered with rocks; the sea breaking ½ mile from the coast.

Point Vilos is 2 miles NNE. from point Lobos; it has given its name to a village built on it.

A small reef called el Desempeño is 200 yards off this point, with which it is connected by a chain of rocks; it uncovers at low water, and during high water it is marked by the surf. About ⅓ mile to the eastward of Desempeño are two banks which are marked by sea-weed. They are just off the village and 400 yards apart; the nearest to the land, which trends SW. and NE. is 400 yards wide, it has 2½ fathoms at its SW. extremity; the depth then increases to the opposite end, where there is 5 fathoms. The other bank has not less than 2¾ fathoms. It is not prudent to anchor on these banks although the bottom is sand.

Finally 164 yards NW. from Desempeño is another shoal in 2¾ fathoms of water.

About 820 yards W. of point Vilos is the islet de los Hu-evos; rocky, barren, and of a yellowish color; its northern extremity is accessible in calm weather; it protects the anchorage of Vilos from the WSW. Islet de los Huevos.

The channel which this islet forms with the mainland is narrow and the sea breaks on the rocks; it should not be passed even by boats when there is any sea, as it fills at once with breakers. At 328 yards S. 61° W. from the southern extremity of the islet there is a rock above water, otherwise it is clear. Other rocks have been reported on the side of the channel, SE. of the same point, but at a short distance from the shore.

The bay of Conchali is comprised between point Vilos and cape Tablas, which is 3½ miles N. 30° W. from the former point. The coast between the two curves to the NE., and forms this extensive bay, which contains two very different anchorages, which have to be chosen according to the season. Bay of Conchali.

There are in this bay two large precipitous rocks, one the isla Blanca, which lies nearest the center, and the other isla Verde, N. 30° W. from the first. Isla Blanca is ¾ mile from point Penitente and 1¼ miles from point Conchas; a little to the eastward of isla Blanca is a smaller rock, remarkable for its black color, called Fantasma; close to the eastward of it there are three other rocks above water. Isla Verde is connected with the continent by a dangerous chain, on which the sea breaks heavily; it is 1,420 yards from point Conchas. At ½ mile SW. from this islet is an uncovered rock, which is the danger farthest from the land.

Vessels can pass between the islands Blanca and Verde; there are 22 to 28 fathoms of water in the channel.

Road de Los Vilos.

At 1¼ miles NE. by N. from the island de los Huevos is point Chungo, whose extremity is foul; the most elevated part is sandy and white; it extends out 500 yards, and is remarkable for the heavy breakers on it with S. winds. In good weather it is possible for boats to land on either side by passing through the rocks, but then only with a pilot.

The roadstead of los Vilos is the space between the island de los Huevos, point Vilos, and point Chungo; this anchorage, which is much frequented by steamers and sailing-vessels, is in latitude 31° 54′ 34″ S., and longitude 71° 39′ 27″ W.; it is sheltered from N. and S. winds, but open to the W. and NW.

The roadstead is quite large and capable of containing many vessels, at single anchor, in its southern part. The SW. swell enters the anchorage through the channel de los Huevos. The depth varies between 3 and 14 fathoms; the best anchorage is in 9 fathoms, bottom sand and shell, N. 61° E. from the summit of the island de los Huevos and N. 16° E. from the signal-mast, near the office of the captain of the port, at the western extremity of the village. Small vessels can approach closer to the coast, but the depth becomes irregular over a bottom of sand and stone.

There is no mole; the landing, which is dangerous for strangers, is in a small creek; the natives land readily on the beach during ordinary SW. winds, but with strong winds, the heavy surf renders all landing impossible, especially at low water, when all landing and discharging is interrupted.

In winter the NW. winds cause a heavy swell and a tremendous surf in the little creek; when the wind is fresh the sea breaks from the island de los Huevos to the island Blanca, when sailing-vessels are in great danger of dragging and being thrown on the beach; it is then better to get under way, and go to Ñagué cove in the NNW. of the bay.

Fresh provisions can be obtained without difficulty, but vegetables only during the summer. Water is drawn from a well back of the village, but it is brackish. In the vicinity, however, are excellent water-courses, at Conchali, Cerro de la Poza de Agua and at the Quebrada Matagorda; these places are too inconvenient to admit of water being obtained at low prices.

The village, which has been in existence 16 years, contains 300 inhabitants; it is symetrically built on uneven ground, and there are about a dozen temporary houses around the custom-house. The surrounding country is capable of cultivation, although the sand of the beach extends for a long distance, and the dry south winds do much harm to vegetation.

Besides the banks around point Vilos there is a shoal, **Directions.** called Chacabuco, in the roadstead; it has 5½ fathoms of water over it and is ⅔ mile N. of the captain of the port's office; it is only dangerous in bad weather.

There is no difficulty in taking the anchorage, as the only hidden dangers are those around point Vilos, and these do not extend farther than 874 yards from the land.

It is water, full and change, at los Vilos 9ʰ 40ᵐ; rise at **Tides.** springs 5½ feet.

To the NNW. of point Chungo and 1⅛ mile from it is point del Penitente, bordered by high cliffs and distinguished by a large rock; it is 233 feet high and overlooked by a hill of 389 feet.

Between the two points are the cove and beach of Agua Amarilla. The cove has an opening of 1¼ miles and is ½ mile deep; it is entirely open to the prevailing winds, and is of no importance as an anchorage.

The beach at its head is of fine sand and is washed by the sea. The extremities of the cove are rocky; in the midst of the rocks to the southward is a small cove called la Ballena, which is only accessible during very fine weather.

Back of the beach is a lagoon, formed by the small river Conchali. The valley to the eastward, which it traverses, is fertile, and contrasts with the barrenness of the mountains in the vicinity.

In the center of the valley is a sandy hill 475 feet high,

to the northward of which opens the fertile gorge of Agua
Amarilla, which has given its name to the beach.

Ñagué cove, which has an opening of about ¼ mile and
about the same depth, is to the westward of point del Penitente; in it the depth varies between 4 and 10 fathoms,
sandy bottom; its north part is shoal.

This cove is too small for sailing-vessels and is open to
the SW. winds and sea, as the islands Verde and Blanca do
not protect it at all from that quarter. The landing is at
the beach near the rocks to the northward of point Conchas,
and is easy when there is no surf.

There is an excellent watering-place and plenty of fish and
shell-fish, but no provisions. It is of some importance, as it
is the only place in these latitudes which is sheltered against
NW. winds, and there is no swell from that quarter.

Point Conchas. Point Conchas closes the cove to the westward; it is
formed by a sandy hill 105 feet high; its shores are rocky
and a dangerous line of reefs makes out toward Verde
island, leaving only a boat-channel.

To the southward of the point is a small cove of the same
name, and farther to the westward another, called Palitos;
but they are only frequented by fishermen during fine
weather, and they cannot in any manner be recommended
as landing-places.

Cape Tablas. The southern part of cape Tablas extends 1¼ miles to the
westward of point Conchas; it is low and rises gradually,
having breakers extending ⅛ mile to the southward.

Cape Tablas is one of the most remarkable projections of
this coast; it is dangerous and is almost cut to a peak; its
latitude is 31° 51′ 24″ S., longitude 71° 41′ 27″ W.; it is 196
feet high and overlooked by a hill of 265 feet. At ¼ mile
SSW. from the cape is an isolated rock, called Morrito del
Pabellon. There are breakers in the immediate vicinity of
the point.

Corales rock. The most serious danger is Corales rock, 1¼ miles S. 34°
W. from cape Tablas. The passage between it and cape
Tablas is deep and free from dangers other than those near
the cape; during rough weather there is a short and high
swell in this passage.

Corales rock is more than 2½ miles from the islet de los
Huevos. From a depth of 10 to 30 fathoms near the islet

the depth increases to more that 44 fathoms near the rock, which is N. by W. from the anchorage of los Vilos. To the SW. of Corales are two small rocks, which are hardly separated from the first; there are from 14 to 30 fathoms around the group.

The coast to the eastward of cape Tablas forms the road- Tablas road-stead of that name by an indentation ⅔ mile in depth, and stead. a little more than 1 mile broad. The best anchorage is in 11¼ fathoms, sandy bottom, S. 16° W. of the eastern part of the island of Lilenes, and N. 73° E. of the Morrito, a small isolated rock north of cape Tablas and 874 yards from the land. The depth of the anchorage varies between 14½ fathoms at the entrance and 4 fathoms near the breakers off the beach. The bottom is of sand or sand mixed with rock; it is good holding-ground, and the roadstead is smooth during the season of SW. winds.

S. by E. from this anchorage is a small, steep point, with a white rock off its extremity. To the eastward of it is the best landing-place, but it is only tolerable, as the beaches are flat and full of rocks and the surf is generally heavy.

The vicinity of the roads is barren; there are no provisions or water, but the latter can be obtained at Sagué, ½ mile distant. There is plenty of game, especially partridges.

Point Pechoñas protects the roadstead of Tablas to the Point las Pecho-northward; it is rocky and broken toward the sea, 255 feet ñas. high, and the breakers extend off it 330 yards.

About 2 miles NE. by N. from cape Tablas is Lilenes island; Lilenes island. it is high and of a greenish color, pointed, and almost circular; its diameter is 984 feet. It is separated from point Pechoñas by a channel 874 yards in width, with from 8 to 12 fathoms of water, bottom sand mixed with stone. The shore of this island is clean, but that of Pechoñas is dangerous, and the sea breaks on it heavily.

Cebollin rock is ½ mile N. 58° W. from the summit of Cebollin rock. Lilenes island, and 2 short miles N. 10° E. from cape Tablas; it is uncovered at low and marked by breakers during high water. At a distance of 164 yards from it there are 15 fathoms, bottom rock; between it and the island there are 24 fathoms, rocky bottom.

Point Loberia is 6½ miles N. 9° E. from cape Tablas; it is Chigua Loco low, and terminated to the SW. by a flat, round hill. The roadstead.

coast between these points curves in a little to the NE. and forms the large roadstead of Chigua Loco, which is entirely open to the SW., and is of no importance.

Boca del Barco cove.   All this coast is bad; the breakers commence about 271 yards from the beach, excepting in a cove called Boca del Barco, situated N. 36° E. from cape Tablas, almost in line with the E. point of Lilenes, and more than 3 miles S. 30° E. from the rock off point Loberia. This small cove is used by coasters and boats; it has a good anchorage off the SW. extremity of its beach, of pebbles. There is a dry rock in the center of the cove, and to the NW. of it two sunken rocks, which must be left to starboard when entering. The depth varies between 4 and 8 fathoms, muddy water. The landing-place is in latitude 31° 47′ 33″ S. and longitude 71° 38′ 17″ W.

If a vessel should wish to load here, she must anchor ⅓ mile to the westward of Salina point, the southern point of the cove, in from 12 to 13 fathoms, bottom sand and rocks. Vessels are completely exposed to the SW. swell, but the anchorage is preferable to that of Chigua Loco.

Chigua Loco cove.   Between Boca del Barco and the elevated portion of point Loberia is a cay called los Bajos de Chigua Loco, ⅔ miles from the land, with which it is connected on the NNE. by a sunken ridge. To the NE. of Bajos is the cove of Chigua Loco; it is useless, and landing is seldom possible without danger. In its center is anchorage, in from 7 to 12 fathoms, sandy bottom, but it is entirely exposed to SW. winds; it is said that point Loberia and the Bajos shelter it from the NW. At the head of the cove are the houses of the hacienda de Chigua Loco, which is in latitude 31° 45′ 26″ S. and longitude 71° 38′ 08″ W.

The coast is generally low, bad, and steep near the beach; about 1½ miles in the interior the mountains rise to 2,000 and 2,600 feet.

Point Mula Muerta lies N. 6° E. of the Bajos; it projects but little, is dark, and bordered by rocks. Between this point and Loberia are two small coves, which are separated by point de las Conchas. The first between las Conchas and Mula Muerta is called Mostaza, and admits of landing in fine weather, but only with an expert. To the SW. by W. the breakers extend ⅔ of a mile to seaward; their extremity is S. 16° W. from point de las Conchas, and S. 24° E. 1 mile from the small hill on point Loberia.

The cove to the NW. of Mostaza is inaccessible and without importance.

Point Loberia is a promontory of moderate height, precipitous, and terminated by a small isolated hill; it is bordered by breakers, which extend 270 yards from the coast; it is overlooked by a chain of mountains from 2,000 to 2,600 feet high. No special description of these mountains, the Andes, is given, as they are so easily confounded during misty weather.

Point Loberia.

The right shore runs N. for 5½ miles, to the Huentelauquen cove, with deep water; ½ mile off shore the depth varies between 14½ and 18 fathoms, bottom rock and sand; at 1½ miles the depth increases to 30 and 35 fathoms, and at 3 miles to 60 and 70 fathoms, muddy bottom.

Huentelauquen cove is tolerably sheltered from the SW. swell, and, although it is only ¼ mile square, it has a special interest for the departments of Illapel and Combarbala, as it is the only passable port known on their coast.

Cove of Huentelauquen.

The south shore is in latitude 31° 58′ 54″ S., longitude 71° 40′ 30″ W.

The depth in the cove varies between 3 and 8 fathoms: in the center it is 6 fathoms. A low and rocky island protects the anchorage from the prevailing winds and sea; the channel which separates it from the main-land is so narrow and so obstructed by rocks that the SW. swell cannot pass through it, and only the chop made by the current is felt. By closing the SW. part of this channel its NE. part would be made into an excellent small harbor for loading small vessels, the isthmus would be enlarged rapidly by the sands, and the anchorage would not suffer from them. The sand is naturally deposited on a coast which makes an angle of 40 degrees with the prevailing winds.

Huentelauquen cove can now admit only two vessels of 300 tons.

The N. shore of the cove is bordered by ravines, and is very irregular; some rocks above water, with plenty of water near them, are 437 yards to seaward of it. The E. beach is of shifting sands; in the NE. are large dunes.

The cove is at present without water or resources, but,

should it become important, all these could be supplied, as the surrounding provinces are prosperous and fertile.

Point Pozo is to the northward of Huentelauquen cove; it is very irregular and steep, and its upper part is level and barren; a reef makes out 400 yards from its foot. Boats can land, during fine weather, in a small creek SE. of the point.

To the northward of Pozo point the coast runs in a little, and forms the unimportant cove of Choapa; its beach is low and sandy, with a heavy surf; the small hill, de los Olivos, about 1 mile from the sea, is a good distinguishing mark.

The river Choapa empties at the southern extremity of the beach; in summer the quantity of its water is much diminished, as it is drawn off to irrigate the valley, which is fertile and rich in products of all kinds.

Point Ventana, in latitude 31° 36′ 40″ S. and longitude 71° 41′ 26″ W., terminates the cove of Choapa, and lies N. 23° W. from point Pozo, N. 2° E. from cape Tablas, and N. 8° W. from point Loberia. It is low, projecting, and dangerous; to seaward of it the sea breaks from time to time; it is overlooked by a chain of sandy hills, which are the northern limit of the valley of Choapa.

Oscuro cove, 10 miles to the northward of point Ventana, is a small bay, running 766 yards northeast and southwest, having a variable width between 350 yards at its mouth and 219 yards at its northeast extremity, where it terminates in a sandy beach, which is ordinarily approachable for the light canoes of the fishermen, but not for larger boats during strong winds.

Its shores are rocky, rugged, lined with numerous dangers, and covered with hills, which rise gradually toward the interior.

Two detached rocks lie off the southern point, but are not of imminent danger; and from the northern point, called Burro, two sunken rocks extend toward the south, over which the sea breaks heavily. The farthest one lies 142 yards from the point.

The depths in the cove are gradual, decreasing from 25 fathoms, sandy bottom, at the entrance, to 18, 16½, 11, and 5½ fathoms, following the center; toward the north and

Point Pozo.

Cove of the river Choapa.

Point Ventana.

Oscuro cove.

south shores the soundings diminish gradually to 11 and 10 fathoms at a short distance from the shores.

Oscuro cove is an anchorage for coasting-vessels and vessels of 200 to 250 tons, anchoring in 11 fathoms, sandy bottom, in the middle of the bay.

The entrance is easy for sailing-vessels, notwithstanding the narrowness of the cove, as they always have a free wind; but going out is difficult; this can be obviated by taking advantage of the morning calms and towing out by ship's boats.

But little swell sets into the anchorage; it will not incommode vessels lying in the cove.

The landing-place is to southward of the eastern sandy beach, and accessible, although not very good on account of the surf.

As there is no wharf, it is not always easy to embark.

The anchorage only admits of two vessels swinging clear.

No provisions or water can be found in the immediate vicinity of Oscuro cove; but at the houses of the Tatoral estate, at a distance of one mile, some articles can be obtained. Between the houses and the cove there is a stream of good water; there is also a small watering-place about 110 yards from the beach. Wood, and some game, consisting of partridges and doves, can be had in S. Abunda ravine. This cove is also called Tatoral.

It is high water, full and change, at 9ʰ; rise, between 4¼ and 6½ feet.

The approximate position is, latitude 31° 27′ 20″ S., longitude 71° 37′ 30″ W.

From Ventana point the coast runs N., with a slight inclination to the westward; it is an almost uninterrupted straight line of cliffs to Maytencillo, which is 21 miles from point Ventana. This small cove is only practicable for balsas; a boat may land sometimes, but there are many hidden rocks. Its position is indicated by a large triangular pyramid of white sand, very regular in form, resting on the cliffs bordering the coast. This landmark remains permanent, thanks to the sand which the wind accumulates on the north side of the cove. *Cove of Maytencillo.*

Ten miles to the northward of Maytencillo are the two points Vano, to the northward of which are some rocks.

All the coast comprised between them and the preceding is composed of rocky cliffs, of a bluish color, about 150 feet high. The land back of them rises to 300 and 400 feet, and, 3 miles in the interior, to 3,000 and 5,000 feet. The deep valley of Arenal is 6 miles to the northward of these points; on the north side of its end near the sea is a sand-hill and at its entrance is a sand-beach.

Limari river. From the cove of Maytencillo the coast is a continuous line for 33 miles, trending N. 7° W. to the mouth of the Limari river, which appears large from seaward, but is inaccessible; the coast near it is rocky and steep. About 2 miles N. of it is a low, rocky point, with a small beach, where boats might land, though there is a heavy surf. The land rises suddenly, forming a chain of mountains 1,000 feet high, which are parallel to the coast to within 2 or 3 miles of the river; their summits to the northward are wooded.

The N. point of the entrance to the Limari river is low and rocky; the southern point, on which a patch of white sand will be distinctly seen, has a rapid descent. The river is about ¼ mile wide at the mouth, but the sea always breaks with violence. In the interior it turns slightly to the NE. and then returns to E. in a deep gorge of the mountains.

Tortoral cove. There is a small bay about 14 miles to the N. of Limari. In its northern part is a sand-beach, but it is always beaten by the surf. The coast to the northward of the bay is rocky, and about 8 or 9 miles farther there is a small peninsula of rocks, in the center of which there is a high pointed rock. To the southward of the promontory a deep cove opens with a sand-beach at its head, but the entrance is so full of sunken rocks and islets that it cannot be entered by even the smallest vessels. During fine weather, however, boats can enter and land in the cove. The outside breaker is not more than 400 yards from the shore; in calm weather the swell sets directly upon it. This cove is called Tortoral de la Lengua de Vaca.

Lengua de Vaca. La Lengua de Vaca is a low, rocky point 8 miles to the north, which gradually rises to a round, flat hill 1,000 feet high, situated about 1 mile to the southward of the point; about 200 yards from it are some rocks just awash, and at 400 yards there are but 5 feet of water.

Bay of Tong-y. From point Lengua de Vaca the coast turns suddenly to

S. 30° E. and forms the bay of Tongoy; it is steep and rocky for 2 miles from the point; there are 14 fathoms about ½ mile from the coast. About 3 miles from the point is a long sand-tongue which extends the whole length of the bay to the peninsula of Tongoy. The southern part of the beach is called Playa de Tanque, and the eastern part Playa de Tongoy.

Off the western extremity of the beach, near Tanque, is an anchorage in 5 to 7 fathoms, about ½ mile from the coast; the bottom is muddy sand, soft in some and hard in other places. With southerly winds it is quite smooth and boats can land without difficulty, but the north winds throw a heavy sea in the bay. This anchorage was formerly visited by American and other whalers. The village called Rincon de Tanque consists of about a dozen ranchos; only brackish water can be obtained here, but about 2½ miles N. 83° E. there is good water some distance from the beach. As landing is generally difficult, it is hard to water. *Anchorage of Tanque.*

There is anchorage all along the bay from Tanque to Tongoy, about 2 miles from the coast, in from 7 to 10 fathoms, sandy bottom.

There is also a good anchorage off the village of Tongoy, which is sheltered from N. winds by the SW. point of the peninsula, in 4 fathoms of water, bottom sand, covering clay; the Lengua bearing WNW.; the smallest vessel must not anchor in a less depth, as the sea breaks violently in shoaler water during fresh N. winds. Large vessels can also find some shelter from N. to NW. winds. *Port of Tongoy.*

With a strong SW. breeze the sea is too heavy for any vessel to remain at the anchorage to the southward of the peninsula, but on its northern part there is a small bay completely sheltered from S. winds. In the southern corner of this bay is a small cove, in which boats can enter during calm weather; this inlet extends in about 1 mile, and at its head is some fresh water which can be taken in boats.

A rock, which is just awash at low tide, lies ½ mile south of the large chimney on the hill; there is a passage between it and the land; it has been marked by an iron buoy surmounted by a wooden ball; a lookout must be had for it, as it is in the route of boats going to the shore and in the position which at first appears to be the best anchorage.

222

FROM CONCEPCION TO COQUIMBO.

Description.   The village of Tongoy was, in 1856, composed of a few
small houses built on a high point, on the south side of the
peninsula; it is much larger at present. The Mexican and
South American Company have a foundery and large store-
houses of mineral ores at this place; when the furnaces are
in operation they will be seen from seaward during the
night. The company have made an embankment of the
scoria of copper, and also a mole, along which the small
coasters load and discharge; the vessels of the company
are loaded and discharged by lighters. A screw-steamer,
which belongs to them, runs from Tongoy to Guyacan, and
is used as a tug. The company ships ores from here to
Guyacan, the United States, England, and Hamburg.
There are some stores, and the communication with Guyacan
is easy. It is, during the summer, frequented for bathing.

The ballast is thrown on the east beach of the bay. There
is a watering-place to the southward of the village.

Tides.   It is high water, full and change, at port Tongoy, at 9^h
10^m; springs rise 5 feet.

Mount Huana-   Mount Huanaquero is about 2 miles NE. of the peninsula
quero.   of Tongoy; it has three peaks, and is 1,850 feet high. The
coast to the westward of this mountain is rugged and rocky,
and has no shelter except for boats. There is a deep bay to
the northward which is sheltered from the S. and W. winds,
but open to the northward. Between it and la Herradura
de Coquimbo, 13 miles, there are no anchorages.

CHAPTER VIII.

FROM COQUIMBO TO THE FRONTIER OF BOLIVIA.

Variation from 14° 13' to 12° 30' easterly, in 1876; increasing annually about 1' 30".

Port Guyacan is a small, land locked harbor, separated Port of Guyacan or Herradura de Coquimbo. from the bay of Coquimbo by an isthmus of about 1 mile in width; it is circular, and has an area of 3¼ square miles; its entrance, between the points Herradura and Miedo, is ¾ mile in width. With a fair wind any vessel can enter, keeping close to the S. coast to avoid a rock about 50 yards off point Miedo; when inside, the anchorage can be chosen in from 3¾ to 20 fathoms, bottom sand and shell, covering a tenacious clay. Should it be desirable to anchor on the side of the Knowsley rock. town of Guyacan, attention must be given to Knowsley rock, named after a vessel which touched on it, which has three heads, with 3.7 feet of water over it and 5 fathoms alongside of it at low-water spring tides. It lies on the north shore of the bay in line with point Miedo and the next point inside it; is S. 71° W. from Cerro Allegre, a round, remarkable hill situated in the NE. corner of the port, and is 163 yards from the nearest land, and 433 yards from the wharf belonging to the smelting-works. It is marked by a red buoy with a cage, moored in 6 fathoms, 40 or 50 feet from its eastern side; the two easternmost chimneys of the smelting-works in line will lead to the southward of it. The N. coast must not be approached closer than 273 yards until Cerro Allegre bears NE.

Off the ancient Herradura, in the SW. corner of the bay, is a perfectly-sheltered spot, in which repairs of all kinds can be undertaken with security; there are from 2½ to 3 fathoms. H. M. S. Beagle made her repairs in this port, remaining several weeks with her crew in camp on the beach.

This land-locked bay would be of still more importance were it farther from the bay of Coquimbo, which is larger but less sheltered; the N. winds, however, which, in winter, raise considerable sea in the latter bay, manage to send an uncomfortable swell into port Guyacan. A disadvantage is that sailing-vessels sometimes find difficulty in leaving the port on account of the narrow entrance, the wind, together with the heavy swell, driving them back into the bay. The favorable wind lasts only a few hours during the morning, and is then generally light and uncertain.

Ballast is thrown on the banks near the river on the east side and to the southward of the village. There are several wharves, but they are in bad condition.

Description.

The Mexican and South American Company established a large furnace for smelting copper-ores at Herradura in 1848, and at the same time founded a town. A bridge of boats was built to Whale rocks, and a mole constructed at which vessels of 300 tons could load and discharge. At present the works at Herradura are deserted, everything having been taken to Guyacan, which is the center of all the factories of the new company, who possess 35 smelting-furnaces, whose draught is furnished by three large chimneys; around the works are the houses of 400 workmen. To the southward is the small town, with about 1,000 inhabitants. The company owns a steamer, which is often used for towing vessels in or out; large vessels are discharged by means of lighters. The furnaces are always in operation, and they give so much light that the vessels of the company can enter during the night. As Guyacan is not a port of entry, vessels before going there to load are obliged to obtain a permit at Coquimbo. The foreign commerce consists in the importation of English coal, brick, clay, iron, &c., and the exports are copper-bars, copper regulus, silver and copper ores, to the United States, England, and Hamburg; banking is done by the company by drafts on England. The water is brackish; but fresh water is brought regularly from Coquimbo; the company has established a distillery, with a reservoir having a capacity of 1,188,810 gallons. There is a large quantity of coal at the works.

Coast.

From point Miedo the coast runs N. 5° W. for 1½ miles. The depth near it is about 16 fathoms, stone and gravel bottom; at this distance from point Miedo is the small precipitous point Finaja, which is clean; to the eastward of it is a small indentation. From the point the coast runs about N. 45° E., and ends in point Tortuga; this stretch of the coast is full of reefs and rocks at the water level, and is therefore inaccessible.

Los Pajaros Niños are small rocky islets, surrounded by reefs, to the NW. of point Tortuga; they form two principal groups; the outside one is about 1¼ miles from the shore, N. 47° W. from the light-house; it is about 130 feet long and 65 wide. About ¾ mile S. 15° E. of it is a large islet, from which a reef of rocks makes out. There is a safe passage ½ mile wide between the two groups, and another between the large islet and the shore; the depth in these is 16 fathoms. Vessels of all sizes can pass through with the fresh SSE. or SW. winds, which generally prevail, but it should not be attempted with light winds, as the current is always strong, and the anchor would have to be dropped on a bottom of rock mixed with sand and shell. *Pajaros Niños.*

Point Tortuga, high and steep, especially at its northern extremity, forms the southern boundary of the entrance to Coquimbo, and is the northern termination of the peninsula, 460 feet high, which separates this bay from that of Guyacan. It has three points; the middle one, near the light-house, is point Tortuga. *Point Tortuga.*

A *fixed white light* varied by *flashes* of 5ˢ duration every 15 seconds, the eclipse lasting 10ˢ, is shown from a square wooden tower, painted white, with the cupola green, and attached to the keeper's dwelling, 200 yards within the extremity of Tortuga point; the building is 25 feet high, and the light, elevated 106 feet above the sea, is visible 12 miles. This light is not visible between the bearings S. 78° W. and N. 75° E. by the south, the land intervening. Coming from the southward the lights from the works in Herradura bay will probably be first seen. It has for some time been proposed by the Chilian government to remove this light to the outermost islet of the Pajaros Niños. *Light : Lat. 29° 56′ 50″ S.; long. 71° 20′ 40″ W.*

Ships approaching Guyacau or Coquimbo bay are signaled from Flag-staff or from Signal hill, there being on each a staff and yard. *Signals.*

Pelicanos rock is around point Tortuga, an isolated rock 26 feet high and 48 yards from the shore; a boat's length from it there are 4½ fathoms. *Pelicanos rock.*

From 40 to 50 yards N. 40° E. from the highest part of Pelicanos rock is a sunken pinnacle rock with 9 feet on it. *Dorsetshire rock.*

15 c

at low-water spring tides, with deep water immediately out-side of it. Vessels going to the anchorage off Coquimbo, on rounding or passing Pelicanos rock, should not approach it inside of 200 yards, or in thick weather not inside of 15 fathoms, but stand to the eastward toward La Serena until the conspicuous church spire, near the town of Coquimbo, comes in line with the extreme of Observation point, bear-ing S. 18° W., when stand in for the anchorage.

Navigating lieutenant A. W. Miller, R. N., of H. B. M. S. Columbine, after sounding for this rock without success, succeeded in finding it by sweeping with a dredge. He states that several ships had struck on an unknown rock, their masters supposing that they were more than 200 yards from Pelicanos rock. In March, 1867, the ship Dorsetshire, of 700 tons, struck on it, and was beached to save her from sinking. Its vicinity has been sounded with care and no other dangers discovered. On the point is a platform with two guns and a house serving as a guard-house, but these cannot be readily distinguished. It was reported that the vessels Chelydra and New Granada touched on a pinnacle rock, situated on a line drawn from Pelicanos to the town of Serena. Captain Harvey, R. N., commanding H. B. M. S. Havannah, searched for it without success, but while sound-ing he found a patch of 6 fathoms, on which the lead would not rest; with 9 and 10 fathoms around it and 12 between it and the shore; its position is 250 yards N. 26° E. of Peli-canos rock. After Captain Harvey and Admirals Cloué and de Lapelin, the vessels of the Chilian squadron searched unsuccessfully for the Chelydra rock, and concluded that it did not exist. It can be assumed, therefore, that these ves-sels, passing near Pelicanos in thick weather, touched on Dorsetshire rock.

The sea around Pelicanos and along this coast is fre-quently covered with a greasy foam of bad odor, which from a distance has the appearance of breakers.

Bay of Co-quimbo. The bay of Coquimbo, which opens to the southward of Pelicanos rock, is situated to the northward of the penin-sula which forms the bay of Guyacan. It is on the limit of the tropical calms and the heavy gales of the higher lati-tudes; vessels can enter at any time during the day.

Between point Tortuga and the Rio Coquimbo it is very

extended, its area being 30 square miles, but it is open to
the northward; there are from 5 to 16 fathoms of water,
with good holding-ground of mud and clay at the anchorage. The maximum depth on a line drawn from the Rio
Coquimbo to point Tortuga is 20 fathoms. The usual anchorage for strangers is in 8 fathoms, with the last point
north of the west coast bearing N. 36° W., the church of
Serena N. 66° E., and the houses near the landing-place S.
65° W. The best anchorage is in the SW. angle of the bay
in 6 fathoms and excellent bottom; but the swell generally
sets in so heavily as to cause a surf, which renders landing,
in all excepting a few sheltered places, very difficult.

The charts of the bay are good; vessels can be taken in
easily by sounding rapidly when approaching the E. coast
or the head of the bay, as the soundings decrease gradually
to a low sandy beach.

Coquimbo is the principal commercial port of northern Description:
Chile; formerly a great disadvantage was its want of fresh
water, which could not be procured without difficulty, as it
had to be brought from a lagoon on the E. shore of the bay;
to remedy this a reservoir of a capacity of about 6,000,000
of gallons was constructed 2 miles SE. of the city, to which
water is taken through pipes; the reservoir being supplied
from the condensation of sea-water. This work was finished in 1865, since which time Coquimbo has been well
supplied with water at a reasonable price. Wood is scarce
and distant from the anchorage. Plenty of fish can be
caught with the seine, but the few fresh provisions procura-
ble are expensive. The town has about 5,000 inhabitants;
there is one government, with several private wharves. A
large store-house has been constructed on one of the moles,
near the government wharf. The most convenient wharf is
the one just to the westward of which is a foundery and a
quay for loading and discharging the copper-ores; the
mountains in the vicinity are very rich in ores.

There are three founderies and one factory of hydraulic
lime at Coquimbo. A railroad 50 miles in length, with
branches to Guyacan and Panucillo, connects it with Serena
and Ovalle. Ballast is thrown on the west coast of the
bay to the northward of the river. The mail steamers
touch here twice a week and the coasting steamer three

times; the Mexican and South American company own a
tug, which can be hired for towing.

The bays of Guyacan and Coquimbo are subject to a
curious phenomenon. On the 24th of April, 1858, at 7.30 a.
m., a sharp shock of earthquake was felt, which was immediately followed at Guyacan by a sudden rising of the
sea of 15 feet. The sea in this port continued to rise and
fall for $1\frac{1}{2}$ hours at intervals of from 3 to 5 minutes, each
flood being a little lower than the preceding. It was four
days before full moon, and the rise took place $1\frac{1}{2}$ hours before high water. Although the sea rose 15 feet, the water
only reached 9 feet above the usual high-water mark. The
bay of Coquimbo was agitated in the same manner; the
sea rose suddenly, covered the quays, and filled the base-
ments of the houses. On one occasion a vessel of 1,400
tons, which was anchored in $4\frac{1}{2}$ fathoms, was left nearly
dry on the first receding of the wave; no serious damage,
however, was done on either occasion to vessels lying in
the port.

Directions.

Vessels bound to Coquimbo should try to make the land
at Lengua de Vaca, which is very salient and sloping toward
the sea, then, steering for point Tortuga, mount Huanaquero will be seen with its three gorges of a dark color.
From there the chain continues to a round isolated hill
1,000 feet high, whose spurs form the port of Guyacan.
This is the Pan de Azucar. All the coast to point Herradura is very dark, but from thence the land is whitish
and can be easily recognized. The views will aid in this, as
also the Signal hill, 478 feet high, which overlooks the
peninsula of Tortuga, and can be seen from seaward. The two
islands, Pajaros Niños, can be seen a long distance; when
they are made, with a steady SE. or SW. wind one of the
channels through them can be taken, but with a light or
variable wind it is best to pass 200 yards from point Tortuga and then steer S. 64° E. until a house on the isthmus
is just open of the cliffs under Signal hill, from whence the
anchorage can be chosen.

Vessels can easily enter during the night by the light
given by the furnaces. In coming from the southward the
coast must be kept at a distance of 3 miles until the lights
of the town of Serena bear S. 87° E.; then steer on this

bearing, which will clear Pajaros Niños, and when the furnaces open from Pelicanos rock, the bay can be entered and anchorage taken in 8¼ fathoms.

When running in with a northerly wind, the Pajaros Niños must not be approached.

Toward evening it is generally calm near point Tortuga, but the swell will drift a vessel in as there is no current, or the boats can be used to tow in. Toward 5 or 6 p. m., however, the cobre, or land-wind, sets in from the NE. or N. In all cases when the wind is dying out near point Tortuga it can be assumed that the land-wind will be strong enough to carry a vessel in.

In approaching this port a vessel must be careful not to be carried to the northward, as the current and the breeze are almost always from the south; for that reason it is recommended to keep close to the Pajaros.

The winds at Coquimbo are generally from the southward and moderate; they are off shore during the greater portion of the year, only interrupted for short intervals in winter by NW. gales. The inhabitants say that the northers are never very heavy at this place, but in winter it is best always to be prudent.

It is high water, full and change, at Coquimbo at $9^h 15^m$; rise 5 feet. Tides.

La Serena, the capital of the province of Coquimbo, of which Coquimbo is the port, founded in 1544, by Francisco de Aguirre, lieutenant of Valdivia, is situated on the NE. shore of the bay; the road and railway connecting the two run along the beach. La Serena.

The houses are mostly built of sun-dried brick, and have but one story on account of the earthquakes. The town and its gardens are provided with water by canals from the river Coquimbo, which is on the N. shore; it has several churches and about 12,300 inhabitants. It was destroyed by the earthquake of 1730.

The landing at Serena can only be accomplished with balsas, on account of the heavy surf; but as Coquimbo is but 6 or 7 miles distant, with every facility for transportation, there are not even balsas to be found at Serena.

The universally good temperature, the agreeable climate, and the clear atmosphere have given the name of Serena to this town. The country, however, often wants rain; in June, 1850, it had not rained a drop for 15 months. In order to give an idea of the value of a shower, the produce caused by one night's rain in the small valley of Huasco, a port farther N., was estimated at $1,180,000 for that district only. This result is really miraculous; before the rain all the valley was as an uncultivated sand desert; a week or ten days afterward the ground was covered with verdure and flowers.

It must be mentioned that the cloudy weather on this coast, to the river Guayaquil, is generally during the winter. The fogs are frequent and sometimes very dense.

Coast.

From the southern extremity of the town of Coquimbo the coast forms a semicircle to Serena and the Rio Coquimbo; from there it runs nearly N. to point Teatinos.

Point Teatinos.

Point Teatinos is the northern extremity of the bay of Coquimbo; it is bold and rugged; the land back of it rises gradually, as it recedes from the coast, to mount Cobre, 6,400 feet high. From Teatinos the coast runs to the north, then to the west, and is terminated by point Poroto, $3\frac{1}{2}$ miles from Teatinos.

About $4\frac{1}{2}$ miles N. of Poroto is port Arrayan or Juan Soldado; it is simply an indentation, completely open to the northward, situated behind a rocky point, in which a boat hardly finds shelter from S. winds.

Mount Soldado.

A little to the northward of mount Cobre is another of the same chain, called Juan Soldado, 3,900 feet high; its northern slope is steep. At its foot is the small bay of Osorno, about $\frac{1}{2}$ mile long, and without shelter. About $\frac{1}{2}$ mile to the northward of the bay is the small village Yerba Buena, which consists of only a few houses.

Tilgo island.

The small island Tilgo is a little to the northward of Yerba Buena, and is separated from the land by a channel 200 yards wide, only practicable for boats. This island has the appearance of an advanced point. There is a large white rock on its western extremity.

Los Pajaros.

The two islets Los Pajaros, 100 to 150 feet high, are separated by a channel 2 miles wide, and lie about 12 miles from the coast. The northern one is much smaller than the other, and, as far as ascertained, there are no dangers between them; but a reef, which sometimes breaks, extends considerably to the southward of the southernmost islet.

Totoralillo is a small bay about 3 miles N. of the island Totoralillo bay Tilgo, opening to the NW.; there are three small islands off its SW. point. The best entrance for small vessels coming from the southward is a channel about 100 yards wide, with 8 to 12 fathoms, which separates the island from the south point; the south swell enters through this channel.

The rock which is above water near the point of the continent must not be approached closer than 100 yards, as there is a sunken rock at about that distance to the westward of it, on which it is proposed to place a buoy. The channel between the island is closed by breakers.

Vessels can anchor ¼ mile from any point of the beach in 6 to 8 fathoms, bottom of fine sand and shell. It is difficult to land at any other place than the mole; the most convenient place is on the rocks near the entrance, but nothing can be shipped from there; the best place for that purpose is the E. extremity of the beach. The land around cape Choros projects enough to the westward to lead to the conclusion that N. winds could not produce much sea at the anchorage, but stormy winds from that direction and from the southward interrupt all loading and discharging.

Fresh water can be obtained from wells near the landing- Description. place. There are two moles, but one in very bad condition; the government mole was destroyed by the sea. Lighters and péons can be hired. Ballast is thrown in two places, on the south shore between the two moles, or on the west shore between the mole belonging to the works Muñoz and the south point of the small creek Temblador.

The village of Totoralillo, with 60 to 80 inhabitants, depends principally on the Mexican and South American company, which has a large dépôt of mineral ores and some small founderies. The ores are shipped for Guyacan, Caldera, the United States, England, and Hamburg. A small steamer which belongs to the company goes to Guyacan from time to time. It is the port of the mines of Higueira, 13 miles to the eastward. There are about 40 mines at this place; it is also the port of El Barco, which has 5 mines. La Higueira is the mining center of the province.

Creek of Temblador. Temblador is a small cove NE. of Totoralillo; it is more difficult to land in than on the other shore, and is less sheltered.

Chungunga island. Chungunga island is about 4½ miles to the northward of Totoralillo, and about 1 mile from the coast. It is a good landmark for the small cove of the same name. Off it is a point of rocks, and a little in the interior a remarkable saddle with a round hillock in the center; on coming from the southward it seems to be the extremity of a mountain-chain which runs to the east of Totoralillo, and attains an elevation of 2,000 to 3,000 feet. A little to the northward of the island Chungunga is a large patch of white sand, which can be distinctly seen from the west; it is at the southern extremity of the beach of Choros; on it there is always a heavy surf.

Cape Choros and point Mar Brava, its NW. extremity, form a small cove, whose shore is strewn with rocks and chains of reefs. It is not very deep, having from 2 to 3 fathoms of water, bottom of fine sand. It is sheltered from N. winds by the island Gaviota, one of the Choros islands, but it is entirely open to the southward.

Choros islands. The three Choros islands are off cape Choros; the inside one, Gaviota, is low, and so close to point Mar Brava that only boats can pass through the channel, which is open to S. and NW. In its narrowest part, 400 yards, is a group of rocks and reefs, on which the sea generally breaks; in its southern portion there are also some reefs, which extend in the direction of cape Choros; there are also three danger-ous rocks, which obstruct the passage.

The best anchorage seems to be N. of the island Gaviota, in 10 fathoms, sandy bottom; but there is always some sea, and the swell often interrupts communication with the shore.

Water can only be found 12 miles in the interior, and is brackish; there is no wood.

The passage between Gaviota and the other two islands is clear; the southernmost island, called las Damas, is the largest, 2 miles long, but it does not protect the anchorage. Its summit is very rugged, and its SW. extremity resembles a castle. Off the south point is a small pyramid, and the breakers extend ¼ mile from the land. The channel between the two outside islands is also clean; but about ½ mile west of the northern island is a rock, which is just awash. It is high water, full and change, at 9ʰ 20ᵐ; rise 5 feet.

Toro reef is 5 miles SSE. of las Damas; it is dangerous, hardly showing above water.

Cape Carrisal is low and rocky, nearly 7 miles NW. by N. of cape Choros, and is surmounted by a remarkable round hill. The intermediate coast is bad; there are from 6 to 12 fathoms at a distance of from 200 to 800 yards.

Apolillado cove opens to the southward of cape Carrisal; small vessels can find shelter in it; it is 5 miles to the north-ward of Gaviota, extends into the land about 500 yards, and is 1,200 yards wide N. and S. It is open to the SW. and NW.; landing in it is impossible and the anchorage unten-able as soon as the winds blow from those directions. There are two small islands off the south point of the cove, but they do not stop the wind or sea.

Vessels which come to these islands for guano prefer to anchor off them rather than enter the cove.

San José, the most important place of the neighborhood, is at least 12 miles from the sea; it has 4,000 inhabitants, and is in the center of the copper-mines, whose products are exported from the ports of Chañeral and Totoralillo.

Carrisal bay is NNE. of cape Carrisal, but it cannot be used by vessels, as heavy breakers commence ½ mile from the shore. A rocky point, surrounded by rocks and break-ers, forms the N. coast of the bay. There is a landing-place near the SE. angle, where the rocky coast joins the beach, but during bad weather the sea breaks there also. There is a rock in the center of the bay.

The bay of Gaviota opens to the eastward of the N. point of the bay of Carrisal; vessels can anchor in it, in an emer-gency, near the point, in 11 fathoms, rocky bottom. There is a heavy surf and the shore is inaccessible.

On the other side of the point forming the northern limit of Gaviota bay is the bay of Logag or port Chañeral, which is well sheltered from N. and S. winds; but the SW. swell, which is always heavy, makes landing difficult; the best place to land is in a small creek to the southward, near the shore. There is another landing-place on the N. side of the bay, but it becomes bad with the least swell; it is called

English cove, and vessels are said to have loaded in it. The shore of the port is always beaten by the surf, which prevents landing, excepting during fine weather. The N. point of the bay is bordered by a reef which extends out at least ½ mile.

When forced to anchor here it is best to do so in 12 fathoms ½ mile S. by E. from the small islands at the head of the bay. This position is ½ mile from the land.

The land around Chañeral bay is low; chains of small hills show some elevated points, their summits are rough and broken, and the soil sandy and barren. Several miles in the interior is a chain of high hills, and between them and the coast are several smaller hills rising from the low land. The village of Chañeral is about 3 miles from the port. It is composed of about 20 houses; there are some near the coast; there is no fresh water within a radius of 10 miles.

Chañeral island.   Chañeral island is about 4 miles W. of the bay of Chañeral; it is flat, excepting its southern extremity, near which there is a remarkable mound surmounted by a round hillock. There are some rocks about ½ mile off the south point of the island. The sea breaks on another rock the same distance from the NW. point. On the N. side of the island is a cove in which boats can land with S. winds; off it there is an anchorage, but in very deep water. An American schooner, which was surprised by a northerly gale, was lost at this anchorage.

Cape Leones.   Cape Leones is about 4 miles N. 18° W. from the west point of Chañeral bay. English cove is just to the eastward of this cape; some rocks and reefs make out ½ mile from it.

Cape Vascuñan.   From cape Leones the coast trends N. 4° W. 4½ miles to point Pajaros, and from there N. 14° E. 4 miles to cape Vascuñan. About 400 yards off this cape is a small rocky islet. The ground rises gradually from the coast, and forms a chain of low hills ½ mile and a higher chain 3 miles in the interior.

The Nicaraguan steamer Delfina was said to have been wrecked on a rock about 2 miles W. of cape Vascuñan; an unsuccessful search was made by captain Montt for this danger. He found no danger of any kind for 3 miles off shore between Vascuñan cape and Pajaros point. He says that the most prominent danger of the coast is a rock off cape

Vascuñan which projects about 200 yards into the sea, with a depth of 10 fathoms between it and the coast, with from 7 to 8 fathoms over it. The coast between cape Vascuñan and Pajaros point is sufficiently clear for a ship to approach it within 220 yards.

The Delfina foundered 546 yards south of cape Vascuñan and 55 yards from the shore.

From cape Vascuñan the coast trends N. 60° E. and forms *Sarco bay.* a small bay, called Sarco, open to northward, but well sheltered from S. winds. There is an anchorage in from 7 to 12 fathoms about ⅛ mile from the shore, but landing is difficult.

Deep Gully bay to the NE. of Sarco bay offers some shel- *Deep Gully bay. (Quebrada honda.)* ter against S. winds. A deep gorge extends into the interior from the SE. angle of the bay. There is a sand-beach in this indentation, and ⅛ mile from it is an anchorage in from 7 to 12 fathoms; landing on the beach is difficult; near it are some huts. The high land to the northward of the bay extends to the coast; the slopes of the hills are covered with yellow sand; the summits are rocks and the coast has a miserable and barren appearance.

Peña Blanca cove is about 4 miles from Deep Gully bay; *Peña Blanca cove.* at the foot of a high chain of hills is a rocky point at whose extremity there is a black sharp peak.

About 2 miles NE. of this is the cove of Peña Blanca. Coming from the westward it appears like a small sandy bay; with exception of a few huts there is little to be seen until close in. This bay is easily confounded with that of Sarco, though it cannot readily be mistaken if point Alcade, 8 miles N. 7° E. of Peña Blanca, is made. It is best to moor in 10 fathoms with the port-anchor to the southward, and the starboard to the westward, and a stern anchor to the eastward, as the swell comes mostly from the westward.

There are two small moles at Peña Blanca belonging to some coal-yards. The landing-place is off the western coal-yard in a small inlet. Water can be procured, but it is scarce and bad.

All the SE. coast of the bay is foul and full of sea-weed; there are two or three rocks above water close to the land.

To the north of Peña Blanca the coast runs N. 15° E., *Coast.* rocky for 6½ miles, when it turns N. 75° W. to point Alcade, forming a deep bay, in whose NE. corner is a small beach, called Tontado.

Point Alcade. Point Alcade is a rocky promontory which forms the extremity on the sea-coast of a spur of the coast chain. Some small detached rocks lie off it, but close to the shore; it rises toward the interior and small masses of rock show above the sand. One of them, higher than the other, has a pointed summit, which can be seen distinctly from the southward. A little back of this summit the ground rises suddenly and connects with the high chain.

Point Huasco. Point Huasco, 6½ mile from point Alcade, is low and abrupt. There are several small islets between it and port Huasco, of which it is the SW. extremity. The only one of these, of any size, is so close to the continent that from seaward it looks like part of it; it can be distinctly seen coming from the southward, but from the northward it is confounded with the rocks back of it. To the SW. of this islet are several other small rocky islets.

Port Huasco. To the eastward of point Huasco is the outer roadstead of port Huasco; the anchorage is not good, the water being very deep and the bottom mostly rock.

A little inland of point Huasco is a small low chain of hills, forming four abrupt peaks which can be seen distinctly from the south and west. Back of these hills the ground falls again for a short distance and then rises suddenly to the high chain which runs east and west, directly to the southward of the anchorage. The last point of this chain forms three round summits, the eastern one having an elevation of 1,900 feet. It is a little higher and the middle one a little lower than the western; they form part of the Cerro de Huasco.

The port Huasco is to the eastward of a second interior point 2 miles ENE. of point Huasco; it has two detached rocks to the NNW. The anchorage is in 5½ fathoms, sandy bottom, 600 yards to the westward of the channel separating the rocks. It is a very inconvenient anchorage and difficult to recognize. The steamers of the P. S. N. Company touch here, as it is the port of Ballenar, a considerable town in the interior. It has a custom-house, a few houses, and some smelting-works about 2½ miles distant. The copper comes from the mines of Huasco and Asiento de Santa Rosa.

On the nights when the steamer is expected a light is exhibited from the jetty, which should be brought to bear S.
30° E., and care taken not to confound it with the lights from
the furnaces, which give a more red and uncertain light.
These would lead a vessel too near the shore and on a rocky
shoal off it.

This port seems to be in a fair way to become prosperous.
The three principal founderies, with their three large chimneys, are a good distinguishing mark. Fresh provisions
are generally abundant. It was formerly hard to water
ships, but this difficulty has been removed by an aqueduct,
which brings the water from the mountains.

The anchorage is completely open to the northward ; but
it is said that northers never blow home here, and that they
occur very seldom.

About 3 miles NE. of the outer harbor is a chain of mountains 1,400 feet high, on whose western slope is a slender
peak. It overlooks a valley in which a small water-course,
the Rio Huasco, runs to the sea ; but this is only drinkable
100 yards from its mouth, off which the sea breaks heavily.
In the valley, but nearer to the port, is a narrow lagoon, or
small brook, with brackish water.

The surrounding country has a more barren aspect than
any other part of the coast ; the soil is everywhere covered
with small stones mixed with sand, from which jagged masses
of rocks emerge. At a short distance in the interior the
rocky soil changes into fine yellow sand, which covers the
bases and slopes of all the surrounding hills ; their summits
are stony and without vegetation. In the low land there
are some few shrubs between the stones, and after a rain,
which seldom occurs, they present a fresher aspect than
could be expected. The river valley then becomes green,
and presents a striking contrast to the other country.

It is high water, full and change, at Huasco at 8ʰ 30ᵐ;
springs rise 5 feet, neaps 4 feet.

Point Lobo is about 10 miles north of Huasco ; it is abrupt, with several small hillocks. To the southward of this
point are several sand-beaches separated by points of rocks,
but the sea breaks with such violence that not even boats
can find any shelter. A little inshore of the point are two
low hills, behind which the ground suddenly rises to a chain
1,000 feet high. There are several small rocks in the bay

to the northward of point Lobo, and about 6 miles from it
is a reef, which Fitz-Roy describes as extending $\frac{1}{2}$ mile from
a low, rocky point; the exterior rock of the reef is high and
detached from the others.

Herradura   de
Carrisal bay.
About 11 miles N. of point Lobo is another rugged point,
with several slender peaks, the highest being 3,050 feet. At
$\frac{1}{2}$ mile from it is the small bay Herradura de Carrisal, which
can hardly be seen unless close-to. Between the rugged
point and that of Herradura, the western point of bay, the
breakers extend $\frac{1}{4}$ mile from the shore.

Off point Herradura is a group of low rocks, which appears,
when coming from the southward, to be off the entrance of
the bay; but the entrance faces the NW., and is between
this group and an island to the NE. of it; there are no dangers more than 100 yards from either side. The bay runs
in about $\frac{3}{4}$ mile to the eastward of the islet; it is sheltered from N. and S. winds, but with a strong N. wind the
swell enters around the island. This bay is small for large
vessels, and they cannot ride at single anchor in the interior
part of the cove; but there is plenty of room to moor $\frac{1}{4}$ mile
outside of the island in 4 fathoms, sandy bottom.

The American vessel Nile, of 420 tons, was moored in this
place during a norther, and was well sheltered. The landing is easier than at any other point from Coquimbo, but
the want of water is a serious drawback. There is a small lagoon about 1 mile from this port, in the valley at the head of
the harbor of Carrisal, but its water is more than brackish;
the péons who load the ores use it, however. A deep valley
which starts at the end of the cove and divides the chain of
high mountains is an excellent landmark for this port. The
part of the mountain south of the valley is highest near the
coast. It can be seen distinctly from the north and south;
on it is a remarkable hillock.

Port   Carrisal
Bajo.
About 1 mile NE. of Herradura is the small port of Carrisal Bajo, well sheltered from S. winds. The anchorage is
good, with sandy bottom; a reef of rocks, which makes out
from the north extremity of the point, at the entrance, and
is terminated by an island, has the same effect as a break-
water; the anchorage is in 5 fathoms, $\frac{2}{3}$ mile W. of the N.
point of the island; it has been marked by a buoy.

It is proposed to build a light-house on the south point
of the entrance of the port, with a light of the 4th order

white, fixed, and flashing, and to be visible 16 miles. There are three wharves and one digue; the ballast is thrown in a small cove, on the N. coast, near the W. extremity of the anchorage. This place is destined to become prosperous. In 1863 it had 1,800 inhabitants and three furnaces. The mines, which are 20 miles distant, and connected with the town by a tramway, employ 3,000 persons. As it is thought more convenient to send the ores from the mines to this place instead of to Huasco, it will probably soon be the principal port for their shipment.

The coast to the northward of Carrisal is rugged and steep; rocks make out for 200 yards from nearly all the points. About 7 miles to the northward is an elevated point, surmounted by a hillock, with some ragged hillocks a little in the interior. To the northward of this point is a cove, in which small vessels can find shelter from S. winds; there is another similar 1 mile farther north. Coast.

The elevated part of the coast is terminated by a high, rocky point, a little to the northward of the second cove; to the northward of it is the small port of Matamores, well sheltered from south winds, where loading is easy; a vessel not drawing more than 9 to 12 feet can moor in the inner harbor, and will be well sheltered from N. winds, in from 3 to 5 fathoms, but there is often a heavy swell. There is an anchorage farther out, under the point, in 8 to 10 fathoms, but the anchor must not be dropped in less than 8 fathoms, as the bottom inside that depth is rocky. This is a good anchorage for small vessels during the summer, but there is no fresh water. Four miles back of the low coast, near Matamores, is a chain of mountains 2,440 feet high; a short distance inland are some rocky hills, of moderate height. Matamores

About 2 miles farther to the N. is point Tortoral, and a little to the northward of it is a small, deep bay, situated at the end of the valley of Tortoral Bajo, which distinguishes it from Tortoral cove, in latitude 30° 22′ S. There is, probably, an anchorage, but Fitz-Roy could not examine it, on account of a heavy surf. To the northward of this the low hills are not rocky, but are covered with yellow sand, except near their summits. Tortoral Bajo cove.

About 6 miles N. 14° E. of Tortoral Bajo is a remarkable rocky point, a little to seaward of which is a white detached P. jonal cove.

rock, and a little inside is a small rounded hill. The small cove of Pajonal is 1½ miles NNE. from it. On coming from the southward it is easily recognized by the hill just mentioned and by a small island off the N. point of the cove, which has a hillock with a square summit in its center. A chain of hills, which is higher than any in the vicinity, lies directly on the N. coast of the cove. In the valley, about 1 mile from the cove, is a chain of steep hills rising out of the low lands. This anchorage is better sheltered from S. winds than any other to the southward, excepting Herradura de Carrisal. There should not be much swell, as the N. point, Cachos, and the island with a square summit project well to the westward. The S. swell is only felt in the entrance, and along the south coast the water is smooth and landing easy. There is a dangerous breaker about ¼ mile N. 87° W. from the extreme southern point; it can only be seen with a heavy swell. The best anchorage is nearly in the middle of the cove, near the S. coast, in 5 fathoms, bottom fine sand. The end of the cove is shoal.

Mineral ores are sometimes shipped from here; there is no drinking-water within 2 miles, and the nearest is bad.

Point Cachos.   Point Cachos, which is nearly 4 miles N. of Pajonal, has an island and several rocks off it. Vessels can pass within ½ mile of this island and the one with a square summit, but there is no passage inside of them.

Salado bay.   At point Cachos the coast turns to the eastward, and forms the spacious bay of Salado, which contains several coves; among them is the large cove of Chasco, which is seen immediately after rounding the point. From a distance this cove appears inviting, but 1 mile from the head of the bay there are only 3 fathoms. Its shores are bordered by rocks, some above and some below water; the latter are not marked by breakers, as they are well sheltered. A little NNE. of the point are two shoals, which are always uncovered.

Middle bay.   Middle bay is another indentation about 1 mile ENE. from these shoals. In its southern angle is a cove, which is well sheltered from S. winds, and has a good anchorage in 7 fathoms, but is open to the northward. With S. winds the water is very smooth, and the swell can only enter with N. winds. At ½ mile from this cove is another small bay,

V

Approaches to Caldera Bay, coming from the South.

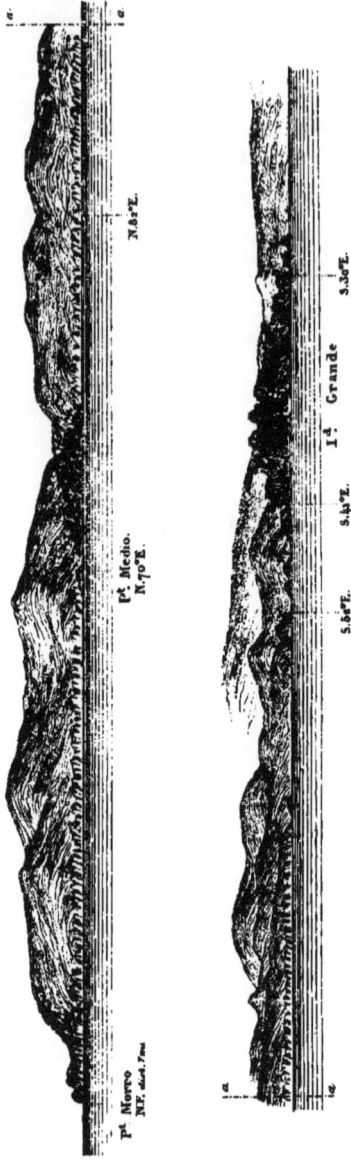

Pt Morro
N.F. dist. 7mi

Pt Medio.
N.70°E.

N.81°E.

a.
a.

Copiapo Bay.

S.84°E.

S.44°E.

Iᵈ Grande
S.30°E.

n which vessels can anchor, but it is not so well sheltered. There are no habitations in the bay of Salado, and no signs of fresh water in the valleys.

The land back of Salado is low, but to the northward of   Point Salado.
the northern bay it rises to a chain of sand-hills running to the eastward and terminating in point Salado, which is rocky and precipitous; off it is a group of islets and rocks, steep-to. The coast to the northward of the point is rocky and indented. Over a space of about 4 miles there are rocks near the shore, then there is an abrupt point, and in the interior a high mountain with a slender summit, showing a double peak to the southward.

Immediately to the northward of this abrupt point is a   Barranquilla de
rocky bay, and alongside of the point a small cove. The Copiapó bay.
Beagle anchored here in 5 fathoms, being only 100 yards from the shore on either side. This bay does not seem of any importance, as it is open to the swell from the northward, though partially protected against the winds from that quarter. Some ores have been shipped from here, but the cove is too small to offer any security, and outside of it the depth increases too rapidly. Boats can land easily in the bay; there is a cove at its head in which no vessel can anchor, but which has a good landing-place; the middle bay of Salado is in every way superior to Barranquilla. The nearest fresh water is in the river Copiapó, 12 miles to the northward.

The coast from Barranquilla to point Dallas, 10 miles   Point Dallas.
NNW., is rocky and irregular; there is no place in which small vessels can find shelter. Point Dallas is of black rock, with a hillock on its extremity; it resembles an island when seen from the southward. The land rises back of it, forming a chain of low sand-hills with summits of rock. One mile west of point Dallas is a shoal awash. The channel between it and the point appears to be large enough for any vessel; a reef runs out far enough from the point to cause, with a heavy sea, breakers $\frac{1}{4}$ mile off shore, but $\frac{1}{4}$ mile farther there are 11 fathoms. When the swell is not very heavy no breakers can be seen off the point.

Nearly 4 miles N. 14° W. from point Dallas are some de-   Port of Copiapó
tached reefs, inside of which, in the bay, is the port of Copiapó; the roadstead is very bad, the swell rolls in heavily,

16 c

and landing is more difficult than in any of the ports to the
southward. The position of the port can be easily recognized
from the Morro, a hill 10 miles to the northward, 850 feet high,
and visible 30 to 35 miles in clear weather. It is remarkable
from being nearly flat at the summit and having two paps
at its extremity; the eastern slope is very rugged. To the
northward is the extremity of another chain of mountains.
To the SW. of the Morro is another hill, whose western side
is steep, and which probably forms a part of the same chain.
In coming from the southward with clear weather, these hills
will be seen before the land in the vicinity of the port. This
port, which was called bad by Fitz Roy, has been abandoned.
The enormous increase of the value of the silver mines of
Copiapó, a city in the interior, and the consequent commerce demanded more facilities, which were happily found
as near the mines as the old port. It is only astonishing that this port was used so long when Caldera and
Ynglés were so near. The principal dangers to be avoided
when entering the port of Copiapó are the shoals of
Caja Grande and Caja Chica. Their vicinity is foul, and
between them and point Dallas are several dangerous rocky
spots.

Caja Grande.   Caja Grande is the northern and outside reef off point
Dallas. It is a submerged bed of rocks extending $\frac{3}{4}$ of a
mile NNE. and SSW., and is $\frac{1}{8}$ mile in breadth. Its position
is apparent, as the sea breaks over it heavily at all times,
whether there is any swell in the bay or not. In 1843,
the captain of la Janequeo reported a rock $\frac{1}{2}$ mile N. 32°
W. of Caja Grande, 11 feet in circumference, with $8\frac{1}{2}$ feet
over it at low water; around it are from 13 to 18 fathoms;
the sea breaks over it at low water, and forms eddies at
high water.

Caja Chica.   Caja Chica is $2\frac{1}{4}$ miles NW. of point Dallas. It is a small
rocky shoal, with a large pointed rock in the center, which
is always uncovered. It is separated from Caja Grande by
a passage 1 mile wide, which seems much less, however, on
account of the breakers of Anacachi and others, which sometimes extend in this channel from the side of Caja Grande.

Anacachi rock.   Anacachi rock, on which a Chilian brig of that name was
lost, is about $\frac{1}{2}$ mile N. 36° W. from Caja Chica and 3 miles

N. 81° W. from the flagstaff of Copiapó. There are only 10 feet of water on it at low tide.

Isla Grande is to seaward of the N. point of Copiapó; it is **Isla Grande.** very easily distinguished, as it has a small hillock on each extremity, the easternmost being the highest. Just to the west of the center of the island is another small round hillock. The middle of the channel between Isla Grande and the continent is clear of danger, but there is such a heavy swell in it that it is seldom taken. A reef runs out 400 yards to the eastward from the N. extremity of the island; 200 yards from the reef there are 8¼ fathoms. There appears to be no danger on the coast of the continent opposite this island. The rocks to the southward of this part of the mainland are inside of the alignment of the points to the north; to the northward of the island are two rocks, one of which is high, but these dangers do not extend out more than ¼ mile.

The nearest anchorage for large vessels is in 5 fathoms, **Directions.** with Caja Chica bearing S. 82° W., the western extremity of Isla Grande N. 23° W., the jetty or landing-place S. 10° W., and the flag-staff, which is 85 feet above the sea, S. 39° W. The Morro of Copiapó kept open of Isla Granda will lead well to the westward of all the dangers off Copiapó.

There are several channels. That between Caja Chica and Caja Grande, though often used, is very dangerous on account of Anacachi rocks. Caja Chica must be kept at a distance of 800 to 1,200 yards, and the flag-staff above Copiapó on the bearing E. 14° S.; but without a steady breeze, one which can be relied on, this passage should not be taken. Neither should a sailing-vessel take the channel between point Dallas and the reefs to the southward; for if the wind should die out, which often happens under the high cliffs, she would be in a dangerous position.

The most natural and best passage is to the northward of Caja Grande. In order to avoid the rocks, when coming from the southward, Isla Grande must be brought to bear N. 59° E., and steer on this bearing until the N. extremity of the sandstone rocks to the northward of the city bears at least S. 64° 30′ E., on which course the harbor can be entered, until the flag-staff above the city bears S. 36° E., when steer for it, and anchor as convenient. In case the flag-staff, which is small, cannot be seen immediately, a house in the city, remarkable for its bright copper roof, can be brought on the same bearing.

In coming from the northward vessels will probably have to beat, when the continent can be approached to within $\frac{1}{2}$ mile, and Isla Grande to within a less distance. When beating to the eastward of Caja Grande, or the other reefs, a line running N. 8° W. and S. 8° E., which joins the W. point of Isla Grande and the mountainous part of point Dallas, should not be passed. With a northerly wind, the anchorage can be approached in 6 to 12 fathoms by keeping the flag-staff S. 25° E.

Vessels must always ride to a long scope in this roadstead, with a second anchor ready, as the sea often gets up without warning. The bottom is hard and holds badly. Soundings are very regular from 12 fathoms to 3 fathoms alongside of the beach; the bottom is principally hard, yellow sand, with some spots of yellow sandstone. Several vessels have been drifted ashore in this bay, having dragged, owing to the heavy sea which sets in suddenly.

Tides.

It is high water, full and change, at Copiapó at 8ʰ 30ᵐ; rise, 5 feet.

Point Medio.

Point Medio, which is very rocky, is on the continent, to the northward of Isla Grande. On its SW. point are two abrupt knobs, and near them several rocks and islets, but there are no dangers to seaward.

From point Medio to the Morro de Copiapó the coast is formed of bluffs, with remarkable patches of white rock on the cliffs to the southward of the point, which is perpendicular and surmounted by abrupt blocks. The Morro rises suddenly a little in the interior.

Ynglós bay.

Passing point Morro, a deep bay, called Ynglés, opens to the SE.; in it are several patches of rock. At the northern limit of a long sand-beach there is a piece of rocky coast, off the extremity of which is a small island.

Port Ynglés.

Port Ynglés is just to the northward of this island, and back of the peninsula of Caldereta, 200 yards off which lies a rock, covered at high water, but it can always be recognized from the breakers. After passing this rock the coast is steep-to, and can be approached to within at least 200 yards. The port embraces several small coves. In the first, which is to starboard of the entrance, small vessels can anchor, but the bottom is of stone; there is a low island to the eastward of this cove. Half way between this island and the east point is the best anchorage during S. winds. Small vessels can go farther into the bight, to the SE. of the island; in it there is a good landing-place. The bay in the NE. angle of the port is well sheltered from N. winds, and the sea never enters it, but landing is difficult. The best place for this is the rocky point at the southern extremity of the NE. beach, or in a small cove among the rocks, where the sea is perfectly calm. This bay, which has from 6 to 9 fathoms, is the best anchorage in the port, but there is no fresh water. The S. coast is too shoal, and a large vessel cannot run in further than off the east point, where there are from 4 to 6 fathoms in midchannel. The bottom in the port is of hard sand; it can easily be seen in 12 fathoms; at the entrance there are 18 fathoms close to either shore.

Port Caldera is 1½ miles from the entrance to port Ynglés, **Port Caldera.** on the other side of an island close to Caldera point, the western extremity of the port.

A light-house stands on a small hillock which overlooks **Light-house:** point Caldera; it is a square, white, wooden tower, 42 feet **Lat. 27° 03′ 00″ N.; long. 70° 53′ 30″ W.** high; opposite to it is a small wooden house, serving as a kitchen. From the lantern is exhibited, 123 feet above the level of the sea, a *white, fixed* light, *flashing every* 90 *seconds*, and visible 15 miles in clear weather. The light cannot be seen until it bears to the eastward of N. 25° E., the Morro intervening. From the light-house, point Cabeza de Vaca bears N. 5° 30′ E., and point Morro S. 54° 20′ E.

The anchorage for large vessels at Caldera is in 9 to 12 **Anchorage.** fathoms, ½ mile from the land and a little to the southward of point Caleta. There are two buoys near the pier, to facilitate hauling alongside. Vessels of 1,000 tons can load and discharge at the pier in 3¼ to 3¾ fathoms.

A sailing-vessel must be careful when approaching with a scant wind, as the current and the swell drift toward the rocks N. of point Francisco, the northern extremity of the bay.

Caldera is a beautiful bay, well sheltered, but more open

* The longitude given of Caldera appears to be 2′ too far to the westward.—Lieut. F. A. Miller, U. S. N.

than port Ynglés. The N. wind, the enemy of all the harbors and roadsteads in Chile, sometimes blows in the bay
and raises considerable sea in its S. corner. But as Caldera is near the northern limit of northers, they are seldom
strong enough to be dangerous to a vessel well anchored.
Point Cabeza de Vaca, 12 miles distant, also protects the
bay a little, and the NE. corner, called Calderillo, is always
smooth.

Description.

The surrounding country, with exception of a few rocks
on the points, is entirely covered with light sand. The land
around the bay is low, but at a short distance in the interior
it rises to a chain of hills, whose summits become higher as
they recede from the coast. To the eastward is a mountain
with a remarkable pointed summit; its slopes are covered
with sand, and near them are two small hummocks. Port
Caldera, the second commercial harbor in Chile, has been
one of the most rapidly-developed places in South America.
A large mole runs out from the rocks in the SE. angle of
the bay, which is used for loading and discharging, and
belongs to the railroad, which is extended to its end, with
every convenience. The railroad runs to the city of Copiapó,
60 miles in the interior, crossing all the mule-paths which
come from the silver-mines. It has branches to Chañarcillo,
San Antonio, and Puquios, in the mining regions. The
road is fitted with excellent American locomotives and cars,
and is under an energetic and capable direction. Trains
run twice a day. There are two other wharves which belong
to the furnaces—Hornos del Norte y Hornos del Sur—in the
eastern part of the port. There is a semaphore station a
little to the southward of point Caleta.

There is a telegraph to Santiago and Valparaiso.

Ballast is thrown to the northward, near the railroad
wharf, on a small chain of rocks.

The town, which is built of wood, rose like magic from
the sand and rocks around the bay. It is laid out in squares,
and has already some fine two-story houses. Fresh provisions and vegetables can be obtained, but no fresh water;
the inhabitants condense it or have it brought by the railroad from Copiapó. During the last few years the railroad
company has started a condenser, which condenses about
7,000 gallons per day. A reservoir containing over 42,000
gallons supplies the city through pipes.

Fish can be caught in the bay, but only with a seine. They are seldom caught along the coast. There are some rock-fish near the outer points of ports Ynglés and Caldera, but there is always a heavy surf.

There are 5 smelting-furnaces at Caldera, the fires of which can be seen at a long distance. The Mexican and South American Company does a large business by importing sil-ver-ores and copper from Chile, Bolivia, and Perú, and ex-porting copper and copper-ores, silver and silver-ores, to the United States, England, and Hamburg.

Beef and mutton can be procured, the animals being brought from the interior by the railroad.

The remarkable point Cabeza de Vaca is about 12 miles to the northward of Caldera. On its extremity are two small hillocks, back of which the land is level for some dis-tance, where there are some low hills which terminate in a long mountain-chain. The coast between this point and Caldera forms several small bays, which are separated by rocky points. There are rocks at short distances to sea-ward of all of them. *Point Cabeza de Vaca.*

There are no dangers more than $\frac{1}{4}$ mile from point Cabeza de Vaca. To the northward of it is a small bay, full of rocks, called Totoralillo, and off the N. point of its entrance is a reef which extends $\frac{1}{4}$ mile from the land and is termi-nated by a high rock. Heavy breakers are observed during rough weather about $\frac{1}{2}$ mile NNW. from this reef.

The coast to the northward of Totoralillo is steep and rocky for three or four miles. A high chain of mountains runs parallel and close to the shore. Obispito cove has a white rock off its south point; further on, the coast is low and rocky, with breakers $\frac{1}{4}$ mile from the shore. About 2 miles N. of Obispito is a point with a small white island off it; to the northward of this the coast runs east, forming the small cove of Obispo, in which will be seen a high sand-hill with a stony summit. The Beagle was anchored in this cove, but it cannot be of much use to vessels, as it is so dif-ficult to land. A little to the northward and inland from Obispo is a higher chain of stony hills, which extends 7 miles. This chain terminates abruptly just inside of a brown point, whose extremity, seen from the southward, seems to have a white patch on it; but the patch is an islet. *Obispito and Obispo coves.*

Port Flamenco·   Port Flamenco is to the northward of this point; it is a good
harbor, well sheltered against S. winds, and better against
those from the N., as the N. point projects enough to keep
the heavy sea from coming in. The land on the N. side of
the bay is very low; the N. point is flat and rocky. A little
in the interior is a detached hill rising out of the low land,
and to the northward is another hill resembling it.

The land at the head of the bay is but slightly elevated ;
back of it is a deep valley which runs between two ranges
of steep hills. All of the hills are covered with yellow sand
from their bases about half-way up; their summits are
stony, and have some stunted bushes on them.

Landing is easy in the SE. corner of the bay, either on
the rocks, or on the beach of a small cove which is in the
middle of a reef of rocks a little farther to the north. In
1835 there were some huts in which two families lived; the
huts were made of the skins of the sea-calf and the guanaco,
and were inferior to the *toldos* of Patagonia. The only
drinking-water these people had was half salt, and found at
some distance from the beach. Their principal occupation
was catching, drying, and salting conger eels, which are
numerous at Flamenco, for the market at Copiapó.

Tides.   It is high water, full and change, at Flamenco at 9ʰ 10ᵐ;
rise 5 feet.

Cove of Las An-   To the northward of Flamenco is a bay without any land-
imas.   ing whatever. It is entirely rocky, with some few scattered
patches of sand, and a heavy surf always rolls on the
beach. The north point of the bay is low, but a short distance in the interior there is another chain of hills, whose
outside slopes are very rugged. To the north of this point
is a small rocky cove called las Animas; it does not appear suitable for vessels, and landing in it is very difficult.
The north point of this bay is formed by a steep rock and a
rounded hill which rises directly out of the water; the sides
of the hills have remarkable black veins running in different directions.

Chañaral de Las   Chañaral bay, much deeper than the cove of Las Animas,
Animas bay.   is to the eastward of its north point. The east and north
shores of this bay are low and sandy, and a heavy surf
breaks constantly on the shore; the southern part is rocky,

off which is the anchorage in 12 fathoms.   There is no sheltered place where vessels can haul in to load.

Mr. James Gales, master of the Florence Nightingale,
says Chañaral is the worst place on the coast, but the following remarks may be of use to strangers: A short distance N. of Las Animas is a clay-colored granite point,
around which there is a deep cove.   About ½ mile NE. of this
point is another point somewhat similar, but whiter toward
the extremity, named Piedra Blanca; 800 yards NE. of
Piedra Blanca is a black, rugged point named Piedra Negra; half-way between these points, and 400 yards from the
shore, is the best anchorage, off the loading-place, in 7 fathoms—black sand.   A dangerous rock, with 8 feet of water
on it at low water, lies 133 yards outside the outermost rock
above water, off Piedra Blanca.   Ships in this bay should
not lie with less than 45 fathoms of cable out, as rollers
frequently set in.

All over Chañaral bay the depth is about 10 fathoms, and
shoal water extends some distance to seaward.   At 5 to 6
miles off the bay, with point Las Animas bearing S. 14° W.,
and Sugar-loaf N. 8° W., there are 38 fathoms.   Vessels
may, therefore, under most circumstances, stand into the
bay with a certainty of finding an anchorage in easy depth,
which would be better than to attempt to tow off if caught
near the shore by the wind dying out.

At the bottom of Chañaral bay is a deep ravine, strongly
resembling the Pisagua in Perú, but without water; this
ravine forms a funnel through which the wind rushes with
great force, from the sea by day and from the land by night,
which does not tend to improve the anchorage.   There is a
mole, but its bad construction and position render it in-
adequate to the wants of commerce.   Drinking-water with
a disagreeable taste can be obtained only after 3 hours' walk
from the village.

The mines are 15 miles from Chañaral.   Ballast is thrown
in a small cove to the westward of the end of the mole.   The
port is regularly visited by English and Chilian steamers.

The north point of the bay is low and rocky; a little in
the interior is a high chain of mountains.   The coast and
hills to the northward of this point are formed of brown and
red rocks; the summits of some of the hills are covered

with bushes. The sandy appearance of the mountains to
the southward ceases here, and the coast seems, if possible,
more barren.

Pan de Azucar cove. The coast between Chañaral and the island Pan de Azucar, 9 miles to the northward, is rocky and offers no shelter.
There is, however, a small bay which affords some protection against N. winds, to the southward of the island; it is
exposed to south winds, and it is difficult to land in it. The
island Pan de Azucar is 600 feet high and about $\frac{1}{2}$ mile
from the land. Coming from the southward, a mountain of
the same form, a little south of the island, may be mistaken
for it, but the latter is lower and its summit more slender.
In the shoalest part of the middle of the channel between
the island and continent there are 5 fathoms. Toward its
northern extremity the sea is calm, and vessels anchor under
its lee, sheltered from south winds, in 6 to 8 fathoms; as
soon as the depth is over $8\frac{1}{4}$ fathoms, the water deepens rapidly to 13 and 20 fathoms, about $\frac{1}{2}$ mile from the island.
To the northward of the channel, on the continent, is a
small bay in which vessels can probably find shelter against
S. winds. Ballast is thrown on the rocky point separating
the two bays.

Fitz Roy observed $\frac{1}{4}$ knot northerly current between this
island and the continent, but it was after a fresh southerly
breeze which had blown for several days. The coasters say
that this current generally attains a velocity of $\frac{1}{2}$ mile.
There are two private moles at the anchorage of Pan de
Azucar.

Point Ballena. Point Ballena is 19 miles N. of the island Pan de Azucar;
it is a salient point, surrounded by several small islets of rock.
The rocky coast between this point and the island Pan de
Azucar is not so high; immediately back of it is a mountain-
chain more than 2,000 feet in height.

To the northward of Point Ballena is a small bay, which
contains a small rocky islet whose white summit is about $\frac{1}{2}$
mile west of the south point; this bay is called Ballenita;
it hardly deserves the name of a port. There are two or
three small sand-beaches among the rocks, on which the
sea breaks heavily, and the hills, which come close to the
sea, have a very rugged appearance.

Lavata bay is about 6 miles to the northward of Ballenita bay; its south point has several spurs of low and abrupt rocks from it to seaward, and inshore the mountains are very precipitous. Directly behind the point is a small cove in which the Beagle anchored, where there is an excellent landing-place. This is the outer cove; there is another inside, which appears to be better, but it is far from the coastline, and there is no information regarding it. The outer harbor appears to offer excellent shelter against south winds, and the water is smooth.

To the northward of Lavata is a point which resembles an island until close aboard; it is connected with the mainland by a low neck of pebbles. The summit of the point is steep, with several rugged peaks. The name of Tortolas has been given to the scattered rocky islets to seaward of this point.

Point San Pedro is $3\frac{1}{2}$ miles from Tortolas islets; it is very abrupt, and a little inshore is a high and rounded hillock. There are several rocks to seaward of this point.

To the eastward of point San Pedro the deep bay of Blanca opens; it is full of rocks, and has no good anchorage. Inside of point San Pedro is a reef which makes out $\frac{1}{2}$ mile. At the head of the bay are several small white islets and two or three sandy coves, but they are too small to offer any shelter to vessels.

About $8\frac{1}{2}$ miles from point San Pedro is point Taltal, which is surrounded for 500 yards from the shore by steep rocky islets, having $8\frac{3}{4}$ fathoms close to them. Behind point Taltal is the fine bay of Taltal, 2 miles wide and 1 deep, with good anchorage in 10 or 11 fathoms 200 yards from the village.

The bay is completely open to the northward, but it has several deep indentations where boats can land easily.

The mountains surrounding the bay are barren, and water has to be brought from a place called los Perales, 6 miles SE. of the anchorage, on the slope of a mountain. The depth increases rapidly from the anchorage; there are 30 fathoms on the line drawn from point Taltal to point Hueso Parado, the northern extremity of the bay. The anchorage is otherwise clear; the sea is smooth, bottom hard fine sand. There are two private wharves off the village. Ballast is thrown on point Hueso Parado. The bay is full of fish and abounds with eatable shell-fish. It is in latitude 25° 25' 30" S.

Bay of Nuestra Señora. The coast between point Taltal and point Grande, 17 miles to the northward, forms a long indentation, which has been called the bay of Nuestra Señora. On the other side of Hueso Parado, 3 or 4 miles from point Taltal, is a white island, on which there are several abrupt hillocks, and a little in the interior is a hill of a lighter color than any in the vicinity.

Point Grande. The coast then continues to the north, projecting to the westward at point Grande, which is 1,570 feet high, and when seen from the southward appears high and rounded; it is terminated by a low and abrupt edge, surmounted by several hummocks; it is surrounded by rocks and breakers for ¼ mile.

Roadstead of Paposo. About 9½ miles from point Grande is point Rincon, near which there is a large white rock. The village of Paposo is situated between these points. It is a poor settlement, of about 200 inhabitants. Near the beach is a chapel. The huts are scattered and hard to distinguish, as they are of the same color as the mountains back of them.

Paposo is an open roadstead, with a very steep bottom and but one landing-place, under point Huanillo, about 2 miles south of the village; it is in latitude 25° 02' 25" S., and longitude 70° 32' 10" W. To the eastward of this small point is an interior cove with 2 fathoms of water. There are a few houses on its banks. The anchorage is in 16 to 20 fathoms, bottom fine sand, about 400 yards WNW. of point Huanillo. The coast between it and the village is full of rocks; the mountains, which are 776 yards from the beach, show some vegetation.

The watering-place of Unquillar, betweeen the chapel and the beach, is about 2 miles from the landing-place. It does not afford very good water, but it is the most abundant and the best between Mexillones de Bolivia and Caldera. Fitz Roy speaks of some wells 2 miles distant. Water is taken in with difficulty except at point Huanillo. The inhabitants fish for conger eels, which they sell to the Indians of the interior at San Pedro de Atacama. Wood can be procured.

A few vessels touch at Paposo to take in dried fish or

VI

Coast from Plata Pt to Grande Pt including Paposa Bay.

Anchorage off Antofagasta.

(Moumt Moreno bearing N.31°W. and Point Moreno N.64°W.)

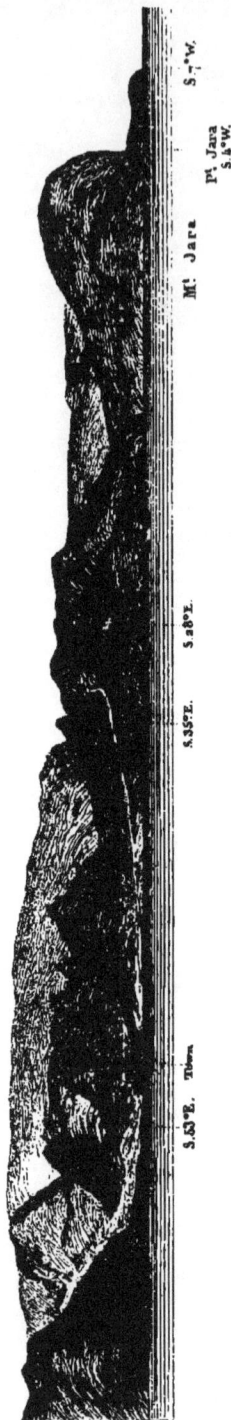

S.53°E. Thm.   S.35°E.   S.28°E.   Mt Jara   Pt Jara   S.⅞°W.
                                                          S.4°W.

copper-ore. The former is abundant; there is but little, however, of the latter, as the mines are from 21 to 24 miles to the SE., and are not much worked.

Vessels bound for Paposo should run on the 25th parallel; **Directions** when within from 6 to 9 miles the white rock off point Rincon will be seen, and soon afterwards the low white point of Huanillo, which should be steered for and the anchor dropped as before indicated.

During clear weather, which is rare, a round hill, higher than the others and directly back of the village, is a good landmark.

Point Plata is 23 miles N. 9° W. from point Grande, and **Point Plata.** resembles it in all particulars. It is 1,670 feet high, and is terminated by a low tongue of land with some rocks at its extremity. These rocks form a small bay to the northward, with irregular bottom covered by 7 to 16¾ fathoms of water.

The coast between points Plata and El Cobre follows in almost a straight line, rocky and steep. It is surmounted by hills from 2,000 to 2,500 feet elevation. There is no shelter whatever.

El Cobre cove, in latitude 24° 14' 50" S. and longitude **El Cobre cove.** 70° 35' 30" W., is about 30 miles N. of point Plata; it is sheltered from the prevailing winds and the bottom is good. Large vessels moor 500 yards N. of some houses on a sand-beach in 16 fathoms of water; fine sand. Small vessels anchor at half that distance in 7 fathoms. Landing is very easy on the sand-point under the shelter of some projecting rocks. The interior point of the cove, off which the anchorage is, is foul. Basse Moreno says that the sea breaks on a shoal 200 yards from the point; it is nearly at the extremity of a bank of rock covered by 2 fathoms.

All the mountains which surround this cove are barren and full of ores. There is no fresh water; it has to be brought from a gorge called Botija, which is to the southward.

This cove can easily be recognized, from the green and crimson color of the mountains and by a small white island 2 miles further to the northward.

Seven miles before arriving at cape Jara, in 24° south latitude, is the coast-boundary between Chile and Bolivia, the neutral territory extending to the 23d degree.

CHAPTER IX.

ISLANDS OFF THE COAST OF CHILE.

Variation easterly, Juan Fernandez 17° 38′ E., San Felix 13° 40′ E.   Annual increase about 1′ 30″.

Juan Fernandez and Mas a Fuera.

The Juan Fernandez group, consisting of two principal islands, belonging to Chile, were discovered by the Spanish pilot Juan Fernandez in 1574. The easternmost of these islands is generally known as Juan Fernandez, but it is also called by the Spaniards Mas a tierra, to distinguish it from the westernmost island, named, from its greater distance from the continent, Mas a fuerra.

Juan Fernandez or Mas a tierra. With the intention of establishing himself on the island Mas a tierra, Juan Fernandez left there a number of goats, which multiplied rapidly and became one of the principal resources of the buccaneers who, from there, plundered the Spanish vessels and settlements. In 1704 the commander of the Cinque ports left on the island the master of the vessel, Alexander Selkirk, famous through Daniel de Foe's romance of Robinson Crusoe.

In 1751 the Chilian government founded a colony upon the island, which was destroyed in the same year by an earthquake. The island then appears to have been deserted until 1819, when the government established there a penal colony. Following a revolt of the criminals this project was abandoned, and the island was farmed out by the Chilian government. Near the landing in San Juan Bautista or Cumberland bay there are at present a few houses with 30 to 40 Chilian settlers, who supply cattle and other requisites to vessels arriving.

The island of Juan Fernandez, about 360 miles west of Valparaiso, extends nearly 10½ miles E. and W., and 5 miles N. and S. It is mountainous and wooded; the NE. part especially is composed of rocky mountains and valleys almost entirely covered with wood; the southern part is comparatively low and flat, and almost barren.

The greatest elevation is the mountain called el Yunque, 3,042 feet, wooded to its summit. From seaward it shows above the other volcanic chains, and from the NE. is a good representation of a blacksmith's anvil. The vegetation is quite rich, though stunted on the flanks of the mountains. The valleys, which are not extensive on the north coast, the only one generally visited by vessels, are watered by numerous torrents.

The wild peach and fig are abundant; the myrtle tree is common; la chonta, a species of palm, as hard but more flexible than ebony, is much esteemed. The principal, if not the only, game is the wild goat. Water is abundant, and easily obtained at San Juan Bautista or Cumberland bay. Fine cattle and goats are raised on the island, with some hogs and poultry, and fruit of indifferent quality. The bay swarms with fish; the cod is taken at Bacalao point, and also at point Loveria; the lobster abounds, and is of excellent quality. At present, this island, as also Mas a fuera, is leased to a Chilian merchant, who keeps up the settlement for supplying vessels and for hunting the seal during the season, the average taken being about one thousand skins in the three months, valued at from $15 to $20 the skin before they are cured. *Productions.*

The climate of Juan Fernandez is much the same as that of Valparaiso. It is more rainy, and though damp from the frequent showers from the clouds, which during the night are arrested by the high mountains, it is considered very healthy. *Climate.*

The fine season is from October to April, often from September to May. During this season the mornings are generally overcast, with occasional rain-squalls. As the breeze springs up the clouds disperse, and the day is clear. Toward evening, and especially during the night, the clouds collect again on the summits, and squalls of wind and rain sweep down to the bay, which are straining to the cables and at times cause vessels to drag.

The bad or rainy season is from May to September, or from April to October. Then it rains usually on alternate days, with calms or strong breezes from the northward.

From October to April the winds prevail from SSW. to SE., their direction being affected by the contour of the valleys. During this season vessels from Valparaiso for Juan Fernandez should hold their wind from the start, to make the island without tacking.

Between the continent and this group the winds blow fresh from S. to SSW., with at times a light breeze from SW. On approaching the islands the prevailing wind is from S. to SE. During the bad season, from April to October, the winds are light and irregular, with calms; winds from the WSW., and at times gales from the NW., which are indicated by the barometer.

The island of Juan Fernandez is easily recognized from whichever side it is made. As there is no shelter to the south, it is usually approached from the northward. El Yunque is a good point of reconnoissance when clear of the morning mists.

From the east point, around the north shore of the island, to the island Santa Clara, the coast is perfectly clear and can be approached within a short distance. It is best, however, not to keep too close to it before approaching the anchorage, on account of the violent gusts from the hills, the variableness of the wind, and the intervals of calm.

Coming from the southward either extremity of the island can be doubled; but for a sailing-vessel it is best to double the east point, to reach more readily the anchorage of San Juan Bautista.

The easternmost point of the island, Guasabullena point, is a high steep cliff; about ⅝ mile S. 74° W. from it is a remarkable islet of a conical form, called Morro Caletas. From point Guasabullena the coast of the island of Juan Fernandez trends about N. 29° W. to point Pescadores, the NE. point of the island, when it inclines more to the westward to point Bacalao, and thence S. 75° W. to point Loveria, the extreme eastern limit of San Juan Bautista or Cumberland bay. This coast is an uninterrupted steep cliff, about 984 feet in height and of a reddish brown color; the points and the coast are free from danger. A small islet lies ½ mile to the S., eastward of Pescadores point but close to the shore. There is a slight indentation in this coast about 1 mile to the northwestward of Guasabullena point, called Puerto Frances; in it there is neither anchorage nor shelter.

VIII

San Felix and San Ambrosio Group.

Cathedral

I. Sⁿ Félix

I. Gonzales

I. Sⁿ Ambrosio.

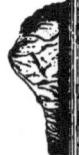

N. 70° E. dist 7 m.

Juan Fernandez Iᵈ (Mas-a-Tierra) S.S.E. dist 18m.

S⁰ Juan Bautista Bay.

St Clara H⁰

Mas-a-Fuera Iᵈ.

S.88°E. dist 42 m.

Cumberland bay is the only anchorage of the island; the depths are considerable, the best anchorage being in about 25 fathoms, bottom fine white sand, about $\frac{1}{4}$ mile from the beach, with the flag-staff on the fort N. 76° W.; the small rock off point San Carlos N. 19° W., and the southernmost of the grottoes S. 83° W., or more to the northward and to seaward, with the southernmost of the grottoes S. 69° W. and the rock off point San Carlos N. 31° W. In the SE. part of the bay the bottom is rock.

Following the coast from the eastward, the bay is easily *Directions.* recognized from the houses which are on its SW. shore, and from seven grottoes situated back of the houses and to the southward of the remains of the fort. As before stated, the coast should not be approached nearer than 1 mile, on account of the gusts which drive down the ravines.

Coming from the N. or NW., West bay may be taken for the anchorage of San Juan Bautista, as from seaward this bay has a better appearance, the land is more level, and the slope toward the mountains more gradual, but by referring to the bearing of El Yunque the mistake cannot be made; besides San Juan Bautista being the only bay from which El Yunque presents its remarkable appearance.

The anchorage should be approached under short sail. On closing the west part of West bay by point San Carlos the depth will be found considerable, but the next sounding will give probably 28 fathoms.

The bay being entirely open to the northward, vessels should not anchor there during the bad season. Although the natives state that the high mountains prevent these winds from blowing home, the same may be said of northerly gales at this anchorage as at that of Valparaiso. Frequently they do no harm, but at times they come suddenly from the N. or NW. and raise a tremendous sea, which of itself is sufficient to force a vessel to leave the anchorage or to drive her on shore.

It is high water, full and change, at Cumberland bay at *Tides.* 9ʰ 55ᵐ; rise 6 feet.

Point San Carlos, which is the western extremity of the *Point San Carlos.* bay of San Juan Bautista, is not so clean as that of Loveria. About 230 feet from it is a rock, the head of which only can be seen at high water. As distances are very deceptive.

17 c

tive under high land, it is best to give this point a good
berth until the rock is made out.

From point San Carlos to West bay, the distance of a
short mile, the coast is formed of abrupt, perpendicular
cliffs, which attain at some points an elevation of 1,250
feet. The bay offers no shelter. It appears to be more fertile than that of San Juan, but the shore cannot be approached, owing to the surf. The only point where landing
can be effected is in a cavity which is seen to the SE. of a
perforated rock, where the beach is of large stones. In this
bay is the grotto where Selkirk lived. A torrent of good
water runs into the bay.

Salinas point forms the NW. entrance to West bay; back
of it is a remarkable mountain, 2,382 feet high, called the
El Pilon de Azúcar. At the foot of the cliff are two land-
slides and a remarkable pointed rock. One-half mile to
the westward of Salinas point is an indentation deeper than
West bay, called Vacaria bay; even landing in it is hardly
practicable, owing to the surf. Off the eastern extremity of
this bay is a rock, under water, lying close to the shore.

Viudo point.   A short ½ mile to the westward of Vacaria bay is Viudo
point, the north point of the island. It is remarkable, the
cliff being cut as the teeth of a saw. About 110 yards off
this point are three dark rocks, steep-to.

From point Viudo the coast runs to the SW. to Los Negros
or Rinion point, which is the NW. point of the island;
about 270 yards to the southward of this point is a high
islet, having the form of a sugar-loaf. The shore here
makes in to the eastward, forming a deep indentation called
Faith bay, which is almost always rough and impracticable.
From the SW. extremity of this bay, Lemos point, the coast
trends uniformly to the SW.

Point Padre   Before arriving at Island point, the SW. point of the
bay.   island, there is a bay, open to the west, called Point Padre
bay; within it are several islands and rocks.

Island point.   Island point is formed of two conical passes; there is a
grotto on it, and off it, close to shore, an isolated rock.

South coast.   From Island point the coast trends to the southeastward
to Higgins point, the south point of the island; close to this
point there are two rocks. From Higgins point the shore,
with several impracticable bays, trends to the northeast-

ward, to Chamelo point; about ¼ mile off shore, toward the middle of this stretch of coast, is an islet called Morro Viñillo, and off Chamelo point, extending to about the same distance off shore, are two rocks, with a rock under water between them. From Chamelo point the coast, called Los Corrales de Molina, trends about E. by S. to the east point of the island.

Santa Clara island, lying south of the western point of Juan Fernandez, and separated from it by a channel over ¾ mile wide, is about 4 miles in circumference and 1,140 feet high. Coming from the northward it has the appearance of a steep cliff; from the southward it presents an irregular aspect of paps and conical rocks; its eastern extremity descends gradually to the sea. It appears to be barren, with a few stunted shrubs on its eastern slope. *Santa Clara island.*

On its NW. point a stream runs along the cliff to the sea. Landing is dangerous, the sea breaking on nearly the whole contour of the island. Off its southwest extremity is a rock, lying close to the island, and there are two others, similarly situated, south of the island. In the channel between this island and Juan Fernandez the sea is usually rough.

Mas a Fuera is about 92 miles to the west of Juan Fernandez; it is about 9 miles long and 2½ miles wide; its summit is 6,036 feet above the sea-level, its northern peak being 4,398 feet high. It is covered with trees, and on its sides are several waterfalls running to the sea, but there is no anchorage. There is a bank on its north coast, but it is steep-to, and the water is so deep that were a vessel to anchor on it she would necessarily be almost touching the land, and exposed to all winds except those from the southward. Cattle have been placed on this island by the farmers of Juan Fernandez. *Mas a Fuera.*

San Félix and San Ambrosio.

The islands of San Félix and San Ambrosio were discovered in 1574 by the Spanish pilot who gave his name to the island of Juan Fernandez. They lie about 500 miles from Copiapó, and are visited solely for the purpose of fishing. They are of volcanic formation and almost entirely barren. Formerly these islands were the resort of multitudes of sea birds and seal, but of late years the seal have almost entirely disappeared. There are no land birds, animals, or insects of any description on these islands, excepting a species of fly. There is no drinkable water on any part of the islands. The mean of many observers places the summit of the island of San Ambrosio in latitude 26° 19′ 52″ S., longitude 79° 57′ 26″ W.

The observations of captain Gormáz places the landing on the NW. side of San Félix, in latitude 26° 16′ 46″ S., longitude 80° 00′ 15″ W.

This group has been visited frequently by vessels of different nationalities from the date of their discovery to the present day. The description here given is taken from the remarks of captain Ramon Vidal Gormáz, who in command of the Chilian war-vessel Cavadonga surveyed these islands in 1874.

San Félix, description.

The island of San Félix extends E. and W. 1¾ miles with an average breadth of about ½ mile, which is reduced to 440 yards to the southward of the cove; the western part, called El Morro Amarillo, is an abrupt projection, almost circular, 600 feet high; from the morro to the southward the island is high and sloping. About 1,000 yards to the eastward of the morro it rises and forms a conical hillock when seen from the north; from this point the island descends gradually to the NE. and terminates in a rocky point, having at low-tide a beach of black sand at its foot.

Detached from the SE. point of San Félix is Gonzalez islet, 549 yards from the island, but united to it by a ridge of rocks awash, over which the sea always breaks heavily; this islet is elliptical, measuring 930 yards NE. and SW., with an average breadth of 438 yards; its height is 432 feet, steep on all sides and inaccessible.

The southern part of San Félix is steep. There are two anchorages off the island, difficult to distinguish, separated by a salient point, opposite which and about 275 yards distant is a line of flat reefs always above water, which extend 328 yards E. and W., leaving a channel with a depth of 2¾ fathoms between them and the island. The western anchorage, or that which is immediately to the SE. of El Morro Amarillo, forms the most of a bight, with a more considerable bank of soundings than the easternmost one. What

may be called the port of the island of San Félix is the triangular space between the cathedral of Peterborough, the
morro Amarillo and the NE. point of the island, but the
anchorage is only the half of this space nearest to San
Félix, within which the soundings vary from 18 fathoms in
the outer part to 4 and 5 fathoms near the land. From the
island to the cathedral the depth varies between 16 and 33
fathoms, bottom black sand with intervals of rock, though
the rock appears to occur less frequently at the anchorage,
where there is a covering of about 1 foot of sand over hard
bottom. The small cove which affords the landing-place of
the island, seen from the north side when near San Félix,
is very remarkable and readily distinguished; it is directly
E. of the morro Amarillo, where the dark lava of the north
steppes of the island joins with the morro. At the bottom of
the bight is an arched grotto formed by various currents of
lava piled above it. The landing is at the entrance of this
grotto; there is no beach, and the only landing is under the
vault of the grotto, where there is from 6 to 12 feet of water
and an excellent place to take the lobster, which abounds and
is equal to those found at Juan Fernandez; the bacalao, a
species of cod-fish, a fine species of eel, called the murena,
and the spotted dog-fish are also abundant.

The rock called the cathedral of Peterborough lies $1\frac{1}{10}$
miles N. 20° W. of the morro Amarillo of San Félix; it is
154 feet in height, having the appearance of columns;
around the cathedral are some high and remarkable rocks.
The depth in its vicinity varies from 16 to 20 fathoms a short
distance from it; the passage between it and San Félix is
deep and clean. There are no hidden dangers in the vicinity of San Félix; the island can be approached from 400 to
600 yards without risk; its shores are all steep, rough, and
inaccessible, except in the cove mentioned and in the vicinity of the NE. point. It must be remembered, however, that
this point is only accessible in a calm and under the most
favorable circumstances.

It is high water, full and change, at San Félix at 9ʰ 40ᵐ; Tides.
rise 6½ feet.

The northern shore of the island of San Ambrosio, seen San Ambrosio.
from the anchorage of San Félix, is in line with the NE.
point of that island, bearing S. 78° E.; the distance between

the islands is 9.6 miles, from the eastern point of San Félix to the western point of San Ambrosio. There is no anchorage around the island. At the distance of from 200 to 400 yards the depth is from 55 to 60 fathoms, the least depths found being 29 and 46 fathoms 74 and 92 yards from the island. The island of San Ambrosio extends 1¾ miles E. and W., with a mean breadth of nearly ½ mile; its form approaches an ellipse; viewed from the sea its summit has the appearance of an incomplete table, without any characteristic peak; it has more verdure than San Félix. There is no indentation of consequence on the shore of the island, though in the middle of its northern part there is a nook which would shelter a boat.

On the western side of San Ambrosio is a small elevated rock separated but a short distance from the island, but sufficiently to see through when looking from the cathedral of Peterborough. To the eastward there are three small, pointed islets, about 402 feet high, pierced in the side of their base toward the island. The most distant of these islets, detached from San Ambrosio toward the E., is about 875 yards; outside the islets there is no hidden danger, and the soundings are deep to their vicinity, but the depths are not so great to the southward as to the northward of the island.

It has been stated that the magnetic influence of these islands affects the compasses of vessels passing. Captain Gormáz states that the volcanic mass of these islands is by no means uniform; the rocks ordinarily magnetic are in some locations positive, in others negative, causing a deviation of the needle from 1° to 3°, according to the prevalence of the masses. At anchor off San Félix, simultaneous observations from ship and shore, employing different stations, show the greatest variation to be 3°, with at times no difference in the angles. In vessels which pass a prudent distance from the islands their magnetic influence could not be sensible.

THE COAST OF BOLIVIA.

CHAPTER X.

FROM CAPE JARA TO THE RIVER LOA.

Variation from 12° 30' to 11° 03' easterly, in 1876, increasing annually about 1' 30".

Bolivia borders the Pacific Ocean for about 250 miles, from cape Jara in latitude 24° S. to the river Loa, in latitude 21° 28' S.; it is bounded to the NW. and N. by Perú, on the E. by Brazil and Paraguay, and to the SE. and S. by the republic of Chile; its extreme length being about 1,100 miles, with a breadth of 800 miles.

The coast of Bolivia is part of the desert of Atacama, which constitutes the province of that name. This province is sparsely populated, containing hardly from 6,000 to 8,000 inhabitants. The coast is generally formed of high sand dunes, leaving but a few feet of beach, which is clean and can be followed at a short distance from the land. Like the coasts of Chile and Perú, it is constantly washed by the southerly swell, and has but few coves or inlets, and those generally bad.

The desert of Atacama extends from Copiapó to the river Loa, and as far as Iquique and 120 miles into the interior, to the spurs of the Cordillera of the Andes. It is barren and uninhabitable; fresh water is unknown, and it never rains; the easterly and southeasterly winds being arrested by the Andes, their snowy peaks condense all the humidity which comes across the plains of the Argentine Republic and Paraguay; and the waters which come down from the Andes are lost in the sands of the desert, which renders them salt. *Desert of Atacama.*

The desert of Atacama is covered by black shifting sands, or dark brown sand varied at times by gravel and stones which are so sharp that the hunters of the guanaco have to shoe their dogs. It is probable that Bolivia was inundated to the foot of the Andes during prehistoric time, and that this part was raised with all this coast of the continent; this is proved by the numerous fossil marine shells, *caracoles*, and large blocks of pure sea-salt, which are found in the ravines of the mountains; the soil itself is full of salts of lime and soda.

The only products which were for a long time taken from these western slopes of the Andes were nitrate of soda and guano; a few years ago, however, some very rich silver-ores were found in the midst of the desert at Caracoles, which is 951 feet above the sea. This place was connected with Mexillones by railroad. The port of Antofagasta was created, and the mines in this desolate part of the earth immediately became a source of strife between Perú, Chile, and Bolivia.

Climate. The climate of the one hundred and fifty miles of coast of Bolivia does not need any special description after the remarks made on that of the northern provinces of Chile. Calms are very frequent, and the trade-winds take the place of local breezes.

The prevailing winds are from SSE. to SSW., modified by the land and sea breeze; the wind is generally light; during the night it is usually calm. There is a thick damp fog from about 9 p. m. to 10 a. m., especially during the months of March, April, and May. There is little movement of the barometer, and no gales of wind; the swell is often heavy and without warning.

Description. The population of Bolivia is stated at about 1,742,352. The capital is Sucre, the ancient Chuquisaca, also called Charcas and la Plata; it has about 24,000 inhabitants.

The republic of Bolivia has for a long time existed on its mines of precious metals; those of Potosi are still celebrated, but they yield less and less. The exportation is much inferior to the importation, and the country becomes poorer and poorer. The civil discords are not calculated to improve this state of affairs. The exportation, which is in the hands of foreigners, consists of guano, niter, copper, and mineral ores. The United States and Europe supply all the necessaries of life.

Formerly Peruvian bark was a source of wealth to Bolivia, but it has been gathered so recklessly and the forests are so impoverished that it has almost ceased to be remunerative.

Cape Jara is a steep rock, rounded on its northern side, 22 miles from el Cobre. The coast between them has the same aspect and direction as that to the southward of the latter. In the northern part of the cape is a small but safe cove, which can be used by small vessels. Vessels which are hunting the seal frequent it, and they leave their boats here to hunt in the vicinity. They are provided with fresh water, and for fuel they use the wrack which grows in abundance on the coast. No provisions are to be found for a considerable distance on either side of the point. Mount Jara, 4 miles to the eastward of the cape, is 3,986 feet high.

Moreno bay commences 4 miles N. 19° E. of cape Jara, and extends to point Las Tetas, the SW. extremity of mount Moreno. The coast between point Jara and the commencement of Moreno bay is high and rocky, and has no remarkable feature other than the Black rock, which lies a little to seaward.

Moreno bay has several anchorages; it commences with Playa Brava, an inhospitable sand-coast bordered by high hills, in the center of which is the port of Antofagasta, about 13 miles from Black rock.

Port Antofagasta is the harbor of export for the nitrate of soda, which is found nine miles from the town, back of the first line of hills, as also of the silver-ores coming from the mines of Caracoles, 114 miles distant, in the desert of Atacama.

The anchorage is to the northward of a bank of stones which extends to seaward for $\frac{1}{4}$ mile. The depth is from 14 to $17\frac{1}{2}$ fathoms; bottom stiff blue mud, covered by sand, coral, and shell. It is $\frac{2}{3}$ of a mile from the moles of the port.

It is necessary to anchor far enough off the bank to have sufficient room to swing with from 45 to 60 fathoms of chain when the fresh wind from SSE. changes to N. Two anchors are always necessary, and in ordinary weather an anchor astern, with the head SSW., to stem the constant swell.

The bank of stones is always washed by the swell. Off its NW. extremity, and a short distance from it, is a rock which is only visible with a moderate sea. It is thought that the Steamer Paita touched on it, and it was necessary to beach her near the mole in the inner harbor. The bank is steep-to, and has about $3\frac{3}{4}$ fathoms alongside.

To the eastward of this bank is a cove with 3 fathoms of

water, with a channel into it 240 yards wide, full of rocks, and from 3¾ to 5 fathoms deep. Two rocks, with 9 to 11 feet of water over them, situated in the line of the northern mole, render it dangerous; the channel must be used only during a calm, and never by large vessels; vessels not drawing more than 13 feet are perfectly secure, and sheltered from the sea by the bank, and can load very easily. At the bottom of the port are two moles, parallel to each other and projecting to the NW. When the weather is at all boisterous the sea not only breaks heavily on the bank, but the entrance of the inner harbor is closed by a bar. This takes place at the change of the moon, and especially during the season of north winds, from April to July; then not even lighters can lie alongside the mole. With the exception of these four months, the wind blows from the south during the day, with fresh land-breezes at night.

Between two hills to the southward of the town is a watering-place, with brackish water, which is used for the animals; the people drink condensed sea-water.

Vessels bound to Antofagasta from the southward should make Jara head, then steer for the head of the bay of Moreno, keeping 4 or 5 miles from the land. When Antofagasta bears NE. by E., a large white anchor will be seen painted on a ridge back of the town. Continue along the land until the anchor bears E. 13° S., when steer for the anchor, and keep the lead going until in 16 fathoms, which is the anchorage. It is best to select a berth outside of other vessels.

When coming from the northward the anchor will be seen as soon as point Las Tetas is doubled; lower, and nearer the beach, the powder-house also will be seen, which is a hut covered with zinc and located above the principal street.

It is advisable not to attempt to enter the anchorage at night, as the lights of the houses mislead, and there is great danger of grounding on the bank of stones.

Antofagasta is an uncomfortable anchorage, owing to the constant swell. It is not sheltered, but neither wind nor swell has been experienced of sufficient strength to cause anxiety for the safety of vessels at anchor. No supplies are to be obtained, owing to the aridity of the country.

A line of railroad from Antofagasta to Salar del Carmen was completed in 1874.

Chimba * bay, 3½ miles N. of Antofagasta, is to the east- ward of an islet 436 yards long, running to the NW., called Oesté, Huanosa, or Dolfin; a short distance to the northward of it are some rocks. The cove is very small, but it is protected from the prevailing winds. The anchorage is 200 yards east of the N. point of the island in 7 or 8 fathoms, bottom fine sand; there are hardly more than 200 yards to swing on every side, except to the NW. The island is tolerably high, and has on it some guano, which was at one time exported, but it has been abandoned; there are also several depots of salt. The channel between the island and the mainland is only accessible to boats. There is no fresh water.

From Chimba bay the coast trends NW., then south, forming the large bay of Jorge, bounded to the westward by mount Moreno.

Mount Moreno was formerly called Jorge. It is the most prominent point on this part of the coast. Its summit is 4,160 feet above the level of the sea, and slopes to the south; but to the northward it ends abruptly in barren plains. It is of a brown color, without the least sign of vegetation; its western slope is cut by a deep ravine. The SW. point of the peninsula of Moreno descends gradually from the summit of the mountain and ends in two hillocks, to which the Spaniards gave the name of Las Tetas. This point is 22 miles N. 15° W. from cape Jara.

Constitucion is a small but convenient ancho:age formed by the continent on the one side and by the island Forsyth on the other. It is situated immediately under mount Moreno, about 5 miles N. of point Las Tetas. A vessel can be hauled up and careened here without being exposed to the heavy swell which is felt in most of the ports on this coast; landing is easy. The best anchorage is to seaward of a sand-spit at the NE. point of the island, in 6 fathoms, muddy bottom. It is best to moor securely, as the sea-breeze sometimes sets in very fresh. Farther out the holding-ground is bad, and on entering, the island or weather side should not be hugged too closely, as there are numerous sunken rocks, but few of which are indicated by sea-weed. It would be

---

* Chimba is the name given by the natives.

best to keep in midchannel, provided the wind admits of reaching the anchorage. There is neither wood nor water in the vicinity.

Esmeralda rock. About 4 miles N. of Forsyth island, opposite a mountain 1,630 feet high, is a bank of islets and rocks, called Lagartos. It is 1½ miles long, trends NNW. and SSE., and is close to the shore. About 1¾ miles from these islets is Esmeralda rock, covered by 9 feet water. On it there is always a heavy swell, if not breakers, at the syzygies. It was discovered by a Chilian corvette of that name. It lies under the following bearings:

Lagartos bank........................ .... N. 83° E.
Mount Mexillones .... N. 25° E.
Mount Moreno .... S. 38° E.

The depth 30 yards from Esmeralda rock is 7½ fathoms, and in the channel between Esmeralda rock and Lagartos islets there are from 8 to 12 fathoms of water.

Point Angamos, or Leading bluff. The chain of plateau which commences at mount Moreno is terminated 12 miles N. 2° E. of Constitucion harbor by a rugged hill called mount Jorgino. On the N. coast of this promontory is the bay of Herradura de Mexillones, a narrow branch of the sea, which makes in to the eastward but affords no shelter.

About 9 miles north of mount Jorgino is point Low, surrounded by sunken rocks, and 5 miles to the NE. is point Angamos, or Leading bluff, a very remarkable promontory, which, with mount Mexillones a few miles to the southward, is the best landmark for all the ports of the vicinity, but especially for Cobija. The bluff of point Angamos is about 1,000 feet high; it faces to the N., and as it is entirely covered with guano it resembles a cliff of chalk.

Abtao rock. About ½ mile NW. of this bluff lies an island which is connected with the point by a reef. No dangers were known outside of it until the steamer Abtao reported several rocks in the vicinity of the large island. The position of the one farthest to seaward, called Abtao, given by Captain Moutt, of the Chilian navy, is 1,620 yards from the nearest land, and on the following bearings:

Point Angamos........... .... S. 55° 51' E.
Mount Mexillones .... S. 06° 51' E.
Point Baja, (low)............ .... S. 34° 09' W.
Little White island, (distant 700 yards)...... S. 10° 21' E.

There are 2½ fathoms of water over it at low water. To clear it give it a berth of 1 mile, and do not change the course until point Angamos bears a little to the southward of SE. by E.

There are two other sunken rocks, one 350 yards to the southward of Abtao rock, covered by 5 fathoms of water, and the other SSE. ½ E., distant 415 yards, with 6 fathoms of water over it.

Abtao rock is on the line of Little White island and a large white patch which will be seen on the slope west of the Morro, bearing S. by E., and also on the line of the first elevation to the northward of mount Mexillones. The rock is not indicated by sea-weed or eddies.

Mount Mexillones is 2,630 feet high, resembles the frustum of a cone, and can be distinguished above the surrounding heights. In clear weather it is a better landmark than point Angamos, but the summits of the mountains on the coast of Bolivia are very frequently covered by thick mist, which renders Leading bluff a better mark, as it cannot be mistaken owing to its chalky appearance, it being the end of the peninsula, and the land suddenly receding to the eastward. The small chain which runs from mount Moreno to mount Mexillones is not connected with the principal chain, and has a form resembling that of a wedge.

In 1862 some important discoveries of guano were made around the base of mount Mexillones, and since 1863 two vessels have carried from it 2,100 tons of guano to Europe. The quality is similar to that of Paquica, the best guano of Bolivia, but it is inferior to the guano of the Chinchas.

The spacious bay of Mexillones de Bolivia opens to the eastward of point Angamos; it is 8 miles wide, and is now much frequented, though there is neither wood nor water. As it borders the great desert of Atacama, its vicinity is perfectly desolate, and it is only the discovery of guano that has given it any importance.

There are two anchorages, both on the W. side of the bay; the one for guano vessels is 3 miles from point Angamos, on a line between two wooden landings, from 200 to 400 yards from the shore, in from 8 to 13 fathoms of water, bottom fine sand; the other is near the village San Luciano. The anchorage is good in from 6 to 12 fathoms, with the flag-staff bearing S. 9° E., or the landing S. 18° W., according to the distance of the berth from shore; vessels can, however, anchor in any part of the cove; the Lamothe Piquet was about ¾ mile distant, in 11 fathoms.

The bay of Mexillones would be a good harbor if it was not for the depth, and the holding-ground being but tolerable. There are in many places of 20 fathoms 600 yards from the land. Violent gusts from the southward come down from the mountains, causing vessels to drag, and it is said that the northerly swell is sometimes felt.

The village of San Luciano is in the SW. angle of the bay; although it sprang up rapidly, it is still a miserable settlement of 500 inhabitants.

Fish and shell-fish are abundant. When distilled water is not used, it is brought from mount Moreno, but it is bad. Chile has for a long time claimed this as within her boundary.

A treaty in 1867 left the guano trade to Bolivia, on the condition that one-half of the proceeds were given to Chile; recently the boundary between Chile and Bolivia has been determined as being near Mexillones bay, mount Mexillones being considered in Chile. A railroad is proposed from this place to the mines of Caracoles.

Gualaguala cove.   Gualaguala cove is 12 miles N. 66° E. from point Angamos; in it there is good anchorage in 7 fathoms, bottom sand and broken shells. Vessels here take in copper-ore, which is brought to the mole by a tramway. At the end of the latter is a shute under which vessels can lie securely.

Bay of Cobija or Puerto la Mar.   From Gualaguala the coast runs nearly N., and there is nothing worthy of mention between Mexillones and the bay of Cobija, which is 30 miles N. 24° E. from Leading bluff. The harbor is protected from south winds by the small point of Cobija, which projects about ½ mile to seaward, and is low and rocky but clean. The anchorage is off the town in 8 to 9 fathoms, E. by N. from the point; bottom sand and broken shells. Four mail-steamers and four coasting-steamers belonging to the P. S. N. Company, touch here every month, making fast to a mooring-buoy which is in 9¾ fathoms ½ mile from the landing.

Light.   A light is hoisted every night on the flag-staff on the point, visible 3 miles, but it must not be relied on.

Cobija Bay.

Coast from Low Point to Teton Point.

This port, from which wools and ores are shipped, has been improved during the last few years; a quay, barracks, and custom-house have been built, and the number of vessels which visit it seem to be increasing. A mole has also been constructed, which renders landing, never very easy, somewhat less difficult; even now during a heavy swell some skill is necessary to take a boat through the narrow channel, formed by the rocks, to the mole. The large water-weeds indicate the isolated rocks. The population is about 2,400, not counting the workmen of the neighboring mines. The towns of Potosi, Chuquisaca, Tupiza, and the other towns in the south of Bolivia get their imported merchandise through this port, and as it is the only port of entry of the republic of Bolivia, vessels wishing to load or discharge in another port must touch here to obtain a license from the custom-house.

The principal articles of export are tin, copper, guano, and ingots of silver.

Good fresh water is scarce, as it never rains. Sometimes a small brook formed of the water condensed by the fog runs in a ravine to the northward of the town; but it is so small that a pipe of the size of a rifle-barrel is large enough to lead it to the reservoir. Condensed water, of which there is always a supply, is generally used. There are some wells, but the water is brackish and cannot be kept in barrels. Fresh provisions can always be obtained at moderate prices; the fruit and vegetables for the inhabitants are brought from Chile and Perú.

The only means of transportation to the interior is by mules; this prevents extension in the exportation of the ores, wools, and other produce of the country. The desert of Atacama commences on the summit of the chain back of the coast, at an elevation of 3,000 feet.

The desert extends 135 miles to the eastward, and no water is found for 90 miles. The mules cross it in three days. It takes 14 days to go to Potosi, a distance of 540 miles; but Indians on foot, having relays, bring a message in 10 days.

It is high water, full and change, at Cobija, at 10^h; the rise, 4.5 feet.

18 c

On the slope of point Cobija is a white stone, which stands out in relief against the black rocks of the land back of it. Generally a Bolivian flag is hoisted on the signal-mast when a vessel is approaching. There is no danger in entering, as the point is steep-to, and can be passed within 200 yards.

The port, however, is not easily recognized. The hills rise directly back of the coast, forming an uninterrupted chain of from 2,000 to 3,000 feet elevation; at their base there is no mark indicating the position of the town. The white, flat rock would be a good mark, were there not another very similar a few miles to the northward. There is, fortunately, a white church on the slope of the mountain, which is an excellent landmark, which can be seen 20 miles in clear weather. It is best to make the land a few miles to the north or south, and then stand along it until the houses are seen. A sailing-vessel should always make the land to the southward of the port.

On coming from the southward, Leading bluff should be made; when steer to make the coast about 9 miles south of Cobija, and coast along it until two islands with white summits are seen off False point; they are 1¼ miles to the southward of the port. The church will prevent Cobija from being confounded with Gatico.

Copper cove is a convenient point for loading copper-ore, as vessels can anchor at a short distance from the land. The cove can be entered easily, and is clear of danger to a short distance from the shore. The best anchorage has the following bearings: Point Cobija, open of the extremity of Rocky point, the western of the cove, S. 35° W., and the jetty S. 38° E. The depth is 14 to 18 fathoms, bottom fine black sand. The anchor must not be dropped in less than 14½ fathoms.

The ores are taken in bags, on balsas, to the vessels, which are anchored near the coast; 50 tons can easily be shipped in this way in a day. All vessels wishing to load here must first obtain a permit at Cobija.

There are no good marks for recognizing Copper cove; the upper part of the mountains is always covered with mist. The white church of Cobija is, however, a good guide during clear weather. When going from Cobija to Copper cove, it is best to steer N. by E., keeping about 800 yards from the coast, until the jetty is opened, then round Rocky point at about 500 yards, and drop the anchor as before directed.

When the church of Cobija has been recognized on coming from the northward, and it bears S. 21° E., steer for it until the huts on the south side of Copper cove are seen; then bring the mole and an isolated house to the northward in line, bearing S. 58° E., which will lead to the anchorage. The huts and mole cannot be seen until 3 or 4 miles from the land. A heavy swell enters the cove, and during calms, or the light prevailing winds from SW., it is difficult for sailing-vessels to get out; they should not attempt it without their boats ahead, as they would run a risk of being drifted ashore.

Huanillo cove, which is 2½ miles from the rocky and dangerous cape of the same name, is 6 miles to the northward of Copper cove. The anchorage is in 15½ fathoms, bottom sand and broken shells. The mining works and furnaces are near the beach, and the metals are taken to the end of the mole by a railway. The former is furnished with a crane, under which vessels can always remain with security. A large condensing apparatus furnishes water to the works and to the people.

Huanillo cove.

From Huanillo cove the coast trends nearly N. 7° E. Between Cobija and Algodon bay, 28 miles, there are some small bays with little depth, and generally sandy, with rocky points, with mountains from 2,000 to 3,000 feet high back of them. About 24 miles from Cobija is Punta Blanca, behind which, to the northward, is a cove in which vessels sometimes load ores.

Punta Blanca.

Algodon bay is small, with a good depth of water; it is sheltered from the S. by point Algodon, off which there is a white islet. The Beagle anchored here ¼ mile from the land, in 11 fathoms, bottom of rock, covered with sand and broken shells.

Algodon bay.

There are three places in this bay in which ores are loaded: Bella Vista, Tocopilla, and Duendas; in their vicinity are important mines. Tocopilla, which is in the south angle of the bay, has about 800 inhabitants, and is a place of some importance. The principal mines and furnaces belong to an English company, which has constructed a wharf and tramway to facilitate the shipping of the metal and the discharge of coal and merchandise. Fresh meat can be obtained at a moderate price. Although there is a spring of good fresh water in the gorge of Mamilla, 7 miles to the northward, condensed water is generally used; there are large condensers, and they furnish the vicinity. Four coasting-steamers of P. S. N. Company stop here monthly.

The spring at Mamilla is 1½ miles from the beach; the inhabitants generally bring the water in pouches made of the skins of the seal calf, which contain from 9 to 10 gallons.

Duendas is in the northern part of the bay, about 1½ miles north of Tocopilla. According to captain Gales, of the Florence Nightingale, there was a prosperous foundery at this place in 1860, the mines only being 1 or 2 miles distant. A mole was constructed out to a depth of 12 feet at low tide, and, although there is always a heavy swell at full and change of the moon, and occasionally at other times, vessels can generally load or discharge. The Florence Nightingale remained here during the two worst months of the year, and, on the average, she did not lose more than one day in the week on account of the swell.

There are two dangerous rocks within the limits of the anchorage of Duendas—Duendas rock, in 15 feet water, and Nightingale rock, in 6 feet. The latter is marked by a large buoy, on which a flag is hoisted when a vessel enters the port. Duendas rock is ⅛ mile N. 55° W. from the mole. The mole is sheltered by a large white rock, with a smaller one off its end. The Florence Nightingale anchored just to the southward of this latter rock.

The bay of Algodon can be recognized by a ravine which runs down to it, and by that of Mamilla to the northward; it has two passes 4,000 feet high on its N. side. The best landmark for a vessel coming from the southward is the broad, light strata on the high land, about 1 mile south of point Algodon.

Cape San Francisco, or Paquica. Cape San Francisco is a promontory about 9 miles N. of Duendas. On its northern slope is a large bed of guano, from which 16,739 tons were taken to Europe in 1862 by 19 vessels. The anchorage is not good; there is generally a heavy swell and heavy breakers on the beach. The guano is placed in bags and hauled to lighters anchored outside

of the surf. Landing is difficult, and often dangerous; vessels moor, head and stern, about 200 yards from the rocks.

Point Arena is low, sandy, and bordered by rocks; it is
16 miles N. 10° E. from cape Paquica, near a remarkable
hill; between these points is a village of fishermen. Vessels can anchor under point Arena in 9¾ fathoms, bottom
fine sand.

The mouth of the river Loa, which is 12 miles N. 21° E.
from point Arena, forms the dividing line on the coast between Bolivia and Perú; it is the eastern point of the west
coast of South America. The Loa is the principal river on
this part of the coast; it runs in a deep gorge and loses
itself in the sand of the beach; its water is bad, from running through a bed of saltpeter, as also from the surrounding copper hills. Bad as it is, the people residing on the
banks have no other; at Chacansi, in the interior, the water
is tolerably good. During the summer the river is about 15
feet broad and 1 foot deep, and it runs with considerable
strength to within a mile of the sea. A chapel on the north
bank, half a mile from the sea, is the only remains of a once
populous village. It is visited occasionally from the interior
for guano, which is abundant.

Point Arena.

Loa River

COAST OF PERÚ.

COAST OF PERU

CHAPTER XI.

GENERAL DESCRIPTION OF THE COAST OF PERÚ.

The entire coast of Perú lies on the Pacific ocean, and is comprised between the parallels of 21° 28′ and 3° 23′ 50″ south latitude. Its western point, which is the westernmost point of South America, is point Pariñas; its easternmost point is the gorge of Loa. Limits.

Perú extends on the seacoast in a SSE. and NNW. direction 1,500 miles, with a breadth varying between 50 and 600 miles. It is divided into 10 departments, with a population of about 3 millions, consisting of the descendants of the Spaniards, of Indians, negroes, and mixed races. General aspect.

The country is naturally divided into three regions: the first between the Andes and the sea, the second composed of the table lands of the Andes, and the third situated to the eastward of these mountains and forming part of the valley of the Amazon.

The country between the Andes and the sea is very contracted, and is traversed by torrents. There are vast sandy plains called *arenales*, which are barren and without irrigation; but if a valley is watered by even the smallest stream, vegetation flourishes. Rain is almost unknown, but is replaced by fogs and dews.

The country is generally healthy, the most common maladies being bilious and inflammatory diarrhea, colic, small-pox, and hydrophobia.

The country is rich in mineral and vegetable products. The mines of precious metals and quicksilver yielded considerably even before the Spanish conquest, and if the production has now diminished it is more especially due to the high price of labor. The mineral kingdom is represented by iron, copper, tin, coal, granite, and porphyry, to which must be added iodine, which is drawn from the mother- Products.

water of saltpeter, nitrate of soda, borate of lime, and sulphate of alumina; which ingredients, found in the department of Tarapaca, have been worked only during the last twenty-five years, and have given this department considerable importance. The nitrate of soda is found between the two chains of the Cordillera around the lake of Titicaca in quantities sufficient to supply the demands of one-half the world. Its origin has been accounted for in two ways : the first is that immense deposits of guano covered the borders of the lake during an antediluvian period and were gradually changed into saltpeter ; the other theory is that when the ground rose the water flowed to the Pacific and disclosed the bottoms of the lakes which covered the plateaux in former times. The exportation of this product has rapidly increased.

The borate of lime is found in the western part, opposite the Cordillera of the Pampa de Tamarugal, under a bed of salt and hard earth from 1 to 3 feet thick, in fine siliceous sand ; like the guano, it is a monopoly of the Peruvian government.

The mineral kingdom is also represented by petroleum, which was found in the department of Piura in the province of Ayavaca, a little to the southward of Tumbez and quite near the sea.

Although there is much to be desired in the culture of the soil, labor is always largely repaid where the country is even slightly watered. In the hot region sugar, rice, tobacco, yams, olives, sweet potatoes, and cocoa-nuts are produced; in the colder climate grapes, wheat, and potatoes. Corn is, and was even during the time of the Incas, the principal nourishment of the inhabitants. Cotton grows well, and is hardly inferior to that of Georgia or Egypt. Clover reaches a height of 3 feet, and is cut five times in a year. Finally, there are many tropical fruits and vegetables.

Probably the most important of all the agricultural products of Perú is the quinquina tree, the bark of which is gathered, dried, and sent to the ports for exportation.

There are a few deer, but the animals from which hides and wool are obtained, as the llama, alpaca, guanaco, vicuña, are numerous, as are also European animals which have become acclimated to Perú.

Guano is found on many parts of the coast, especially on Guano. the southern portion, and all the islands along the coast are more or less covered with it. When the foreign trade in guano commenced it is estimated that the stock was in the vicinity of thirty million tons.

The Peruvian government monopolized the trade, and at first only opened the Chincha islands to foreign commerce; a few years since it opened the Guañape and the Macavi, and later the Ballesta islands. The Chincha islands are exhausted, and the Guañape and Macavi will soon be. The Lobos, which were not opened in 1871, and the deposits along the coasts are all that remain; but on the southern coast it is computed that about eight million tons remain. At first the government showed great negligence in the working of this article, so that at least one-seventh or eighth was washed away and lost. Later it has been more careful of these deposits, which formed one of its best sources of revenue.

The quality of this guano is not the same everywhere; that of the southern deposits is the best, as it contains the most ammonia, which is due to the absence of rains. The guano of the Chinchas is therefore much better than that of Guañape, that of Pabellon de Pica alone compares with that of the Chinchas, but that is reserved for the agriculture of the country. That of Lobos is of inferior quality.

The coast is generally straight and precipitous, and is Coast. beaten, especially in winter, by a heavy SW. to SSW. swell. There are but few coves in which small vessels or boats can find shelter from the waves.

In many places ships boats' cannot be used, and have to be replaced by those of the country—*balsas, caballitos*, etc.

The swell is less heavy in the summer and the communication with the shore therefore easier; ten or fifteen miles to seaward this swell is not experienced. The lines of cliffs are sometimes broken by sand-beaches of great extent, especially to the northward, but the surf is equally violent.

Immediately back of the coast is the Cordillera of the Andes; its proximity to the coast and its elevation change all the water-courses into torrents.

The rivers are generally full in the summer months, and often overflow in February and March; they fall in the winter months, and are nearly dry during August and September. As it seldom if ever rains, it is difficult for the inhabitants and for vessels to procure fresh water. None is found within a radius of 40 miles of Iquique. The wells on the shore often give brackish water; generally, nearly all the water used is distilled, and consequently expensive.

Winds.   The peculiar atmospheric and hydrographic properties of the coast of Perú are worthy the attention of the navigator. There is no other place in the world where the conditions of the climate are so agreeable and uniform. Winds from SSW. to ESE., blow regularly during the day and night, known as the *terral*, or land-wind, and the *virazon*, or sea-breeze, according as they are more to the eastward or to the southward; all winds from the SE. to ESE. are called the terral, those from SE. to SSW. or from seaward the virazon. The virazon sets in about 10 or 11 a.m. and blows till sunset, when it dies out and the breeze gradually comes from the land; it is always cool, damp and lighter than the virazon, and it lasts until morning.

The virazon is lighter in proportion as it comes up later, or as it blows more directly from seaward; it then dies out earlier. The land and sea breeze are often separated by an interval of calm. When blowing from SE. to SSE. these winds acquire the greatest force and steadiness, but they never exceed a very fresh breeze.

The coast of Perú is never visited by storms or hurricanes. The barometric variation is insignificant; there is no thunder or lightning; the rains which take place from June to August are so inconsiderable that they hardly deserve the name of showers. When sailing a short distance from the coast, the sky and horizon have often an appearance so dark and threatening as to alarm those who do not know the want of significance of such signs in these latitudes; the most violent squalls under these circumstances do not necessitate taking in the top-sails or courses. When the sea-breeze is only a little fresh it always scatters these dark mists. It is very seldom that more than one reef has to be taken in the top-sails, and that only when to the south of the 16th parallel, more than 100 miles from the coast, or in the vicinity of cape Blanco, or point Nazca.

To the northward of Callao the wind can be more relied

on; the sea-breeze is more regular and of greater force than farther south.

It must also be mentioned that though the winds are always moderate, very violent gusts often sweep down the mountains, even after the sea-breeze has set in; it is always best to reduce sail before running into any of the small anchorages.

Fresh winds from N. to W. occur during the morning from April to August, at a distance of 3 to 4 miles from the land, but they very seldom last during an entire day, and usually not more than 5 or 6 hours; at times these winds extend farther to seaward.

Winds from S. to SSE. are felt at various distances from the shore; they blow with more regularity than the virazon and terral. Off Arica they extend 90 to 120 miles from the land, and they get nearer to the shore as the distance from the equator is decreased; in the latitude of Lima they are 30 to 36 miles from the land. Farther to the northward they are at about the same distance. To the southward of Arica the same takes place, with the difference that these winds become less regular as the line of variable winds is approached. The winds increase in force as the distance from the shore is increased. At first it will be noticed that the sea-breeze lasts longer; then that the trend of the coast has less influence on the direction of the wind; later, again, that it is stronger during the night than during the day; and, finally, the distance where it has its greatest force is arrived at; this distance probably varies in the same proportion as the belt of variable winds which is along the coast, as the distance is the greatest off Arica, and diminishes as the width of the belt of variable winds, in proportion to the distance north or south of this port. It appears, however, that the chain of high mountains, which extends the whole length of the coast of Perú, intercepts the prevailing winds, and that its influence is felt at a great distance, as the prevailing winds, with a constant force and direction, are only found 520 miles from the coast. The winds which blow from the S. and E. within this limit are generally very variable in both force and direction.

Calms are very frequent on the coast of Perú, especially from November to March. Near the land and in the harbors it is generally calm until 10 or 11 a. m., at which time, as before stated, the virazon sets in. Sometimes, however, the calms last for two or three consecutive days in these months.

The zone comprised between Chala head and the port of Iquique is most subject to them, and there they are of the longest duration.

Fogs are common, from December to the beginning of May, along the entire coast, from Chala head to point Aguja. They are often so dense that nothing can be distinguished at a ship's length. It is therefore necessary to adopt all precautionary measures known to navigators when sailing along the shore or off a harbor; to sound frequently, and to listen for the sound of the surf on the beach; more especially on the coast comprised between Guañape and Lambayeque, as here the water is shoal, the coast dangerous, and the noise of the breakers can be heard from 6 to 8 miles from the shore. It often happens that these fogs last for 24 to 36 hours.

The fogs are less dense on the northern coast of Perú. In the province of Piura (Paita) it rains occasionally.

The dews at night in this part of Perú and its vicinity are noted; they are caused by the land-winds which sweep along the perpetual snows of the Andes.

The action of the tide is feeble along the whole coast, the rise and fall being at no place more than 7 feet.

A general current, known under the name of the Humboldt current, runs along the coast of Perú, leaving it at cape Blanco, to the northward of Paita. It follows the general direction of the coast; being N. between Loa and Arica, it then turns to the WNW., and from Pisco NW. From cape Blanco it runs to the Galápagos islands, to the WNW., and then west. Its breadth varies on leaving the coast between 120 and 180 miles, but it increases to 500 when off the Galápagos islands.

This current has some anomalies; at certain periods a feeble current to the southward is experienced, but it is very irregular, and of short duration, and close to the coast; the season of these cannot be determined. Fitz Roy often

observed these southerly currents immediately before or
during north winds, but this is not always the case, and no
general rule can be deduced. These irregular currents may
have some relation with the causes which produce remarkable eddies in the water in the vicinity of the Galápagos
islands.

The Humboldt is a cold current; its waters are of a lower
temperature than those surrounding it. The temperature
increases, however, in going to the north; on the parallel
of Coquimbo it is 57° Fahrenheit; on that of Cobija, 64°; of
Arica, 64°.6; of Pisco, 66°; of Lima, 65°.3; of Trujillo,
69°; and finally, on that of cape Blanco, 74°.3. A few hours
after leaving this current, to the north of cape Blanco, the
temperature is 77°.9.

The velocity of this current is variable; the minimum
along the coast observed by the La Bonite was 3 miles in
24 hours; that observed by the Venus was 6 miles; the
maximum observed by these vessels was 26 miles. Fitz
Roy speaks of 50 miles between Paita and the Galápagos,
where it has its greatest strength. Humboldt gives 15 miles
a day. The coasters, who navigate entirely by dead-reckoning, always add 5 miles to their westing to obtain the
longitude.

From these results it can be concluded that the force of
the current is at least ⅓ mile per hour in a NW. direction;
the general average is 10 miles per day. When no observations have been taken on account of fog, it is prudent to
take this into consideration.

The four seasons follow each other in this part of the Temperature.
globe without any considerable change in the temperature.
The winter hardly deserves the name; it cannot ever be
said that it is really cold. The air is sometimes cool and
damp, and the sky is low and charged with thick fog; then
a European needs his cloth clothes. The summer is more
sensibly felt; it is generally warm from the morning until
2 or 3 p. m. The thermometer then varies between 69°.8
and 77°.9 in the shade, on board ship. The nights are cool
and agreeable. The spring and fall pass unnoticed. In
general the temperature is equable the year round.

The coast of Perú, as well as that of Chile, Bolivia, Ecua- Rising of the
dor, and New Granada, shows evident signs of rising. The soil.

western part of the desert of Atacama is covered with shells and saline matter, which seem to have come from the ocean during a comparatively recent period. The sea has retired from Iquique as well as from Cobija, Coquimbo, and other places whose shores are washed by the Pacific. At Arica it is estimated that the sea has receded 475 feet during the last forty years, necessitating the extension of the landings. There is no doubt that the island of San Lorenzo and perhaps the neighboring coasts have been raised 82 feet within the period of the known history of this country. It appears, however, that the ground on which the city of Callao is built has fallen, as the site of the old town is partly under water. This is probably local, and does not affect the general rising of the coast, as farther to the northward, at Colon, Santa Martha, and in many places on the coast of New Granada, the land is visibly higher than when first visited by Europeans.

Earthquakes.   The coast of Perú is subject to frequent earthquakes, though some of them are light and hardly felt on ship-board; others have been fearful, especially the memorable earthquake of October 28, 1746, and of August 13, 1868. At the former date, at 10 p. m., many houses and public buildings were destroyed; the sea receded from its ordinary bed in the bay of Callao, and half an hour afterward the bay was filled again, the water coming in with extraordinary velocity from the NW.; it inundated the city, which was then nearer the point, drowned the inhabitants, and swept everything before it for ½ mile to the interior. The four largest vessels which were anchored in the bay broke from their anchors, and two were thrown high and dry above the presidio to the SE.; another was thrown in the plaza, and a third, heavily loaded with wheat, on the shore. One of the former was the Spanish frigate San Femin, carrying 30 guns. The waves tore down all walls, houses, and churches, and hardly any traces of the brick floors of the latter could be found. It threw 24-pounder guns, which were in battery, outside of the walls, and scattered guns of other calibers around the platform. According to reliable statements, at least 5,000 people perished and only 30 survived.

The earthquake of August 13, the memory of which is still fresh, passed with more or less force over all the coast of Perú comprised between Iquique and Pacasmayo; it was also slightly felt in some of the ports of Chile and to the northward as far as California, and caused a general disturbance in the currents and forces of the tides.

Arica was the center of this phenomenon; its destructive effects were formidable and complete, and in a few hours this industrious and flourishing town disappeared altogether. At 5 p. m. its buildings were destroyed by the shock, and at 7 the ruins were inundated, and the traces of their old position almost obliterated; the waters receded several times in a few minutes, exactly as in Callao during the last century.

The vessels at anchor in the bay were thrown on the beach after having been for a time the playthings of the currents and eddies; some vessels were thrown as far as 550 yards inland and completely destroyed. Among them was the United States gunboat Wateree, the United States store-ship Fredonia, the Peruvian corvette America, the French brig Dourregalan, the English bark Chanaral, and the Central American bark Rosa Rivera, and all the coasters in the port.

At Iquique the sea inundated the richest part of the city and destroyed all the buildings near the shore, including the custom-house with all the merchandise it contained, and all the distilleries of fresh water by which the city was supplied. Most of the vessels rode it out well, but some dragged for long distances, and they were constantly fouling each other. An English bark which was loading saltpeter in Junin cove, was thrown on the beach and totally lost. The town of Pisagua and Mejillones was destroyed; of the three vessels which were at anchor at Mejillones one was destroyed and the others much injured by colliding.

At Ilo the sea ran up the gorge for ½ mile, tearing down before it all the olive trees and houses of the village, and breaking in pieces a schooner which was anchored at Pacocha, where the sea ran up the beach for more than 450 yards; Tambo, Mollendo, Quilca, Camana, and Ocoña, as also all the valleys of this province which were near the sea, experienced similar inundations.

The towns of Arequipa and Moquegua, and the surround-

19 c

ing villages, were almost entirely destroyed by this earthquake.

The English steamer Santiago, which was anchored at Chala, having steam up, was saved by going to sea. The inundation destroyed the lower rows of houses of that port. The shock was also felt at the same hour at Callao, but with much less intensity than farther south. Its duration was four minutes; the ground had a prolonged undulating motion, like a slight wave on the water, accompanied by a slight noise.

In the bay of Callao very irregular and almost gyratory currents were felt from eight o'clock in the evening; the sea rose with great rapidity 7 or 8 feet, and in some places even 10 and 12 feet, above its ordinary level, and fell again in the same proportion, or even lower, all in the space of a few minutes, returning to its ordinary condition a little after one o'clock in the morning. No damage was done excepting to vessels anchored too close together. The same phenomenon was observed in the inlets and coves as far north as Pacasmayo.

The wind on that day had been, along the whole coast, the usual light SSE. to SE. wind. The sea was calm, and the weather clear and fine. The barometer showed no remarkable change, and the thermometer stood at its usual height. No sign or indication gave the least warning. It is estimated that in all Perú 4,000 persons perished, and 74 millions of dollars of property was destroyed.

Passages. The passages made along the coasts of Perú are distinguished as down and up—*bajada y subida.* The down passages are those from southern ports to the N. This name is given to them as the prevailing winds blow from S. to ESE., and the coast trends from S. to N. and SE. to NW.; vessels sailing to the northward have a free wind. In making these passages, the following rules should be observed: 1st. Always make a direct course when it is possible. 2d. Sail along the coast at a distance not less 6 nor more than 15 miles from land. 3d. The above precaution is indispensable, as fogs are frequent, not only to prevent accidents, but to make the port of destination with ease. 4th. As many days pass without being able to take observations, and inasmuch as the unequal currents cause a loss of confidence in the dead-reckoning, it is necessary to sail in this manner, as by so doing points on the coast easy to recognize are always in view; for, as will be seen in their description, they are numerous and not easily mistaken. Thus it is easy to ascertain with certainty the position of the ship.

The passages from ports north to those south are made in a very different manner. Owing to the geographical configuration of the coast and the winds which prevail, in going from a northern to a southern port the direct course cannot be made.

Under the head of "Winds" it was stated that day and night they blow between SSW. and ESE.; following these changes, it will be seen that each determines the tack on which the vessel should sail. Therefore, at the commencement of the virazon, or when the wind is, for instance, SE. by S., take the starboard tack, or stand toward the land, and keep it as long as possible, or until the wind comes from SE. by E., which is off the land. Then tack and stand off the land; keep on this tack, unless the ship falls off from SW. by S. If far from land, however, it is well to go about before.

By working in this manner the coasters, which are generally poor vessels, make their passages more speedily. They usually make two tacks in a day; one in-shore, which begins before noon, and the other off-shore, which is taken at eight or nine o'clock at night. This is the method generally adopted; but a commander of a ship must take advantage of every alteration in the wind, and tack accordingly. In general, a day's work may be considered good when progress has been made not only toward the south but also toward the east, or at least without getting to the west. It is a bad day's work to gain southing without lessening the longitude; and this is because the greater the distance between the ship and the coast the greater is the change in the winds, which come from the east, so that the sea-breeze is even lost, and therefore it becomes more difficult to make the land. To avoid these inconveniences, a tack should be made for several hours in-shore, even if the ship does not head better than E. by N. or ENE.; the navigator must not be deceived by heading well to the south.

This is the mode of navigating along the coast suggested by experience, and which has been followed many years by those engaged in the coasting-trade. It must be noted that in the passage up it is not well to approach the land within less than 5 or 6 miles, as nearer the current is stronger and the calms of longer duration. This has been often proved.

Paita to Callao. To illustrate the foregoing explanations, which may serve as a guide to navigators, a description is given of two passages, one from Paita to Callao, and the other from Callao to Arica. The best hour for leaving Paita, as well as all other northern ports, is at night; but the custom-house regulations forbid sailing after sunset. Starting at the close of the day, care must be taken to hug the points to windward at a safe distance. On leaving the bay, the coast trends to the SE., and as at this time of day the wind is SSE. or S., it will be necessary to keep on the port-tack to gain as much to windward as possible. In this way, night finds the ship near the coast; and in this position she gets the benefit of the land-breeze, which is felt between eight and ten o'clock, and she continues all night on the same tack. The land-breeze is very free in the vicinity of Paita, and to such an extent that it often gives one a free course to S. by E. When the wind begins to change to the S., which will be before noon, tack in-shore. In this manner keep on until Aguja point is rounded, making short tacks, if necessary, on account of being near land. On arriving at the latitude of the Lobos de Afuera island, great care must be taken to gain more easting and southing until within the meridian of the island.

The part of the coast where the longest and freshest sea-breezes are experienced is that between Cape Blanco and Lobos de Afuera; a free course is often obtained to ESE. and SE. by E.

Between the meridian of the islands of Lobos de Afuera and that of 80° longitude a vessel may sail 1½ or 2° southwardly, and then be very careful that on reaching the parallel of 11° latitude she is not beyond the 79° of longitude; then work on between this meridian and that of 78° longitude to the parallel of 12° latitude, crossing which she will be to windward of the Hormigas rocks. A good lookout should be kept for these rocks, as they are low and scattered over a circuit of two miles. They are readily recognized, as the sea which sets upon them breaks heavily, and the sound of the breakers can be heard at a great distance.

It is safest to pass to windward of the Hormigas from 3 to 5 miles; in this manner the navigator may be sure that at the commencement of the sea-breeze he will be able to steer for the head of the island of San Lorenzo. This island need cause no fear, not even when one is standing in at night, as its shadow is very perceptible, even in dark nights; in clear weather the light-house on the northwest head will be seen.

In passages from Callao to the ports of Iquique, Arica, Callao to Iquique, &c. and Islay, work inshore as far as Chala Morro, as within that distance a fresh wind is almost always experienced, especially near the San Gallan, the Infiernillos, and Nazca point; the last named is the most windy place on the Peruvian coast. Therefore, in this vicinity the tacks should be short and very close to the land, in order to have the benefit of the fresh sea-breeze, and at nightfall be near the coast to take advantage of the land-breeze, which rarely fails.

Between San Lorenzo and San Gallan islands every care should be taken not to get embayed at Asia island or Cerro-Azul, as it would be very difficult to work out on account of the calms and counter-currents. The surest way is not to pass toward the land of a line drawn from San Lorenzo island to the North Chincha island.

From Chala Morro to the southward the same occurs, and the ship's course should not be less than thirty nor over one hundred miles from the coast, as beyond that distance the winds may set from ESE. or E., in which case it would be difficult to again make land.

When on the parallel of the port of destination every care should be taken to near the coast without losing in latitude, for it is difficult to gain it near the land.

By standing for the coast some miles to windward the ship will have the advantage of fetching her port, even when the weather is calm, as it may be done with the aid of the current and the light variable winds.

If the passage is only from Callao to Pisco or to the Chincha islands, an offing of twenty-five to forty miles should be made until southwest from Cerro-Azul, when work inshore, even if it be to within 6 miles; for from that locality to Pisco light variable winds from the north are always experienced in the morning, veering round gradually to west and south, with which much progress may be made. It frequently happens that although a vessel may be at break of day in front of Cerro-Azul in a calm, she succeeds in anchoring at Pisco in the afternoon. On this part of the coast the current runs with considerable strength to the WNW.

Above all, the special cautions given for each port throughout this book should be kept in mind.

In the passages to Iquique, Arica, and Islay from ports north of Callao, beginning with Huacho, another route should be followed, the off-shore, as the extent of coast to be passed is comparatively too great for the above-described course to prove advantageous. It has been observed that, notwithstanding the greater distance sailed, the off-shore passages are the shortest. The course then to be pursued is, on leaving any of the northern ports, keep the port-tack, on which the ship's head comes more to the south as the longitude and latitude increase, coming up to S. by E. on the parallel of 20°. Continue on this tack until the variable winds, generally met with south of the tropic, are gained, which may obviate the necessity of going, which is often the case, as far as 28° or 30°. With the variable winds keep on an easterly course until sure of making a direct course with the prevailing winds from S. to ESE., which will be met with more to the northward; then steer in for the coast, several miles to windward of the port of destination.

The experience of many captains gives preference to the off-shore route in going from ports north of Callao to those south. Following these instructions in different months of the year, two passages were made from Lambayeque to Arica, one in twenty-seven and the other in thirty-one days; and a third from Pacasmayo to Iquique in twenty-one days. While making these passages, it happened that two vessels left Lambayeque together and preferred the in-shore route; one was forty-eight days in going to Islay and the other seventy to Arica.

It should be remembered that from Arica to Coles point

the current sets toward the land, and that it is therefore dangerous to get too near the coast.

It happens in this route, that when the winds south of the tropic change they frequently do so from E. to N., and then to W., where, or at SW., they remain fixed ; this circum-stance is noted that the proper precautions may be taken. It also happens, when sailing with a light wind from N. or NE., that it changes violently to the SW. or S., with heavy squalls.

From a port of Perú to Chile a vessel when clear of the port should steer between south and west, keeping within $6\frac{1}{2}$ to 7 points of the wind, and endeavor not to brace up too sharp, in order to make the best possible progress. This is advisable, as the wind becomes free as the ship separates herself from the coast going south.

Many are in the habit of trimming with a fore-topmast stud-ding sail, and steering by it ; and this is advisable. Keep on until in the variable winds which are generally met south of the tropic, between it and the parallels of 23° or 30° ; these winds always blow strong from the SW., S., or NW., and are accompanied by rain-squalls. With them the ship may continue her course to the S. and E., out of sight of land till on the parallel of the port of destination, or, in the summer season, a little to the south ; when running in for the port, the coast should be made to windward. Should the vessels be bound to Valparaiso alone, on meeting the variable winds steer for the island of Juan Fernandez; passing in sight of it on the north, keep to the east; if in summer, until point Curaumilla is recognized but in winter work for the port itself, and if possible do not enter if when standing in-shore the wind is northerly and the barometer falling, for in such case the anchorage is dangerous, and it is preferable to lie to outside until the wind goes to the west, when the vessel may enter without risk.

The winter months on the Chilian coast are from May to August, during which season the northers prevail. Heavy storms from that direction and from the NW. occur and sudden changes to the SW. with terrible gusts of hail, espe-cially south of the parallel of 33°. The latter is the most dangerous because the heavy seas occasion disasters among the shipping.

During the rest of the year the prevailing winds are from SE. to SW., and strong southerly winds are frequent with perfectly clear weather. Storms from these points are also experienced, particularly on the coast of Chile. The fogs on this part of the coast are almost constant, and the rains extend through nine months of the year.*

---

* For routes in the Pacific Ocean see publications Nos. 5 and 38 of the United States Hydrographic Office.

CHAPTER XII.

FROM THE RIVER LOA TO CAPE NAZCA.

Variation from 12° 03' to 10° 13' easterly in 1876, increasing about 1' annually.

The gorge of the Loa is the southern limit of Perú; it is in latitude 21° 28' south, and is formed of cliffs of high and barren hills, which from the coast extend almost parallel to each other to the interior, for a short distance, and terminate in a gloomy-looking mountain; there are guano-beds at several points. Gorge of the Loa.

The Loa is but a rivulet running through the bottom of the gorge just described, which containing niter renders its waters bitter and unhealthy; it filters through the sand at the beach, not having enough force to open a channel. Tolerably good water can be obtained in the interior, at Chacausi. River Loa.

The low and sandy beach forms Loa cove in front of the gorge. Vessels can anchor in from 7 to 10 fathoms, about ¼ mile from the land. The best landing is on the beach to the southward; at a short distance there are some fishermen's huts and an abandoned church. The place is wholly destitute of resources. Loa cove.

Fitz Roy mentions an anchorage a little farther to seaward, in from 8 to 12 fathoms, muddy bottom, ½ mile from the land, the chapel bearing N. 20° E. This cove can be easily recognized from seaward; it is in the deepest part of the bay, formed by point Lobo to the northward and point Arena to the southward; the mountains to the southward are nearly of equal height, while those to the northward are higher and irregular.

The coast remains high and continues its northerly direction, projecting to the westward, for 5 miles, where it forms false point Chipana, which is clean; to leeward of it is a very regular cove, with anchorage in 6 to 8¾ fathoms, very near the land. It is not inhabited. False point Chipana.

Landing is easier under the shelter of the point, near its extremity, than in the bay of Chipana, although there is a good deal of sea at the syzygies.

Chipana point. Chipana point is to the NE. of Chipana cove; off it there are some scattered rocks on which the sea always breaks, the highest being about ½ mile to the northward; they must be passed to leeward.

Chipana bay. Chipana bay is to the eastward of Chipana point. The beach is of sand, and the depth from 8¾ to 11 fathoms about 600 yards from the land. There are a few inhabitants, but no resources.

Running for this point, the land should be made in the latitude of the Loa, where a double white patch will be seen on a hill near the shore, and a similar one a little farther to the northward; these marks, which are visible 6 to 8 miles, being made out, the course can be shaped for False point, where is the anchorage in 7 fathoms, sand and broken shells, sheltered by low, level ground. There is no danger in entering; although the land is low, it may be approached within ½ mile in from 10 to 6 fathoms. The anchorage inside the long kelp-covered reef, which extends 800 yards to the northwestward of the north point of the cove, might perhaps be preferred, but the landing there is not so good.

Huanillo point. The high and slightly projecting point of Huanillo is about 6½ miles to the north. There is here an abundant deposit of guano, but the constant swell makes it very dangerous to reach it. It can, however, be shipped on days when the sea is quiet by anchoring close to the point, in a depth of from 9 to 12 fathoms. It is necessary to have very good ground-tackle, else the vessel will be in great danger. In this manner a vessel was loaded in 1858. There are no permanent inhabitants.

Chomache point and bay. Chomache point, 5 miles to the northward of Huanillo point, is terminated by several rocks and small islets. making out about 1 mile to the westward, over which the sea always breaks. The coast to the northward of the point forms Chomache bay, in which there is anchorage in 9 to 12 fathoms, very near the land. There are here some deposits of guano; the locality is uninhabited.

Los Pajaros islets. The two small islets of Los Pajaros are 5 miles to the northward and about ⅓ mile from the coast. They are clean and of average height; their white color is due to the guano with which they are covered. They are the rendezvous of birds and seal.

IX

Views off the Coast of Perú, between the Lat.^{s} 22° and 19° S.

Point Lobos or Blanca is 1½ miles north of these islands. It is formed by high mountains, which, directed from the interior toward the coast, unite and descend to the sea; its surface is full of white and yellowish patches, by which it can easily be recognized. It is also the southern limit of the large cove or small gulf which opens to the northward and eastward, and contains many remarkable points, which will be hereafter described.

Point Lobos or Blanca is important on account of the large guano deposits which are on it, as its quality differs little from that of the Chincha. All the guano which is found here is covered by a thick crust of material which the natives call *Caliche;*[*] it has to be broken with picks in order to uncover the guano, after which it is easily removed.

Around the point, and to leeward of it, is the anchorage, opposite some houses on the shore; there are from 9 to 13 fathoms close to the rocks. There is another good anchorage to the eastward of a small islet. The breakers generally render landing very inconvenient; but they are not so heavy as on other points of the coast, as the point offers some protection.

Small vessels generally take the guano for the use of the country. Loading is easy during fine weather, as boats can run alongside of the rocks in several places. There are no resources whatever, and the few inhabitants are occupied in loading vessels.

The gorge of Pica, opens 2 miles to the northward; it is formed by very high mountains, which run into the interior so close together as not even to form a valley; some rocks make out from the coast to the northward of the gorge; several of these have a whitish color from the guano which covers them; from a distance they look like vessels under sail.

The Pabellon de Pica is about 5 miles from the gorge, at the head and almost in the center of the bay; it is a most remarkable point, 1,100 feet high; a sort of promontory, with a semicircular base, which runs into the sea; it contrasts strongly with the surrounding hills, as they are brown, naked, and burnt by the sun; its shape is that of a tent, to

---

[*] The *Caliche* is impure nitrate of soda, or common salt, mixed with other substances.

which probably it owes its name; its surface is covered with yellow and white patches. The mountains of Carrasco to the eastward are 5,516 feet high.

The Pabellon de Pica contains a large deposit of guano, though large quantities have been taken from it from time immemorial.

The best anchorage for loading is opposite and to the westward of the Pabellon; there are from 12 to 14 fathoms within 300 or 400 yards of the shore. Shipping is easy by means of a "messenger" from the rocks to the sea; small boats can receive the bags directly from above.

Cove of Pabellon.

Pabellon cove, having some few houses on its shores, is to the northward of the hill; it is the best landing place, as boats can run alongside of the rocks. The vessels which load guano anchor in the cove in from 9 to 10 fathoms, about 400 yards from the shore. The inhabitants are fishermen: they load the guano in the small boats.

The exportation of this guano, as also that of point Lobos, is prohibited. That which is used in the country must be taken by vessels of the country having a license from the custom-houses of Iquique, Arica, or Islay.

Patache point.

From the Pabellon de Pica the coast runs N. for some miles, afterwards NW., and then W., terminating in Patache point, which, formed of low, projecting hillocks, ends in two small islets. This point and Lobos, 14½ miles apart, form the extremities of the cove just described.

Large patches of guano are found on several parts of the intermediate coast.

In this indentation the depth does not exceed 50 fathoms; to the northward of point Patache there is good anchorage in from 7 to 10 fathoms, near the land.

Cove and islets of Patillos.

About 4½ miles to the northward are the three small islets of Patillos, which are very white, being covered by a thin bed of guano, and can be seen from a great distance; they are only separated from the land a few hundred yards. To leeward of them is the cove of Patillos, having a depth of from 7 to 10 fathoms, ⅓ mile from the shore. On the beach are several store-houses, painted white, which formerly contained saltpeter, which was loaded in the cove; they are now abandoned.

The small point of Yapes is 2½ miles N. of Patillos islets; Cove and islets of Yapes. it is high and projects but little. At a short distance are some high rocks, which form the group called Yapes. To the NW. of the point is a small cove of the same name, with a depth of from 7 to 10 fathoms, 400 yards to leeward of the islands. The islands must be passed very close to avoid several rocks which are ⅛ mile to the northward of the cove. Saltpeter has often been shipped from this place.

In the northern part of this stretch of the coast is the cove Leña cove. of Leña, with a depth of from 4 to 6 fathoms. The same precautions have to be taken on entering it as for Yapes. The landing is difficult nearly the whole year, on account of the surf.

Caramucho cove is 1¼ miles north of Yapes; it affords a Caramucho cove. bad anchorage in 9 or 10 fathoms, without shelter; scattered rocks make out ½ mile from the shore, and it is frequently exposed to a heavy swell.

Chucumata cove is 11 miles N. of Patillos and WSW. of Chucumata cove. mount Oyarvide. It is somewhat sheltered from the southward by a small point; the anchorage is near the rock in from 7 to 10 fathoms; there are no inhabitants.

Ligate cove is 1 mile N. of Chucumata cove; it offers no Ligate cove. shelter or convenience of any kind.

The coast runs N. and S., with some slight curves to the Oyarvide heights and mountain. westward, for 28½ miles from point Patache. It is a curtain of high hills, called the heights of Oyarvide; its summit, which is 5,800 feet above the sea, is a little back to the NE.

It is dangerous to approach this coast nearer than 3 miles, on account of the calms and heavy swell.

Tarapacá mountain is the northern termination of Oyar- Tarapacá mountain. vide heights; its summit is an inclined plane, with the greatest elevation to the northward.

Grueso point is a spur of Tarapacá mountain. It is formed Grueso point. of high dark rocks, which advance into the sea in a convex form. There are several isolated rocks off the point, some extending out ½ mile; but all are visible.

On the northern part of point Grueso there are some remarkable white points and three hills, which can serve as a landmark for Iquique. This is important, as all this coast presents the aspect of a wall rising abruptly from the sea, and the points do not appear to project, especially at sunrise.

Chiquinata cove.   From Grueso point the coast runs 4½ miles NE., still keeping the same appearance, after which it resumes its northerly direction and forms Chiquinata cove, which abounds in deposits of guano, covered, like that of point Lobos, by a bed of caliche and sand.

Molle cove.   Molle cove, in the bight formed by this part of the coast, can be easily recognized. Toward its end and on the hills which surround it is the road leading up the mountain, which can be seen for a long distance. The depth is from 5 to 9 fathoms very close to the shore. There is a private wharf, alongside of which vessels of 600 tons can moor; large shipments of saltpeter are made from here. There is no fresh water in this cove or those to the southward, but condensed water can be bought at a moderate price.

Coast of Iquique.   At the foot of the mountain, to the northward of Molle cove, is a beach of light sand several miles long; at its foot is the excellent port of Iquique.

Iquique islet.   Iquique islet is a very short distance from the shore; the channel between can only be used by boats. Its surface is low and level, with the exception of a hillock 46 feet high; it is surrounded by rocks, and those to westward of it should be specially avoided, as some of them are covered at high water and are 400 yards from the islet; those to the northward of the island extend only 200 yards. The company of English steamers show a light from this island; but it is only lighted during the nights when the steamers of that line pass, and it cannot be relied upon. It should be at the western extremity of the reefs instead of on the island.

ort of Iquique.   The port of Iquique is to the northward, and to the leeward of the islet, which keeps off the southern swell. It is convenient, safe, and spacious, extending over an area of 4 miles. The bottom is of sand and rock, with a depth of from 5 to 10 fathoms. Fitz Roy gives for the anchorage in 5½ fathoms the following bearings: Point Piedras, N. 1° E.; the western point of the island, S. 68° 15′ E.; and the church steeple, S. 2° E. One anchor with 30 fathoms of chain is sufficient. The principal landing is in the little cove off the wall of the temporary custom-house.

There are many low rocks off the coast on the side of the channel which separates it from the island; they are covered at high tide, and are always marked by large masses of sea-weed. Going to the steps of the custom-house, this channel must be passed, heading for the street to the southward of the custom-house, and between the two rocks known as el Toro and la Vaca; these must be avoided, as boats may be endangered by them, especially during the night or during the heavy swells from June to August. A permanent mole is being built by which this will be avoided.

There is no safe anchorage to the southward of the island on account of the many scattered rocks, and the want of shelter.

Vessels bound to this port should be careful to make the land to the southward, especially the white patches of point Grueso, and they should not approach the shore nearer than 4 miles before making out the bight of Iquique. Directions.

The course should then be shaped for the dune or the most northern white patch of point Grueso; the port will then be easily reached by heading for the island and the church-steeple, as soon as they are made out. Vessels are often becalmed near the island, in which case they must be towed by their boats, as they would otherwise be carried to leeward of the port by the force of the current.

It is high water, full and change, at Iquique, at $8^h 45^m$, rise 5 feet. *Tides.*

From the southward, Iquique is the first of the larger ports of Perú, or of those which have custom-houses and are ports of entry. *Description.*

There is no vegetation and no fresh water in the vicinity of Iquique, but all kinds of provisions can be obtained, though at high prices; water is condensed from sea-water, which makes it dear, but the price is not unreasonable. There is always plenty of coal, and steamers can purchase it at reasonable rates.

Iquique is the principal dépôt of saltpeter and borax. These products, which are found in many parts of the province of Tarapacá, to which this port belongs, are exported in large quantities.

A railroad has been built from this point to la Noria, 70 miles, the principal center of the saltpeter factories; the station is near the beach at the northern extremity of the town; it runs up the hills to the northward, turns to the south, still ascending, and disappears in the interior amongst the heights of Molle. It can be easily seen from seaward.

The ancient and celebrated silver-mines of Huantajalla are at present almost abandoned, though near to this port. They will undoubtedly be worked again when the population and commerce increase, but at present labor is too expensive.

In the port of Iquique, as in nearly all those on this coast, a peculiar kind of *balsa* is used for fishing and loading saltpeter, when boats cannot be beached; they consist of two separate parts, each of which is composed of two skins of the seal; they are sewed perfectly tight and pointed at the ends, and are inflated by means of a tube with a mouthpiece of bone; the two parts are lashed together with rope made from skins, and a platform of planks is placed on it; the latter supports the cargo and the man who steers the balsa, who, on his knees, propels the float with a double paddle of cane with a skin stretched on each extremity. The natives make passages on them from the southern cove to the ports of Arica and Islay, but then they are furnished with a mast and light sail of *tocuyo*—a coarse and solid cotton fabric of the country.

Point Piedras.   The roadstead of Iquique is bounded on the north by point Piedras, which is high and rocky, and projects but little. It derives its name from the rocks, which lie close to it and are visible.

From point Piedras the coast runs N. with a slight inclination to the westward; it is formed of very high and steep mountains, with but few beaches, and can be approached as close as desired, as the depths are considerable and the current sets to seaward, even during calm weather. The passage from Iquique to the western ports should be made as stated in the preceding chapter. It often takes eight or ten days to go from Iquique to Pisagua when such a course is not followed.

Colorada point and cove.   Colorada point is 11½ miles to the northward of point Piedras, the cove of the same name being immediately to leeward of it; it affords no shelter, but is clean, with a depth of from 10 to 13 fathoms very near the land. The road to the interior can be seen on the mountains.

Two small islets, called Cololue, are some hundred yards off Colorada point. They are whitened by a thin bed of guano.

About 5 miles to northward the coast is cut by Aurora gorge; it is small, and extends but a short distance into the interior. It is inaccessible.

Little hill and point Mejillones extend about 1½ miles to leeward of the gorge. The former is an isolated hillock detached from the coast and forming a peninsula of moderate height, but lower than the coast. It can be easily recognized from this and its position, and besides by a road leading to the mountains, which is seen to the southward. Some small islets and rocks extend from the hill 100 yards to the SW.; its northern point, or that of the peninsula, is black, and in passing it the detached rocks which extend out about 200 yards to the northward must be looked out for.

Mejillones cove is to the northward of Little hill; it has from 7 to 11 fathoms of water close to the shore. The best anchorage is in the middle, about 200 yards from the shore, in 10 or 11 fathoms; it is narrow and inconvenient. When there are more than three or four large vessels, it becomes absolutely necessary to moor head and stern. Large quantities of saltpeter are shipped from this cove, where there are large store-houses and always vessels loading. The town is small and without resources. On the hill and islets are thin beds of guano.

It is prudent always to have an anchor ready when approaching the cove, as the calms and currents often necessitate towing by the boats.

When bound to Mejillones from seaward, the land should be made at Iquique, and the peninsula, which at first appears as an island, will soon be seen, and after it the road to the mines. The land should then be gradually closed, so as to pass about 300 yards to the northward of the peninsula. Enough room must be allowed at the anchorage to swing clear of the rocks off the N. point during calms.

About 9 miles to the northward is Junin cove, whose anchorage, from 12 to 19 fathoms, is not sheltered, and is subject to a heavy swell.

A very small point runs out a few yards to the southward of the cove and gives a protection to the landing, for which

20 c

there is a commodious wharf. Vessels should always have a stern mooring, in order to stem the constant long south swell. The road down the mountains, which is clearly marked on their flank and forms an acute angle with the summit to the north, is a good landmark; another is a white tower situated on the upper profile of the cliffs above the anchorage.

The saltpeter is brought down to the store-houses, which are on the small points, by carts. The only fresh water is condensed.

Pichalo point.   Point Pichalo is the most remarkable point on this coast. It is 15 miles N. of Mejillones, and projects nearly 2 miles perpendicular to the coast, showing several hills, and, descending gradually to its extremity, it forms almost right angles with the coast to the northward and southward.

Guaina-Pisagua bay.   In doubling the point just described attention must be paid to a rock lying off it 200 yards; Fitz Roy gives its distance 100 yards. After this is passed the point is clear, and can be approached closely, when the spacious bay of Guaina-Pisagua opens. It is surrounded by very high mountains, and has little or no beach; the anchorage is in its eastern or central part, in from 9 to 15 fathoms, very near the land and off the town of the same name. A hidden rock, on which the sea does not always break, must be avoided; it lies 200 yards off the middle of the town. It is best to anchor off the western extremity of the town until acquainted with the position of the rock, after which the vessel can be hauled into any convenient position. Fitz Roy places this rock 400 yards from the shore, with $4\frac{1}{2}$ feet of water over it. He indicates the anchorage as $\frac{1}{4}$ mile from the village in 8 fathoms, point Pisagua bearing N. 3° E.

This bay is often subject to very heavy gusts, which sweep down the mountains from 10 to 11 a. m. to the first hours of the night, causing vessels to drag into deep water, and rendering it necessary to make sail. These gusts shift quickly and with great violence from one point to another, between east and west by the south. Vessels should double the point under short sail, and this precaution should not be disregarded even if it be calm outside.

The port, one of entry, has but a small population; water and provisions of all kinds can be obtained, but at high

prices, as everything is imported. After Iquique, the greatest quantity of saltpeter is shipped from this port, and there
are always several large vessels at anchor. Small quantities of coal can be obtained at moderate prices.

A railroad is being constructed to the factories called de
Sal de Obispo, which can be plainly seen on the mountains
to the northward, on which it forms an acute angle in wind-
ing back to the southward to attain the heights.

Point Pisagua is the northern limit of the bay of Guaina- *Gorge and point Pisagua.*
Pisagua for a distance of 2 miles. The gorge is immedi-
ately to leeward. A very small rivulet of good water runs
through it during the summer. The people, who have their
huts on the shore, are principally occupied in fishing and
transporting water to the town. During the winter the
brook dries entirely and shallow wells are dug to reach the
water.

The Pisagua river, like the Loa, has not sufficient strength
to reach the sea. Fitz Roy says, in contradiction to Garcia,
that it is dangerous and that landing is difficult.

Pisagua bay is to leeward of the point and opposite the *Pisagua bay.*
gorge of the same name. The depth near the point is between 6 and 10 fathoms; there are some few inhabitants,
but no resources of any kind. Guano mixed with sand is
found at several points in its vicinity.

Gorda point is 16 miles from point Pisagua, and offers *Point Gorda.*
nothing remarkable. Some rocks above water run out from
it 1 mile.

The gorge of Camarones is 6½ miles to the northward; it *Camarones gorge.*
makes a perpendicular cut in the cliffs, which are from 2,000
to 4,000 feet high. At its opening it is about 1 mile wide;
in front of it is a cove, with an anchorage in from 9 to 11
fathoms near the shore, but the heavy surf leaves no safe
point for landing; the sides of the gorge are covered with
bushes. On the southern side are some white stones which
resemble houses when seen from a distance. In the summer months there is generally a small stream of good water.
The coast south of the gorge to point Gorda, and north to
the gorge of Vitor, has no beaches whatever; the sea beats
against the foot of the mountains, on many parts of which
are patches of guano. This coast is steeper and higher than
that in its vicinity.

Point Madrid.
The name of point Madrid is given to a small point about 12½ miles to the northward of Camarones gorge. It is clean, but of no importance.

Cape Lobos.
Cape Lobos is a very remarkable point, 27 miles to the northward of Camerones; it projects but little, is of convex form, and rises in steps. It is quite clean, and of a dark color at its base; toward the top there are some very bright white patches; thin beds of guano. It is said that small vessels can anchor under its lee.

Vitor gorge.
The coast is cut by Vitor gorge 2½ miles north of cape Lobos; it is deep, resembling that of Camarones, and about ¾ mile wide. The valley extends several miles into the interior, and has but little vegetation. During the summer months a small stream of water penetrates to the sea.

The low and sandy beach in front of the gorge forms a regular cove, with from 6 to 9 fathoms close to the shore. Boats cannot always be beached, as the sea often breaks heavily.

La Capilla.
The coast to the southward of Vitor gorge is formed of steep, high mountains, surmounted by a sort of table-land, averaging from 2,000 to 4,000 feet in height; that to the northward, as the preceding, trends almost north and south for 11½ miles; at this distance a slight indentation commences to the east, called La Capilla, surrounded by hills lower and more inclined.

From cape Lobos to the north there are white patches and bands, indicating guano.

La Licera.
To the northward of this indentation the coast makes a slight elbow, called La Licera. Troops have been disembarked at this point, but landing is always attended with danger and the loss of men, as there is no time when the surf is not heavy.

Mount Gordo.
Mount Gordo lies to the northward and near this beach; it is 878 feet above the sea, and terminates in a bluff.

El Morro de Arica.
The Morro de Arica forms the southern extremity of the port of that name. It is a large mass of rocks, which rises almost perpendicularly to the height of 530 feet, and is crowned by an almost horizontal platform, with a slight inclination toward the interior. The part toward the sea has on it several patches of guano, which gives it a silver appearance from a distance; it can be seen for 20 or 25 mile

Views off the Coast of Peru between the Lat.s 19° and 17° S.

Mt Ilay
3900

Mt Ilay
3900

Mt Ilay
2900

in clear weather. It would be difficult to mistake it for another point, but it is more readily recognized when coming from the southward than when coming from the westward.

The small island Alacran is separated from the Morro by a very narrow channel. This passage is dangerous and impracticable; it is strewn with shoals and filled with large masses of sea-weed. The island is low and uninhabited. A fort was being constructed on it and was well advanced, when it was destroyed by the earthquake and inundation of August 13, 1868. There are some rocks around the island, but very near it; the most distant, those to the westward and northward, being less than 100 yards distant. The point for landing is in the Morro channel, avoiding the rocks until a little creek, almost in the middle of the channel, is open, where there is a small wharf, but it is often preferable to beach the boats. *Alacran island.*

The port of Arica opens to the northward of the island of Alacran, which shelters it from the southward; it is formed by the coast, which runs out from the Morro and first curves to NE. and then to NW. The port is large and convenient, having a depth of from 5 to 10 fathoms near the land; bottom coarse sand. A vessel is almost always safe with one anchor and 45 fathoms of chain. There is some rocky bottom to leeward which would endanger the anchor. *Port of Arica.*

The best anchorage is to the northward of the islet, a little toward the channel, at a distance of from 400 to 600 yards. This is the most sheltered place against the swell which often sets in from the SW. It is best to moor head and stern, head to the SW., with a second anchor ready for letting go. During the months from June to August there is often such a sea-way as to interrupt all work, but by mooring as above indicated there is no danger.

Vessels bound to this port should make the land to windward and approach it after passing Vitor gorge. When from 9 to 12 miles from this gorge the Morro de Arica will be seen, resembling a steep, white hill, and also the rounded hill called Monte Gordo farther inland; and, on a nearer approach, the island of Alacran, which cannot be mistaken for the Morro de Arica, as it is low. The bight of Arica is subject to frequent calms or to light variable winds, and the *Directions.*

currents are strong, so that vessels heading directly for the port instead of to the southward of it are carried from it. Vessels to the southward, on the contrary, have only to drift, and need only the lightest breeze. The calms near the land are not dangerous, as the coast is clear and the current sets to seaward. Should it become dark or calm when approaching the port, or when not wishing to enter, it is best to anchor, or drop a kedge as soon as the lead shows a good depth for that purpose, otherwise a vessel would find herself to leeward of the port and probably be unable to make it during the day.

Arica is amply provided with provisions of all sorts; water is taken in by casks, which are rolled to the wells. It is the third great port of entry from the southward, and the second in point of importance. It is united with Tacna, the capital of the department of Moquegua, in which it is situated, by 45 miles of railroad. There is also telegraphic communication.

This city was but a short time since one of the most flourishing on the coast, but was completely destroyed by the earthquake and inundation of 1868; it will probably be soon rebuilt, as the amount necessary for that purpose has been granted by the government. The water consumed in the port is drawn from the valley of Azapa, where the wells give excellent water. It is said that there is much intermittent fever in the town, and it is certain that the Morro interrupts the sea-breeze.

The wharf, which was destroyed by the earthquake, was replaced in 1870 by a small landing, which is not always convenient, as it is dry at low water, when it is better to run on the shingle-beach, inside of the landing.

The bottom having altered since the earthquake of 1868, it is recommended not to anchor to the eastward of a NE. $\frac{1}{2}$ E. line passing over the western summit of Alacran island, nor more than $\frac{1}{2}$ mile to the northward of the parallel of the Morro, in from 5 to 9 fathoms.

The ground back of Arica rises gradually to the snow-covered peaks of the Andes. Among them the volcano of Arequipa can be seen at a distance of 90 miles. During clear weather Tacna, which is distant 20 miles in a straight line, will be seen. There are few views so majestic as that f the Andes between Arica and cape Sama; some of these mountains have a height 19,630 to 22,960 feet.

The principal commerce of the port of Arica consists in the importation of foreign merchandise for its department and in transit for Bolivia, and in the export of barilla, tin, wool, hides, cotton, and precious metals from Perú, as also from Bolivia. Small quantities of coal can generally be obtained.

It is high water, full and change, at Arica at 8ʰ 50ᵐ; rise, 5.3 feet. *Tides.*

The valley of Azapa is at the extremity of the city of Arica; it rises gradually but sensibly to the interior; it has very little water, but abundant and excellent fruit is raised; the olives in particular have a great reputation. The plans for increasing the amount of water in this valley appear favorable to the enterprise. *Valley of Azapa.*

The valley of Chacayuta is alongside of that of Azapa, and extends to the northward of Arica. In the interior are seen the mountains of Tacora covered with perpetual snow. The shores in this port are sandy and low; the sea always breaks and makes landing dangerous. The coast of the continent, which has thus far run nearly N. and S., changes its general direction abruptly to NW. *Valley of Chacayuta.*

The coast keeps its low and sandy aspect for 28 miles, running nearly N. 50° W. to the heights of Juan Diaz, which are close to the sea, and of moderate height. The soundings to seaward are regular; from 2 to 3 miles from the coast there are 14 to 28 fathoms, muddy bottom. *Heights of Juan Diaz.*

Quiaca point, a low tongue of land, is 5 miles to the NW. of Juan Diaz; it unites near the beach with the cliffs which extend to the northward. *Quiaca point.*

Sama head is remarkable for its elevation, 3,890 feet; it is 42½ miles NW. of Arica; seen from a distance, it has a dark-blue color. It can be easily recognized 30 or 40 miles, as it is the highest land in the vicinity of this part of the coast. It seems to project outside of the coast-line, and is the landmark for Ilo. The rocks which run out from it are close to and above water. *Sama head.*

Sama point is formed by a spur running out from the hill toward the sea. *Sama point.*

Sama cove.   To the northward of Sama point is Sama cove; it is clean,
with a depth of from 9 to 13 fathoms near the shore. On
the mountains surrounding it the path can be seen which is
used by the muleteers who come to load guano. The vessels loading the guano for the neighboring valleys are the
only ones which visit this small cove. There are only two or
three straw huts, in which fishermen and the guards over the
guano live; landing can only be effected with balsas.

Locumba river.   Locumba river empties 5 miles to the NW., after irrigating
the valley of the same name. It is full of water from February to June, and almost dry the remainder of the year.

Ticke point.   Ticke point is 4 miles NW. of Locumba river; it is stony
and in close proximity to the mountains of the interior.

Picata point.   Following the coast to the northward, Picata point is 3½
miles from Ticke point. Both are small, and without importance. All the neighboring coast is clear.

The soundings to seaward are regular; 2 miles from the
coast, between Locumba and Coles points, there are from
14 to 20 fathoms.

Point Coles.   Point Coles is 32½ miles WNW. of Sama hill; it is very
salient and low at its western extremity; it rises toward the
interior, and the ground becomes rugged; it can be easily
recognized, as its western extremity is formed by several
islands and high rocks which extend out, and can be seen at
a good distance; from a distance it resembles an island.

After doubling this point, the coast trends nearly N. 28°
E. to the gorge of Ilo, forming several small coves, known
as Inglés, Calienta, Negro, Pacocha, and la Picuda coves.
After doubling the point, the land must not be approached
to the southward closer than 1 mile, as it is strewn with
rocks above water, called Tortuga, Leones, Tres Hermanos,
and others, which are submerged and run out to the distance
of ½ mile.

Inglés cove is the best landing-place in the bay of Ilo, but
it has been closed to prevent smuggling. Back of it is a
mountain, remarkable for its plateau, which joins the mountains to the NE. of Ilo.

Pacocha cove.   Following the shore of the point at the distance above
mentioned, the land can be approached as soon as the Hermanos bear to the southward of east, when Pacocha cove,
the most important anchorage in Ilo bay, will be open. The

anchorage is good near the shore in from 12 to 14 fathoms, stony bottom. There is a wharf which affords all conveniences for landing; there is no fresh water, the inhabitants bringing it from the river Ilo.

Pacocha is the point of the terminus of the railroad uniting the coast with the rich valley of Moquequa; it has without doubt great conveniences for a good port. At present there are but few houses and inhabitants. There are no resources, with the exception of small quantities of fresh provisions. The exports are wines and olive oil, the principal products of the surrounding valleys.

The anchorage is subject to strong winds and heavy gusts from about noon to early in the night. These do not, however, raise much sea, owing to the proximity of the land to windward; a very heavy swell is frequent from June to September.

The port of Ilo is at the head of the cove opposite the gorge of the same name. The village was completely destroyed by the inundation which followed the earthquake of August, 1868, when some of the wooden houses were actually thrown up on the heights. The anchorage is farther to the southward in from 8 to 10 fathoms; but the land should never be approached closer than 400 yards. There is always a heavy cross-swell in this roadstead, for which reason vessels generally prefer the anchorage of Pacocha. The road along the shore is short and good. There are many rocky places in the bottom of the anchorage at Ilo, rendering it dangerous to vessels, and causing loss of anchors. It is best to anchor with one anchor and a stern line. *Port of Ilo.*

The principal commerce of Ilo consists in olive oil. All the inhabitants of the gorge are employed in its preparation. The distance from Ilo to Moquequa is 54 miles.

Ilo gorge is formed by the slopes of the mountainse immediately to the north and south of it, the vegetation beginning at the edge of the sea. It is watered by a brook which runs to the ocean and gives an abundant supply from February to May. There are some scattered rocks at a distance of ½ mile to the westward of the gorge. *Ilo gorge.*

Sopladera point is 5 miles NNW. of Ilo gorge. There are some few rocks off it, above water. *Sopladera point*

Chuza cove.

Chuza cove is to leeward of Sopladera point. It can be easily recognized, as the olive trees grow down to near the sea. The depth is from 9 to 10 fathoms 600 yards from the shore. There is fresh water from natural falls, and the landing is generally easy, by beaching on the north side of the point.

Yerba Buena point.

The coast to the NW. becomes a little more salient, and forms Yerba Buena point, 3½ miles from Chuza cove.

Cove and gorge of Yerba Buena.

Yerba Buena cove is immediately to the northward of the point, and the gorge of the same name is 1 mile farther to leeward. There is an anchorage in from 8 to 10 fathoms, 400 yards to leeward of the point; landing is generally difficult, on account of the surf, but it can be effected by running on the sand-beach to leeward of the point, where there is fresh water. The cove is easily recognized, from the gorge and the olive trees which it incloses. The gorge is formed by two parallel cliffs, between which the vegetation can be seen. The north coast, as the south, is formed of high and barren mountains. All the coast, as far as Tambo, is from 196 to 392 feet high.

Point Pacay.

Point Pacay is 13½ miles N. 40° W. from Yerba Buena gorge; it is high and rocky, with a ravine at its western extremity.

Pacay cove.

Pacay cove is to the northward of the point. The bottom is rocky, with a depth of from 18 to 20 fathoms 800 yards from the land. There are ordinarily masses of guano on the beach, brought by vessels for the agriculture of the country. There are a few huts occupied by fishermen and watchmen. This place is devoid of resources, and often rendered difficult of access on account of the surf.

Islet of Jesus.

The coast, going to the northward, continues clear. It is bordered by some islets near the land, the most remarkable of which is the islet of Jesus, which is 3½ miles from Pacay and ½ mile from the land; it is high, clean, and covered by a light bed of white guano, by which it can be easily recognized.

Cocotea cove.

Cocotea cove runs into the land to the northward of Jesus islet. It has mountains at its extremities and a gorge at its head. The best anchorage is inside, between the islet of Jesus and the north shore, in from 8 to 13 fathoms. The swell is heavy. Vessels remaining but a few days let go one anchor, haul into 5 fathoms off the huts, and moor head and stern, as the cove is very narrow. This place, devoid of resources, is only visited by vessels bringing guano for the country, and by small vessels bringing provisions for the inhabitants of the valley of Tambo.

Cape Peje-Perro is a small promontory which runs out from the coast 3 miles NW. of Cocotea; it offers no shelter, and landing is difficult on account of the continual surf. Cape Peje-Perro.

Following the coast for 9 miles, which trends to the WNW., the large and fertile valley of Tambo, in the deep gorge of the same name, is seen. It is the only place on this part of the coast where there is extensive cultivation. The gorge is wide toward the sea, and contracts toward the interior. The slopes of the mountains and the ravines are perfectly barren, making a strong contrast with the luxuriant vegetation of the plain, which is watered by a river. This vegetation commences close by the sea and continues for a short distance to the north and south in front of the ravines which terminate the gorge. The valley abounds in the necessaries of life. Its richest products are rice and sugar-cane. Valley of Tambo.

The beach in front of the gorge extends to seaward and forms Méjico point, which is low, sandy, and almost inaccessible, owing to the heavy surf. Méjico point.

There is no safe anchorage on this part of the coast; there are but from 5 to 6 fathoms of water 1 mile from the land, and from 19 to 28 fathoms, with muddy bottom, at a distance of 3 or 4 miles.

In clear weather the Misti, or volcano of Arequipa, can be seen through the valley of Tambo, though a long distance in the interior. It has a conical form, and is covered by perpetual snow. It is 20,290 feet high, and at sunrise, in clear weather, it can be seen 100 miles. Volcano of Misti.

From the gorge of Tambo the coast runs NW. by W., with a low sandy beach and high cliffs, close together, forming an opening called la Ensenada. The appearance of the coast does not change for 6 miles, when it makes a short turn, forming the cove of Mejia, with an anchorage in from 9 to 10 fathoms 1,000 or 1,200 yards from the land. This cove gives no shelter, and the surf is heavy and constant. Mejia cove.

A small quantity of fresh water can be obtained from a very small gorge in the neighboring mountains, called Chule.

There was formerly a creek in front of this gorge, but it is closed up with sand. It was proposed to make Mejia the terminus of the railroad to Arequipa, and the landing of the materials had commenced, but the difficulty of this operation decided the engineers to abandon it, and the terminus was established at Mollendo.

The coast from 2 miles to the southward of Mollendo is covered with ashes or very fine white sand.

Port of Mollendo.

The port of Mollendo is 9 miles NW. of Mejia cove, and nearly 5 miles to the eastward of point Islay; there are some scattered rocks to the N. and S., but they are all near the shore and above water. The bottom is of coarse sand; the depth near the shore is 12 fathoms, with 22 fathoms at 600 yards. A vessel rode out here the tidal wave of 1868.

This place was adopted for landing all the materials for the railroad to Arequipa, which is now extended to the city of Puno, 13,900 feet above the sea level. The workshops and houses of the employés are on top of the cliffs; they are visible from seaward, and are the best mark for the anchorage, which is clear, but without shelter against the constant SW. swell. Vessels generally moor head and stern, with the head to the southward, stemming the swell. The stern moorings must be strong on account of the strong current setting to the WNW. The mail-boats have two buoys for this purpose moored N. and S., in 20 fathoms, 800 yards from the rocks. An island to the southward of the upper buildings has been united to the coast, and the railroad track starts from this junction; this peninsula separates the two inner coves. In the northernmost is a temporary wharf, which is inside of the small point and the rocks off it. The water is brought through pipes from a neighboring gorge, and condensers are also used.

Light.

A *white light* 147 feet above the sea, *visible* 3 miles, is shown from the flag-staff on the peninsula.

As before stated, there is a strong current in the port, always setting to seaward. There is often considerable sea, which, though not dangerous, often interrupts communication with the shore for two or three days. This is especially the case from June to September. The breakers off the point of the island often extend to the land.

Directions.

Vessels bound to Mollendo should make the land off the

valley of Tambo, and then hug the coast for 3 miles. Sailing-vessels must bear in mind that if they are becalmed in the evening there is no other anchorage excepting that off Méjico point; farther north the land has to be foo closely approached for anchorage.

It is high water, full and change, at Mollendo at $8^h\ 00^m$; rise, 5.2 feet.

Chiguas cove is a small indentation 3 miles to the westward of Mollendo, where boats can land during fine weather.

Guerrero gorge is the largest gorge in the mountains of the interior. The railroad to Arequipa can be easily seen following its declivities.

From Mollendo the coast trends nearly E. and W. for 5 miles. All the coast between the valley of Tambo and point Cornejo to the north can be easily recognized; high and barren mountains descend gradually to the sea without leaving any beach; all of them are covered with large white patches, caused by a kind of cinder or white earth; it is thought to be the product of some volcanic eruption which took place long since.

Below the White mountains, to the westward of Mollendo, is a dark band a little above the sea, called point Islay. It is impossible to mistake it when attention is given to this difference in color. Rocky islets are scattered around the point, but they are high and near to it; there is a rock awash ½ mile south of the point, or 437 yards S. by E. of the southernmost rock, and S. 45° E. of the western Alvizuri islet; the sea breaks almost constantly over it, so that it can be readily seen; it is called San Malo; to clear it the westernmost Alvizuri islet should not be brought to bear to the northward of N. 20° W. while to the southward of the rocks off point Islay.

The three islets called Alvizuri or White islets are off the north point, and mark the entrance to the port of Islay. They are clean, of moderate height, white and rugged. They are generally passed to the northward as close to as desirable, as there are 20 fathoms at the foot of the western rock. On coming from the southward a channel will be seen between the islands and rocks off the point; it is perfectly clear, the depth being from 11 to 17 fathoms. It can be taken when wishing to enter and take a better anchorage

Tides.

Chiguas cove.

Guerrero gorge.

Coast of Islay

Point Islay.

Alvizuri islets.

to windward; it is not advisable, however, for vessels visiting the port for the first time.

The port of
Islay. The port of Islay is distinguished from all others on the
coast of Perú by its configuration. It is a large basin, surrounded by dark, rocky, and almost perpendicular cliffs,
which leave no beach. The bottom, which is very steep, is
of rock; 100 yards from the shore there are from 11 to 13
fathoms, and the depth increases to 26 and 30 fathoms in the
center of the port, and 600 yards from the land the depth is
between 30 and 40 fathoms. Vessels should therefore keep
as close to the southern shore as possible; they will be close
to the wharf on that side, and can run their stern fasts to
the rocks.

There is a pier, with an iron frame and a wooden platform, constructed on some islets and the shore. An inclined
plane commences at this point, about 295 feet in length,
which has a tramway and stationary steam-engine, by means
of which the merchandise is taken to the custom-house.
This building is situated at the shore end of the road, and
the town commences near it.

Off the small point of the redoubt, in the middle of the
south shore, is a shoal on which the sea always breaks.
Landing is often difficult, even alongside of the wharf,
and experienced men are necessary when the swell breaks
heavily on the shore, which is generally the case at the
syzygies, and especially at the equinoxes.

From the southward Islay is the fourth port of entry of
Perú, and the principal one of the department of Arequipa.
The departments of Cuzco and Puno are also to a great
extent provided through it. Large quantities of foreign
merchandise for the use of these departments are imported,
and large quantities of wool and other products of the country are exported. Provisions can be obtained in the town,
but fresh supplies are not abundant. Water is taken from
the iron pipes along the wharf. The population is about
2,000.

The city of Arequipa is connected with Arica by a railroad
90 miles long. There is also a telegraph. This city, of 35,000
inhabitants, is, in a straight line, N. 45° E., 50 miles from
Islay. It is 7,580 feet above the sea, on the plain of Quilca

It is surrounded by snow-covered peaks, above which the volcano of Arequipa rises to a height of 20,200 feet.

Vessels coming from the southward should make the land at point Méjico, off the valley of Tambo, which is so remarkable that, with the description given, it cannot be mistaken. After making this point, which can be seen at a distance of from 9 to 18 miles, according to the weather, head for a cut in the mountains to the westward.

The railroad to Arequipa runs through this cut (Guerrero gorge) after following the foot of the hill from Islay. On approaching the coast, running to the northward, the white patch before mentioned will be seen, and soon after the dark belt which forms point Islay; on drawing closer in, the reef off it will be distinguished, with the white islets of Alvizuri, after which head for the western one; when, passing close to leeward of it, steer for the anchorage, taking care not to let go the anchor until as near the south coast as prudent.

Some give a small olive-wood as a landmark for the southern anchorage, which is on the slope of the hummocks, about 6 miles N. by W. from point Islay. It is a dark, lozenge-shaped spot, of a dark-green color, which contrasts strongly with the glittering sand around it.

With a free wind and a knowledge of the coast, it is best to steer through the passage between the eastern Alvizuri island and the island next toward the land, as the wind will be ahead when doubling the outside or western island.

For this purpose, Flat rock, a small islet off the western point of the cove of Islay, must be kept just open to the northward of the town, or a vessel can go to seaward of this anchorage and drop her anchor as soon as the town is well open, with Flat rock to the southward and the point to the northward of the town, bearing N. 56° E. The best anchorage is just inside of the point of Flat rock, off the landing, in from 10 to 12 fathoms. The mail-boats anchor as soon as the second island is covered by Flat rock.

When coming from the westward, the land should be approached on the parallel of 17° 5′, making the land 3 miles south of point Islay. When not quite sure of the longitude, point Cornejo should be recognized in passing. It is very remarkable, and will be easily seen in clear weather by remaining on that parallel. The valley o' Quilca, the first green land west of Tambo, can be recognized; otherwise look out for point Cornejo, and, when off it, point Islay will be seen bearing east, and resembling two islands off a steep point. During favorable weather a mountain 3,340 feet high, with a bell-shaped summit, will be seen NE. of the town, and afterward the town itself, having the appearance of black spots on a white ground. Then shape a course for the anchorage under the Alvizuri islets.

This part of the coast is subject to frequent calms, for which reason the vessel's head must be kept to the southward of the port, to avoid being drifted to leeward. The current sets to the westward at the rate of from ½ to 1 mile an hour. If a vessel should be becalmed near the islands, boats must be used to tow in.

Tides.   It is high water, full and change, at Islay at 8^h 50^m, rise 7½ feet.

De la Fuente islet.   De la Fuente is a high islet some yards to the northward of the wharf, and separates the port from Matarani cove.

Matarani cove.   Matarani cove is in front of a small gorge of the same name which opens to the northward of the port, in the same indentation. This cove is bordered by a small sand-beach, but it is inconvenient for vessels which have to discharge at the mole, on account of the distance. The depth is from 11 to 14 fathoms near the land. Vessels loaded with guano for Arequipa and its valleys anchor in this cove, where the guano is unloaded at steps which run up the heights. There is a small isolated rock in the center of the cove.

Mount Islay.   Mount Islay is toward the interior, NNE. of the port; it is the highest peak of the mountains running along this coast, being 3,340 feet above the level of the sea; its summit is dark, of a conical form, with white patches on all sides.

Mollendito gorge.   From Matarini the coast continues NW. for 3 miles, steep, with rocky cliffs, which are cut by the small gorge of Mollendito, off which there is a narrow beach with anchorage for small craft. Fishermen often reside here temporarily.

Santa Anna cove.   The appearance of the coast continuing the same, about 9 miles NW. is the small cove of Santa Anna, without shelter.

The coast runs east and west for 2½ miles from Santa Anna cove; it is rocky, high, covered with white patches, and

Views off the Coast of Peru. between the Lat.s 17° and 16 ° S.

terminates at this distance in Cornejo point, which is of a reddish color, its western extremity being composed of low isolated rocks, on which the sea always breaks. This point is often confounded with that of Islay, but this mistake cannot be made when the preceding remarks are consulted.

Nonato gorge is seen at the distance of 1 mile, in the northern part of the point. Off it is a very small cove, without convenient anchorage or shelter. *Nonato gorge.*

Vessels can, however, moor at the head of this cove without feeling any swell, and as there is sufficient depth, urgent repairs can be made. For this purpose the anchor is dropped in 27 fathoms, 400 yards from the entrance, and the vessel hauled in. There are no resources, but an abundance of fish in this and the following cove.

The coast for 2 miles to the northward continues high and stony, when a slight indentation forms Guata cove; it gives no shelter and can only serve as anchorage for small coasters. At the head of the gorge is a well of brackish water. *Guata cove.*

Arauta cove is 3 miles NW. from Guata cove; it is the best of the three just mentioned. To the southward of it is a small white islet which can be easily recognized. It was intended to establish a port here instead of at Islay, but it offers no important advantages and has serious inconveniencies, such as, a heavy swell, no shelter for large vessels, and the steepness of the mountains, which rise almost perpendicularly out of the sea. The bottom is of stone, and the depth from 19 to 21 fathoms, 400 or 500 yards from the shore. This port is sometimes used when the inundations of the river Quilca interrupt the communication between Quilca and Arequipa. *Arauta cove.*

From Arauta cove the coast, rocky and steep, trends N. 50° W. for 7 miles, when the cliffs are cut by the beautiful and fertile valley of Quilca, on the plain in the gorge which is inclosed by steep and lofty mountains running from the shore to the interior. The cultivated spots in the center of the valley are seen between the mountains and offer a splendid view from seaward. The valley contains a river with an abundance of water. To the southward is a steep and clean islet. *Gorge and valley of Quilca.*

Quilca cove is seen to the northward when near the opening of the gorge. Its entrance is narrow, but there is a very *Quilca cove.*

21 c

good anchorage for small coasters and a fine landing-place. The depth is from 6 to 9 fathoms, the best anchorage for large vessels being to the southward, between the cove and the gorge, when the eye is the apex of a right-angle whose sides run toward the church-door and the island to the southward. The English steamship company has generally at the anchorage a small red buoy to indicate the best position.

Lartigue gives the following anchorage: church N. 5° E., and the islets S. 70° W.; bottom fine gray sand. It must be remembered that the water deepens rapidly outside of the plateau. Vessels must anchor with the head to the SSW., with a good stern-line, as the current often sets strongly to the SE. The bottom varies greatly, being either of rock, mud, sand, or gravel. During freshets, the water of the river is very muddy, the current is strong, and small mud-bars are formed on the rocky bottom.

Tides.   It is high water, full and change, at the Quilca river at 8ʰ 00ᵐ; rise, 6 feet.

Description.   Quilca was the principal port of Arequipa during its colonization, but it was abandoned on account of the heavy swell which often sets into the cove, the distance and little shelter of the anchorage, and finally the greater advantages of Islay.

The town is very small and has no resources, excepting some fresh provisions.

Landing off the valley is dangerous, and great difficulty is experienced in taking in water from the river. Quilca is visited principally by small vessels trading in oil and provisions. Boats must run in to the head of the cove and beach, but this is often difficult and dangerous on account of the surf.

Landing is also possible in the small cove of Miélo, 1 mile NW. of that of Quilca, but its entrance is full of rocks.

Directions.   Point Cornejo, the landmark for Quilca, can be recognized by its reddish color; the difference in height of the coast to the north and south also serves as a guide. About 2 miles NW. of the point, the mountains forming the valley will begin to show, and afterward the city itself.

The coast to the NW. of Quilca is abrupt, with some few sand-beaches.

Point Pano is 9½ miles WNW. of Quilca cove; all the Point Pano.
intervening coast is bold; soundings commence from the
N. part of the point toward the valley of Camaná.

The coast trends W. by N., 6 miles from point Pano, to Fuerte mount-
ain.
mount Fuerte, which is close to the sea, and has the appearance of a curtain of a fort; it is a very remarkable point,
and can be easily distinguished when to the southward of
Camaná valley.

The large and fertile valley of Camaná opens to the NW. Camaná valley.
of mount Fuerte; it is from 2 to 3 miles wide near the sea,
and its vegetation, interspersed with some white houses,
can be seen from a long distance. There is good depth
for anchoring throughout, but when it was intended to
establish it as a port, vessels were accustomed to anchor in
from 7 to 11 fathoms, muddy bottom, from 1⅛ to 1½ miles
from the shore on which the town is situated, nearly due
south of Fuerte mountain. It is necessary to anchor as
indicated, as the sea breaks a long distance from the beach.
A heavy swell, no shelter, and a strong current from the
southward cause constant anxiety and risk. It is never
prudent to attempt landing in ships' boats.

Camaná has a large commerce in olives and olive oil with
the interior and the ports along the coast. The other products are consumed by the inhabitants. The river Camaná
has a good quantity of water throughout the year.

From this valley the barren coast still continues to trend Coast.
to the west and north, with beaches of sand backed by high
mountains, for 17 miles, when steep cliffs 600 feet in height
commence. These are marked by white patches, and close
to them lie some rocky islets.

The deep gorge and valley of Ocoña opens about 2 miles Valley and
gorge of Ocoña.
from the NW. extremity of the cliffs; traversing the valley
is a river of good water. There is no safe anchorage or
good landing-place. The Clorinde anchored in 19 fathoms,
sandy bottom, 2 miles SSW. from the middle of the valley.
The gorge is formed of high and barren mountains to the
N. and S. The principal cultivation is the olive and the
vine, which afford some commerce.

Pescadores point is 11 miles WNW. of Ocoña gorge; it Pescadores
point.
is formed of high hills and steep cliffs, of a dark color, which
descend gradually to its extremity. On its surface are

some patches of guano; there are some rocks close to it. About 1⅓ miles south of the point is a sunken rock on which the sea does not usually break at high water and in calm weather. To the northward of the point is a sheltered cove, convenient and clean, but as there is no inhabited place in its vicinity it is without importance. The bottom is of rock, with a depth of from 5 to 11 fathoms near the shore.

Gorge and valley of Atico. The chain of barren and high mountains which follows the coast continues WNW. without interruption for 21 miles, where it is cut by the gorge of Atico. The valley is cultivated; there is a sufficient quantity of fresh water, and a settlement at the entrance. There is neither anchorage nor safe landing; the sea is rough, and the breakers commence a long distance from the shore.

Point Atico. From the gorge the coast runs nearly east and west for 4 miles, and is terminated by point Atico or Blanca; its entire extent is covered with peaks and hillocks which from a distance resemble islands, but they form a peninsula connected with the coast by a low, sandy rock. Their surface is covered with white and yellow patches of guano, which is taken in small vessels for the agriculture of the neighboring valley.

Roadstead of Atico. In the northern part of the point is the excellent roadstead of Atico; it is sheltered and clean, with from 9 to 11 fathoms near the shore; boats can be beached easily. There are a few houses, and a small traffic has been lately commenced. During the war of independence it was used for landing and embarking troops designated for operations in the provinces to the southward. Caraveli is the most important town of the vicinity. The south coast must not be closely approached after doubling the point, as some rocks above water are detached from it, but at short distances.

Loboso point. Loboso point is 8 miles NW. of the roadstead of Atico. It is low, stony, and projects but little; its western extremity consist of black rocks. It is covered with guano.

Saguas cove. The general direction of the coast for 21 miles is NW. to the small cove of Saguas, which has a depth of from 8 to 14 fathoms 600 yards from the land. It is not sheltered from the heavy SW. swell, which often renders the beach inaccessible. A small gorge or dry bed of a torrent is seen in the ravine.

Port of Chala. From Saguas cove the coast inclines more to the westward for 12¾ miles, to the new port of Chala, which is a small cove with some rocks to windward of it. It is subject to a continual swell, with a heavy surf on the shore. From October to March calms are frequent and of some duration. The bottom is rocky, with from 14 to 20 fathoms of water 1¼ miles from the landing. There is no shelter against the heavy SW. swell, and the holding-ground is bad. There are no resources, and fresh water is so far removed that it is expensive.

Through this port there is a traffic with some of the provinces of the departments of Ayacucho, Cuzco, and Arequipa. It is the nearest port to the city of Cuzco.

In the vicinity are some good mines of copper; some have been worked. Metal is shipped to foreign countries without duty. According to the statements of competent persons these veins are good and abundant, but they have been almost abandoned on account of the want of capital and workmen.

About 8 miles WNW. of port Chala is the high and rocky point of Chala, projecting a little to the westward. It is formed by one of the hills which run out from Chala head, and is terminated by a conical hillock. Chala point.

Chala head, remarkable for its height, 3,740 feet, is near the beach, is of a light color and convex form. Seen from the southward, it resembles large steps descending to the sea. It is visible from a great distance, and its position in regard to the remainder of the chain makes it appear isolated. During the rainy season its slopes are covered with vegetation. At its foot is the valley of Chala. Chala head.

From point Chala the coast runs to the northward, with less inclination to the W., for 7 miles, to Tanaca cove, in which the depth is 7 to 8 fathoms 700 to 800 yards from the land. This cove offers no shelter, and the sea always breaks on the beach. Fresh squalls are often experienced. Tanaca cove.

Atiquipa gorge opens 3½ miles WNW. of Tanaca cove. It contains the river of the same name, and is cultivated to the sea. The shore in front of it is low and sandy, and offers neither shelter nor landing, it being constantly beaten by the surf. The inhabitants live on their own produce and the trade with the interior. The climate is considered very healthy, and its pastures, which are celebrated, are favored with springs of excellent water and abound with cattle. Atiquipa gorge.

Ocopa cove.   From Atiquipa the coast trends WNW. for 2 miles, to Ocopa cove, which has a rocky bottom, with from 7 to 9 fathoms ½ mile from the shore, where there is some cultivation. The heavy swell renders the approach of the land dangerous, and there is no shelter for vessels.

Valley and gorge of Lomas or Chaviña.   Lomas gorge, which opens 6 miles WNW. of Atiquipa, is broad, and the plain, which is inclosed by high mountains, is covered with vegetation. A rivulet runs through this valley. There is no anchorage or safe landing. The inhabitants live by agriculture; there are some few exportations from the neighboring provinces. This gorge and the preceding can be seen from a long distance.

Point Paquija.   Point Paquija is 2 miles WNW. of Lomas gorge. It is high, of rocks, and projects but little. A reef, over which the sea breaks, runs out ⅔ mile to the westward of its extremity.

Point Lomas.   The coast trends WNW. 10 miles to point Lomas; a low chain of hillocks forms the intermediate coast. Point Lomas is low at its junction with the land, and high and rocky to seaward; it resembles an island from a distance, and is surrounded by scattered rocks above water.

Port Lomas.   On the N. side of point Lomas is Lomas road, the port of Acarí; it affords a safe anchorage in its eastern indentation, 500 to 600 yards from the land, in 7 to 11 fathoms of water, sandy bottom; the SW. swell comes in, and the cove is subject to fresh gusts, rendering it necessary to enter under short sail and to anchor with a good scope of chain. There are some huts and store-houses on the shore. In the latter are stowed the produce for exportation, consisting of cotton, sugar, chancaca, rum, and aguardiente. In certain seasons of the year vessels come here to hunt the otter. It is a watering-place for the families of the town of Acarí, 27 miles distant. Families bring all their provisions, including water, with them; the water of the wells being so brackish that it can hardly be used. The important hacienda of Chocavento is 26 miles from the port, and gives rise to most of the traffic.

The Andes are seen along all this part of the coast, in a continuous chain from 3,000 to 5,000 feet high.

Point Lobos or Sombrero.   N. 56° W. of port Lomas is the small point of Sombrero, composed of high mountains running into the sea; some high rocks are off it and the coast to the northward. They

Views off the Coast of Perú between the Lat.ᵈ 16° and 15° S.

are all near the land, and the sea breaks constantly over them.

Direction bluff, the termination of a range of table-land, is parallel and close to the shore, 8 miles WNW. from Sombrero point; all its upper part is a plateau which is most prominent to the northward; inclined cliffs descend from the edge of this plateau to the sea. Off this coast are some dark-colored islets and rocks about ½ mile from the shore, extending as far as San Juan point. One of them is 1 mile from the land, and lies S. 34° E., 2 miles from San Juan point. *Direction bluff.*

Point San Juan, 7 miles NW. by W. from Direction bluff, is very projecting, and is surmounted by two low hillocks with broad bases; the outer one is the largest. When to the north and south of them, and near the land, they appear like islands, as the land which connects them with the continent is low. *Point San Juan.*

The point is surrounded by rocks, some of which are nearly awash, but the sea does not break over them; the most distant are above water; they bear WSW. from the point, and extend about 1 mile to seaward.

Point San Juan shelters the southern portion of the excellent and large port of San Juan. The anchorage is good in 8 to 14 fathoms, muddy bottom, near the SE. shore, where boats can land with ease. Vessels can be hove down here, but everything would have to be brought, even wood and water, as none is found here, and the place is uninhabited. *Port of San Juan.*

In order to enter the port of San Juan the reef should be passed at least 200 yards to the westward, and the land should not be approached closer until well inside of point San Juan, to the northward of which, at a short distance, is a sunken rock. Haul close to the wind, or tack, if necessary, to reach the anchorage. The north shore can be approached closely, as it is steep to. This shore is formed of irregular broken cliffs, with a sandy plain at the head of the bay. The port can be recognized by the Morro de Acarí. *Directions.*

The Morro de Acarí is N. 18° E. of point San Juan, in the middle of the shore of the cove. It is very remarkable and can be readily recognized. There are several low mountains, to the northward of which is this isolated mountain; it is 1,650 feet high, and terminates in a sharp point. It has the form of a stone used for filtering water, with the opening *Morro de Acarí.*

underneath. A less elevated peak detaches itself to the northward, in which direction are high mountains in the interior. It is a splendid landmark for the port of San Juan or the bay of San Nicolas.

Point San Nicolas. Point San Nicolas separates the port of San Juan from San Nicolas bay, and is N. 41° W. 8 miles from point San Juan; close to its western extremity is a small island; white and yellow spots, caused by thin coverings of guano, will be seen on the black rocks forming the point, which, with the island, is surrounded by low and dangerous rocks, the farthest to seaward being about ½ mile to the northward, which necessitates caution when doubling this point to enter the bay of San Nicolas.

Bay of San Nicolas. The point just described shelters the fine bay of San Nicolas from the southward. There is a well-sheltered anchorage near its south shore, in 7 to 12 fathoms, where the landing is good. As soon as the island off the point is cleared the coast can be hugged, as everything is clear inside. There is no permanent population, and the bay is only visited by vessels which come to load cotton, cochineal, and other produce of the farms in the vicinity. There are nearly always piles of cotton bales on the shore ready for exportation. There is no water or other resources either at the port or for a considerable distance. Landing is not so easy as in port San Juan, as there is no beach in the sheltered part.

Tides. It is high water, full and change, at San Nicolas at 5ʰ 30ᵐ; rise, 3 feet.

Point Beware. Point Beware is the northern limit of the bay. It is N. 49° W. from point San Nicolas. It is high, steep, of a dark color, and surrounded by rocks above water.

Changuillo gorge. Changuillo gorge is 11 miles NW. of point Beware. In it there are some cultivated spots, and also a brook of fresh water. There is no anchorage or shelter in front of it, and there is nearly always a heavy swell and a surf on the shore. The brig Hector, which took this for port Caballos, to which she was bound, anchored with two anchors and 80 fathoms on each chain, but the swell was so heavy that she was thrown on the rocks and totally lost.

On all this coast from Atico to the bay of Independencia there are 50 fathoms 2 miles from the land.

CHAPTER XIII.

FROM CAPE NAZCA TO THE BAY OF CASMA.

Variation from 10° 13' to 9° 45' easterly in 1876. Increasing annually about 1 .

Cape Nazca lies 4 miles NW. by W. from the gorge of **Cape Nazca.** Changuillo, and 1,020 feet above the level of the sea. At its foot are two hillocks of sand, one higher than the other, which terminate in a point. The high land of the promontory is of a dark color, which makes it appear as if a cap was thrown over it. The point can be closely approached, all the rocks off it being above water.

Doubling to the N. of Cape Nazca there are two small **Port Caballos or** points inside of a cove, which opens to the eastward; the **Nazca.** lookout station situated on a hillock will also be seen; after passing the inside point, and when the mast at the lookout bears S. 61° 30' W. distant 300 yards, the anchor can be dropped in 6 fathoms, bottom coarse sand. This is the most sheltered place and the best for shipping cotton; the best landing is on the neighboring beach. near the rocks farthest to the eastward, on the weather side.

It is desirable to anchor in port Caballos before 10 a. m., as the variable winds before that hour are favorable for entering; later the fresh southerly winds commence. If the land should be made later, the port should be approached under short sail, and it will be prudent to reef the topsails, as the gusts in the cove are so heavy that care is necessary even with these precautions; the fine breeze near point Nazca must not be trusted; it is quite different at the anchorage. The wind which springs up every day between 10 and 11 a. m. blows more or less strong until 8 or 9 p. m.; it then commences to die out, and is calm about midnight. These are the best hours for loading and discharging. During the conjunction and opposition of the moon the wind often lasts through the night, when all communication with the shore is difficult and dangerous. The lookout mast must not be relied upon as a landmark, as it is a light stick, and may be carried away by any squall; the position it occupies on a small hill, separating the beaches of sand and rock, makes it a good distinguishing mark.

Vessels should moor with a good scope on each chain, as the wind blows in gusts sufficiently heavy to try the best ground tackle.

The port is uninhabited, and is without resources. One hut and the ruins of an old store-house are the only buildings on the south shore. The bales of cotton are thrown on the shore and left without a permanent guard. This port is only frequented by vessels which export the cotton, cochineal, sweet wines, and other produce from the farms in the vicinity.

Ica river.　Ica river empties 8 miles NW. of point Nazca, forming near the coast a narrow and tortuous opening, which can only be recognized at a short distance. Some green patches extend down to the sea; it is inaccessible.

Olleros point.　The coast trends WNW. for 8 miles from the mouth of Ica river to Olleros point, which is low and sandy, with two small islets to the westward; some few vessels which come to load the produce of the farms in the valley of Ica have anchored off its northern part in 7 to 9 fathoms, but the anchorage offers no security, and the heavy swell and surf have caused it to be abandoned.

Table of Doña Maria.　The table of Doña Maria, 2,150 feet high, is more than 1 mile inland and 4½ miles from point Doña Maria. It is one of the most remarkable points on this coast, rising above a chain of mountains which trend NW. and SE. near the coast; it is of a conical form, with a flat summit.

Point Doña Maria.　Point Doña Maria is N. 79º W. from the mountain and N. 53º W. from point Nazca. It is high, of rock, and of a dark color, with white guano beds at its foot. Its S.'and W. sides are surrounded by dark, pointed rocks, some of the westernmost of which lie 1 mile from the land.

Los Infiernillos.　Los Infiernillos is the name given to all the rocks off point Doña Maria. Among these is a small sugar-loaf shaped islet, the farthest to seaward from the point, and 54 feet above the level of the sea. There is no danger to seaward of the Infiernillos; there are 53 fathoms 2 miles outside of them. The steamer Santiago, of the English company, ran bows on these rocks, and was only saved by her water-tight compartments and braces.

Point Azua.　The coast trends N. for several miles, and then, turning to the NW., forms point Azua, 10 miles from point Santa Ma.

XIII

Views off the Coast of Perú between the Lat.s 15° and 14° S.

ria. There is a steep hill at its end, and off it are several rocks. On the high mountains in the interior, which extend toward the sea, is a light belt, probably of guano, from point Santa Maria to a little to the northward of Azua.

The height of Morro Quemado, 2,070 feet, and its prox- *Morro Que- mado.* imity to the beach, render it very remarkable. After form- ing a small point which projects to the northward, it pre- sents an apparently plain black surface which is in reality inclined; it is surmounted by a comb, which is its greatest elevation. It is the termination of the high, light-colored shore to the southward.

The low, level, whitish islands of Santa Rosa are to the *Islands of Santa Rosa.* northward of Morro Quemado. There are some rocks and small islets near the shores of these islands, as also that of the main land. Should it be desirable to pass between them and the continent the south coast should be kept closer aboard than the northern one.

The Serrate channel is ¾ mile wide, and separates the *Serrate chan- nel.* islands of Santa Rosa from Morro Quemado. This passage is clear and safe, having from 14 to 20 fathoms of water. It leads to Independencia bay.

Las Viejas island is ⅔ mile NW. of the northern extremity *Las Viejas isl- and.* of the northern Santa Rosa island, and is 3½ miles long SE. and NW. The summit of this high island is on its south- ern part, from which it gradually descends toward the N. Between it and the Santa Rosa islands is a channel ⅔ mile wide, but so full of reefs as to leave no clear passage. The northern part of the island is clean, and can be approached to a short distance. A very remarkable conical hillock, separated from the high land, will be seen on the southern part.

There is an excellent anchorage on the NE. coast in a small cove, protected from the northward by a small point. The depth is from 5 to 7 fathoms, 600 yards from the shore. Large quantities of guano are found on the hill off this an- chorage; the beds are increased by the birds and marine animals which have abandoned the Chincha islands. It is not inhabited.

The islands Viejas and Santa Rosa, and the heights of *Independencia bay.* the Morro Quemado, form the SSE. and E. boundaries of the large bay Independencia, which extends NW. and SE.

12½ miles; with a width NE. and SW. of 2½ to 4 miles. The bottom is of rock everywhere excepting in its southern part, where it is of coarse sand; the depth is from 8 to 20 fathoms. The best anchorages for protection against the strong winds are in the bight in its southern part, and on the NE. shore of Las Viejas.

Until the year 1825 this bay was unknown; it was discovered by two vessels, the Dardo and the Trujillana, with troops for Pisco, entering it by mistake; the vessels were wrecked and many perished. This mistake and error in the reckoning was very probably due to a southerly current setting through the Boqueron de San Gallan.

The bay has two entrances, one by the Serrate channel, and the other, called Trujillana, to the northward of the island las Viejas; when coming from the southward the first channel is the shortest, but it is also the narrowest; the Trujillana is 4½ miles wide; both are equally clean. In entering the bay the same precaution should be taken as on entering the bay of Caballos, as it is equally subject to violent gusts. The best and widest passage is close along the north coast of the island Viejas. There are generally a few huts of fishermen, who come here temporarily, on the east side, at a village called Tungo; they bring all their provisions, and even water, from Ica, the capital of the province The bay owes its name to its being the first anchorage of the transports carrying the united army, which, under command of general San Martin, proclaimed the independence of Perú.

Tides.   It is high water, full and change, in Independencia bay at 9ʰ 50ᵐ; rise, 4½ feet.

Mountains of Carrasco.   The mountains of Carrasco, 3,000 feet high, rise toward the eastern part of the bay.

Carretas mountains.   The Carretas mountains, forming the northern boundary of Independencia bay, are 1,410 feet high; they present the appearance of a rocky promontory which descends to the southward in steps, and terminates in a point with an island a short distance off it. The entrance, called la Trujillana, opens between this point and the island Viejas. When to the eastward of this group of mountains, a dark-colored detached hill, very abrupt, is seen, which increases in height to the northward; at its highest point it has a rectangular-shaped cut; on it there are bright patches of guano. The junction of these mountains forms a peninsula which con-nects with the continent to the northward.

The coast continues high, abrupt, and of a dark color for 6 miles in a northerly direction. Mount Wilson, 1,420 feet high, lies close to the coast, 4½ miles north of Carretas head; here the coast commences to fall, forming a spacious cove, which terminates 10 miles north of mount Wilson. Mount Wilson.

Zárate islet is in the middle of this coast, about 1 mile from the shore; its upper part is almost level, and its sides nearly perpendicular; its color is a dark yellow with black patches; on it there are some beds of guano. Zárate islet.

Salinillo cove is to the northward of a small point on the same parallel, and to the eastward of Zárate island; it is without shelter, has a depth of from 7 to 11 fathoms ⅓ of a mile from the land, and bottom of rock and coarse sand. The best anchorage is to the eastward of the small islands which lie to the northward of Zárate. Salinillo cove.

The beach is continually beaten by the surf, and it is not prudent to use ships' boats for landing. It is the harbor of export for the salt mines of Pisco, about 2 miles in the in-terior. The salt is brought alongside in lighters. There is no permanent population, the people who ship the salt com-ing with it.

From Salinillo cove the coast runs north for 4 miles, then west, and is terminated by point Huacas, which is nearly black, high and abrupt; its upper outline is an obtuse angle whose shorter side forms the western limit of the coast de-scribed; the other side is prolonged, and descends gradually to the SE. Point Huacas.

Lechuza mountain, a short distance in the interior, is 1,580 feet high; it is of a light color, with a pointed summit. Lechuza mount-ain.

From point Huacas to Paracas point the coast trends to the north a little easterly. All this part is high, formed by high mountains of a dark color. A vessel can approach it as near as desirable. There are a few rocks above water close to this point, and to the eastward of it is a cove with an anchorage in 6 to 11 fathoms. Paracas point.

From this cove the coast continues high and trends to the eastward, terminating in point Ripio. From thence it trends to the southward for 3½ miles. Point Ripio.

On the inclined plain which forms the hills between the two last-described points are three enormous crosses, perfectly executed ; their origin is unknown. They are formed of a large wall of white stone, following the inclination of the mountain from near the coast to its summit. When opposite to it it has the appearance of a chandelier with three branches and a foot. They stand out uniformly and perfectly against the dingy tinge of the mountains.

The peninsula of Paracas is the high land and mountains described, comprised between the cove east of point Huacas and the western coast of Paracas cove. It is united with the continent to the SE. by a sand plain 4 miles wide.

San Gallan island, 1,365 feet high, is 2½ miles from the peninsula of Paracas; it extends NW. and SE. 2½ miles. It is high, barren, and of a light color. Its profile is a convex curve whose extremities and center are most prominent and the most convex. This part is generally covered by a horizontal belt of fog, which is dispersed as soon as the wind called the *Paraca*, of which it is considered the precursor, sets in.

The island is surrounded by some small islets, some of ·which are a short distance from its northern shore, one of them resembling a ten-pin. Guano is found on different points of the island.

San Gallan is nearly always the land-fall of vessels bound for Callao, after doubling cape Horn, coming from Australia or the coast of Chile.

Piñeiro rock lies 1½ miles S. 4° E. from the S. extreme of San Gallan island ; it is just awash, and with a smooth sea can be seen, but when blowing hard, with a weather tide, the confused cross-sea fills the channel with foam and renders it difficult to distinguish the rock. It is much in the way of vessels entering through the Boqueron de San Gallan.

The Boqueron de San Gallan separates the island of San Gallan from the peninsula of Paracas. The depth varies from 20 to 30 fathoms near the shore. Vessels coming from the S. and wishing to pass through this channel must keep close to point Huacas, keeping nearer to this coast than to that of San Gallan until Piñeiro rock is passed, when the middle of the channel can be taken or the island approached in order to avoid the calms caused by the high lands to the southward.

Lieut. A. Miller, U. S. N., remarks that, from the experience on board the U. S. S. Richmond and Omaha, a constant southerly current seems to set through the Boqueron de San Gallan, and that this is also the experience of several of the captains of the P. S. N. Co.'s steamers. *Current.*

The name of Tres Marias islets is given to three islets in the northern part of the Boqueron. The southern one is about 3 miles from point Paracas; their general direction is nearly N. and S. *Tres Marias islets.*

The island of Ballesta, 1¼ miles to the northward of the Tres Marias, is of moderate height, and pierced at its southern extremity, forming a natural bridge. Two islets of the same height lie near it. All are covered with a thick bed of guano, which is now being taken away. There is an anchorage on their NE. side, in 18 to 20 fathoms, 300 yards distant. Landing is generally difficult. *Island and islets of Ballesta.*

Salcedo rock, 3 feet under water at low tide, is 1 mile S. ¼ E. from the SE. extremity of the southern Ballesta island. Near its base the summit of the northern Ballesta is seen between the other islands, whose tops are a little more indented. The sea seldom breaks over this rock. *Salcedo rock.*

At 4½ miles to the eastward of Ballesta is Blanca island, also called Novillo, of moderate height, and of a whitish color, from the beds of guano. Near it is a small islet of the same height. All these islands are clear, excepting Salcedo rock. The depth is not less than 13 fathoms near them, and 800 yards to leeward of Ballesta there are from 25 to 30 fathoms. *Blanca island.*

The spacious and excellent bay of Paracas is to the SE. of point Ripio. The bight runs 4 miles S., and is 2½ miles wide, with from 6 to 11 fathoms at the entrance, diminishing inside; the bottom is mud. It is one of the most convenient anchorages of this part of the coast. There are some huts inhabited by fishermen on the east shore. The best anchorage is off these, ⅔ mile from the beach, in 4¼ to 5 fathoms; near the shore the water is shoal. In the vicinity are some wells of fresh water, which can be taken aboard in casks or by small crafts having a long hose and pump. There are no resources, and the inhabitants live by fishing *Paracas bay.*

and watering vessels. There is no tide, heavy swell, or surf in this bay.

The low and sandy shores have some wooded patches. The army under General San Martin effected its first landing at this place.

Vessels wishing ballast can procure it at point Ripio, as they anchor close to, and take it aboard in their boats.

San Andres. The small fishing village of San Andres is 4½ miles to the northward of Paracas; it is visited as a watering-place during the summer.

Port of Pisco. All of this bay embraced between the island San Gallan, the peninsula of Paracas, and the coast running from it to the N. and E., constitute the bay of Pisco; but the port of Pisco is 5½ miles N. 76° E. from Blanca island. It affords no shelter whatever against the almost continual SW. swell. The best anchorage and most convenient place for loading and discharging is to the SW. of the wharf, 400 yards from its extremity, in 4 fathoms. Vessels drawing more than 19 feet must anchor a little farther out, with the church bearing N. 87° E.

From 11 o'clock a. m. until sunset there is a regular and fresh wind, known as the *Paracas*, as it comes from the bay of that name; then all work has to be suspended, as the boats cannot pass to and from the shore. Vessels should not anchor with less than 45 fathoms of chain.

The space between the commencement of the breakers and the shore is called the *tasca;* in this port, as also in the others on the coast of Perú, with a heavy surf it is more than 500 yards.

Pisco has a fine pier, built of iron, with a wooden platform running out 733 yards into 3¾ fathoms of water.

Boats can land alongside of this pier in all weather, as the sea never breaks at the steps. In the afternoon, however, landing is difficult for small boats on account of the heavy swell which sets through the piles. On either side of the mole is the custom-house and the office of the captain of the port.

Light. On the end of the mole is a harbor-light 46 feet above the level of the sea. It is *fixed, red,* visible 3 miles, and serves to indicate the landing and the anchorage.

Vessels coming from the southward bound to the bay of Pisco and the Chincha islands should make the land in the vicinity of cape and mount Carretas, keeping account of the westerly current which sets off the coast at the rate of about 15 miles a day; then steer for the Boqueron de San Gallan, keeping a lookout for the Piñeiro rock, in the vicinity of which there is an irregular sea and eddies. Point Paracas should not be approached too close, as some shoals have been reported to the northward of it. It should be kept distant 1 short mile, and steer for Blanca island; pass close to southward of it, and steer for the church.

The coast can be easily recognized from seaward by the island of San Gallan, with the peninsula of Paracas back of it; the latter also appears as an island from the difference in height compared with the land near it, which cannot be seen at so great a distance. When closer, the Chincha and Ballesta islands come in sight, after which any of the channels, excepting that between the Ballestas and Tres Marias, can be taken.

When coming from the north there is no danger to be avoided. After doubling the Chincha islands a course can be steered for the anchorage, always remembering that the depth rapidly decreases toward Blanca island, but there is no danger.

Loading and discharging the produce of the country is done by means of boats, which are hauled on the beaches off the store-houses on the south side; they are held by a hawser, one end of which is anchored and the other fast on shore.

The town of Pisco is on an elevation $\frac{2}{3}$ mile from the sea. The principal church, with two white towers, can be easily distinguished, and is an excellent mark for the port. Provisions and fruits can be had in abundance. Water can be taken from an iron pipe which is on one side of the dock. On the beach there are a few warehouses and stores.

Pisco is the principal port of the coast province of Ica, whose capital is 42 miles distant. A railroad has been built between the two places. The telegraph communication extends to Lima.

Large exportations are made from this port, consisting of spirits of wine in jars of baked earth called *botijas* and

22 c

piscos, sugar of different qualities, wine, cotton, beans, dates, and other produce.

The department of Ayacucho and part of that of Huancavelica export from Pisco the wool of the sheep and vicuña, precious metals, and other articles of industry.

The reputation of the wine of Pisco is gradually increasing in the United States and in Europe. The modern processes of making wine are used with great success in the large establishments of Pisco and Ica. This branch of industry has increased from year to year, and it will undoubtedly become the largest of this province. The most commonly known wines are the Falconi, Ledos, Latorre, Elias, &c. The production of sugar and cotton increases yearly.

This port is the principal one for the importation of foreign merchandise into the province Ica, Ayacucho, and Huancavelica. A railroad is being built from Lima to connect with that of Ica at Pisco, which will cross the rich valleys of Cañete and Chincha.

Pisco river.   About 2½ miles N. of the store-houses at Pisco a watercourse or torrent, called Pisco river, empties. It has an abundance of water in the summer, but is completely dry in the winter.

Heights of Caucato.   The heights of Caucato rise from the beach to the northward of the river Pisco; they are the only ones near the sea in this latitude.

Port of Caucato.   The name of port Caucato is given to a small bight in the coast to the northward of the heights. It is generally used for shipping the produce from the farms of Caucato, and for landing the machinery and guano used on them. The anchorage is without shelter; a heavy swell sets in, and the surf hinders boats from landing. The loading and discharging is generally done by lighters from Pisco; there are no inhabitants. The depth, 1 mile from the land, is 4 to 5 fathoms.

Valleys of Condor and Chincha.   The fine valleys of Condor and Chincha commence beyond the Caucato heights. Their numerous fine farms can be seen from the ocean. The principal produce is sugar-cane, cotton, the vine, vegetables of all descriptions, and fruits.

Chincha river.   From the heights of Caucato the coast continues to the north, low, with a sandy beach, for 9 miles, to Chincha river, which is generally well supplied with water from January

XIV

Views off the Coast of Perú between the Lat.s 14° and 12 ½° S

The Chincha Islands seen from the NW. at a distance of 6 miles.

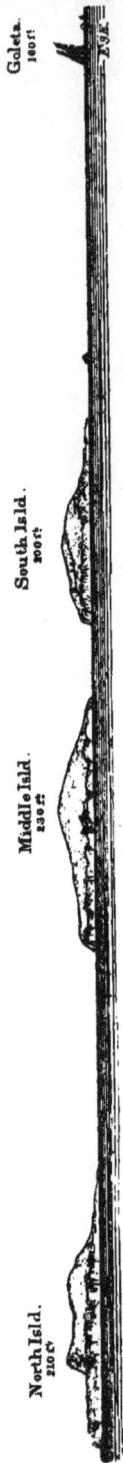

North Isld.
210 ft

Middle Isld.
230 ft

South Isld.
200 ft

Galeta.
180 ft

I.S.E.

R.Caleta
N.½W.

Green Land
NW¾N.

to May, but is nearly dry at its mouth during the remainder of the year.

Port Tambo de Mora, a new port, is immediately to the northward of the mouth of the river Chincha. It is used for shipping wine, cotton, sugar, vegetables, and other produce from the valley of Chincha. *Port Tambo de Mora.*

The best landmark for this port is the limit of the valley or low beach. It is 1 mile south of the cliffs; its depth is $3\frac{1}{4}$ to 5 fathoms, over muddy bottom, $\frac{1}{2}$ mile from the land. There is no shelter against the SW. swell. It is not prudent to use ships' boats without knowing the *tasca* or the state of the bar. There are always some small vessels at the anchorage taking in provisions and fruits for Callao. The houses of the settlement on the beach will be seen; there is a telegraph station connected with the capital.

The valley of Chincha abounds in all kinds of fowl, fruits, and fresh provisions, which can be had at low prices, but water is expensive and hard to take on board.

Many accidents have taken place in this port on account of the imprudent use of ships' boats.

The Chincha group of islands takes its name from the valley on whose parallel it is. It is composed of three principal islands, named North, Middle, and South, according to their position. *Chincha islands.*

These islands were uninhabited and without importance, when the national government comprehended the great value of the guano they contained and commenced to take advantage of it. Since then the guano beds have been the principal source of wealth of the state, and the guano has been sold for the benefit of the government. The great superiority of the guano of the Chinchas is due to the total absence of rain, the ammonia, one of its principal elements, being thus preserved in all its strength.

The beds of the North and Middle islands are exhausted, and only some thousands of tons of guano were left on the South island in 1870. In 1872 the shipment of guano to Europe was stopped.

The guano used in the country is only taken away by Peruvian vessels, speculators having no other remuneration than the cost of the freight. The vessels must get the necessary permission at Pisco and take the officer, who accompanies the cargo to its destination, on board.

It is also prohibited to foreign vessels to go directly to these islands. They must get permission and the order to load at Callao. After having loaded, they must return to this port to get their papers for the port of destination.

The vessels are loaded in their turns as they arrive by means of hose which lead from the cliffs to the holds of the vessels. The vessels whose first turn it is to load must furnish water to the island, as they have none.

The islands are commanded by a special governor appointed by the general government. There is also a captain of the port and some lower officials. Most of the guano was gotten out by Chinamen.

Directions. In going to the Chinchas from Callao it is recommended to stand off the land during the night and near it during the day, until to the southward of the 13th degree of latitude, and then 4 to 5 miles from the coast to Pisco. In the autumn Captain Harvey, R. N., recommends running 26 hours to seaward and 22 toward the land; at the end of 48 hours the vessel will be to windward of San Gallan.

The currents around the Chinchas are very uncertain, and often set to the northward $1\frac{1}{2}$ miles an hour.

North island. North island is in latitude 13° 38′ 12″ S., and longitude 76° 22′ 55″ W., and 11 miles N. 64° W. from the port of Pisco. The island is formed of rocks, whose surfaces were entirely covered by guano. The beds were in some places 100 feet thick. The island is 108 feet high, about 1,592 yards long, and from 700 to 800 yards wide.

The principal anchorage is in $11\frac{1}{2}$ to 26 fathoms near the land; rocky bottom. It is necessary to approach as near as possible, so as not to be in too great a depth. Men-of-war had best anchor on the east side, in order to escape the immense quantity of dust which is blown to leeward by the *Paraca* during working-hours. As before stated, this wind is very fresh, and blows every day from ten or eleven a. m. to sunset, and sometimes, though rarely, until nine or ten o'clock in the evening.

Some sunken rocks extend 300 yards from the two points which form the extremities of the principal cove on the N. coast. The NW. rock, said to be covered by 6 feet of water, was searched for by Commander Marq-Saint-Hilaire, of the French navy, but not found, even with the aid of the pilots;

the SE. rock is covered by 7 feet of water. Near the first
is a shoal in 2¾ fathoms, which was found by the Commander St. Hilaire. These have been marked by red buoys;
the danger will be cleared by keeping the east point of
Middle island open of the SE. point of North island until
the NW. rock commences to open from the latter island.

The relative bearing of these two shoals is S. 50° E. and
N. 50° W., 284 yards apart.

The buildings on the island are the dwellings of the
authorities.

Numerous coasters from Pisco and Tambo de Mora trade
here in fruits and fresh provisions, which are sold at moderate prices; but water is expensive, as it has to be brought
from Paracas.

Middle island is separated from North island by a channel Middle island.
of an average breadth of ½ mile. There is a good anchorage
in the entire channel, the holding-ground being the best in
the vicinity. The bottom is of rock, and the depth near
the shores from 9 to 14 fathoms. Vessels can enter or leave
by either end of the channel. The rocks are visible and
near the shores.

In 18 to 19 fathoms there is a bottom of white sand and
shells. A vessel must always anchor either a little east or
west of the hose, to avoid the dust brought by the *Paraca*.

South island is separated from Middle island by a chan- South island.
nel ¼ of a mile broad. This channel is a bad anchorage,
and has in it some scattered rocks. In its center it is shoal,
and cannot, therefore, be used by large vessels. Attention
must be paid to a reef of rocks which makes out from the
E. side of this island 400 yards; the sea does not always
break over it. The reef consists of two parts, having a passage between them.

Lieutenant Janet, French navy, discovered another reef,
with 2 fathoms of water over it, 1,120 yards S. 7° W. from
the east point of the middle Chincha, and 164 yards N. 40°
E. from the east breaker of South island.

The best anchorage is at the E. entrance of the channel.
The western end is contracted, as its center is obstructed
by rocks, leaving room for only a small number of vessels.
This island is at present the only one containing guano; it
is loaded by means of lighters, which receive their guano

from the wharf or hose, at different points. There are some few small, high islets to the westward. The SW. part is so full of rocks that it cannot be approached.

Goleta islet. Goleta islet belongs to this group, and is 1,526 yards SW. of South island. It owes its name to its resemblance to a schooner under sail. It is 170 feet high, and surrounded by some scattered rocks on which the sea breaks. There is another small islet 872 yards to the SSE. of it.

Cañete valley. Passing Chincha valley, the barren coast runs north with a little inclination to the westward, with cliffs from 425 to 530 feet high, with elevated mountains in the interior. One of the best cultivated valleys of the coast of Perú, called Cañeta, is 24 miles from Chincha valley. The high mountains divide here, and show the plantations extending from the sea to the horizon in the interior; in the midst of the varied colors of the vegetation are houses and factories. The principal production is the sugar-cane, which yields the best sugar of Perú, as also large quantities of chancaca and rum. Of these there is a considerable export, especially to Chile, California, and Australia. The valley abounds in vegetables, fruits, fowl, and cattle.

Cañete river. Cañeta river empties near the southern limit of the valley. From January to May, it has plenty of water; during the remainder of the year there is but little at its mouth, it being used in irrigating the farms. The surf is too dangerous to allow landing opposite the valley.

Point Fraile. Close to the northern limit of the valley is a small isolated mountain of moderate height; its inland part forms an elevated hillock, while that to seaward is a steep, rocky hill, covered with a white deposit; its prolongation is point Fraile or Cerro Azul, which can be easily recognized by the blue color of the small hill.

Port of Cerro Azul. The coast to the northward of point Fraile forms a cove called the port of Cerro Azul. The anchorage is insecure; the bottom is of rock, and the SW. swell is always felt. Vessels can anchor on an E. and W. bearing, ¼ mile distant from the hill, in from 5 to 7 fathoms. Fitz Roy gives another anchorage farther out in 7 fathoms, ¾ mile NW. by W. of the hills.

Boats must land on the beach in front of the store-houses, to leeward of the point. There is a wooden dock which offers some facilities. It belongs to the proprietor of the factories of la Quebrada and Casa Blanca, and was constructed by him, as was also the railroad to the different farms and the facilities for the exportation of the produce.

The town at the head of the cove is small. Fresh provisions and fruits can be obtained there, or at the settlement of Cañete, 5 miles distant. The principal commerce consists in the exportation of sugar, chancaca, liquor, rum, alcohol, cotton, and other less important produce of the valley; there is telegraphic communication with the capital.

The town of Cañete, the capital of the province of the same name, and its commercial center, is 5 miles in the interior; it has an extensive traffic with the towns in the vicinity. Cañete.

Point Loberia is 5½ miles to the northward of Cerro Azul. It projects but little, and has three rocks above water off its extremity, which form a small cove to the northward of no importance. Point Loberia.

Point Malpaso de Asia is 9 miles to the northward of point Loberia; it is small, steep, and extends out from a sand-beach. Point Malpaso de Asia.

Asia island, toward the NW., about 1 mile from the coast, is in shape of a tent, and of a light color, due to a thin covering of guano. It is surrounded by small islets, the northernmost of which reach to the coast. Asia island.

Chocalla point is 3 miles to the northward of Asia island; it is of black rock with white spots of guano, and projects but little. To leeward of it is a cove, in which there is no convenient landing on account of the surf. Chocalla point.

This cove is limited to the northward by the height of Salzar, which is near the coast and 582 feet above the sea. Height of Salzar.

The river Mala, which is but a rivulet, after watering a small valley, discharges 1½ miles N. of the height of Salzar. Landing here is dangerous on account of the surf. River and valley of Mala.

The coast for 8 miles to the NW. is formed by low cliffs bordered by sand-beaches as far as the isolated hill Calavera, which is of black rock with white patches; its western side is nearly perpendicular, and terminates in a sharp point. Calavera head.

This coast runs nearly WNW. to a point of rocks which Point Chilca.

descends to the sea gradually, and is terminated by a small hillock, 298 feet high, named Chilca.

Port Chilca. From point Chilca the coast runs to the northward 2 miles, and forms the port of Chilca. An island, which is to the north and close to the coast, is free from dangers. The port is small, the best anchorage being between the islet and the east shore, in 5½ to 14 fathoms. Some rocks above water extend 200 yards from the east point, which is within the entrance of the port. The water is generally shoal to a distance of 140 to 160 yards.

The settlement of Chilca is 1 mile from the anchorage; it is small and supplies are scarce; water is brought from wells at some distance, and most of the inhabitants are employed in the salt mines, whose products are sold in the valleys of Lima. From January to May they cultivate land toward the interior, and raise sufficient to last them through the year.

Tides. It is high water, full and change, at Chilca at $5^h\ 30^m$; rise, 4 feet.

Lurin. The settlement of Lurin is 12 miles to the northward and ⅓ of a mile from the coast. There is anchoring ground in all this bight of the coast 1 mile from the land, in 7 to 10 fathoms. Vessels can anchor the same distance off the town in 5 fathoms, but there is no shelter, a heavy SW. swell, and a surf that often renders landing impossible. This settlement possesses a river with an abundance of water in the summer; during the remainder of the year but little flows to the sea, it being used for irrigating the farms. Vegetables, cotton, and sugar-cane are raised; fruits, cattle, and fowl are abundant. As the temperature of Lurin is excellent and healthy, it serves as one of the watering places of Lima.

Pachacamac island. Pachacamac island is ½ mile long, 200 yards wide, and 400 feet high; it is 1⅔ miles from the shore and in the center of Lurin cove. Its profile is regular, and it is covered with thin beds of guano. It lies NW. and SE.

Islets of San Francisco and Sauce. The two small islets, San Francisco and Sauce, are to the southward of Pachacamac island and close to it. They are rendered inaccessible by the heavy breakers. One of them resembles a rounded sugar-loaf.

Corcovado reef Corcovado reef is 1 mile south of these islets. It is about 1 mile long and formed of scattered rocks, over which the

Views off the Coast of Perú between the Lat.^{s} 12½° and 11¾° S.

Approach to Callao Roads from the Southward.

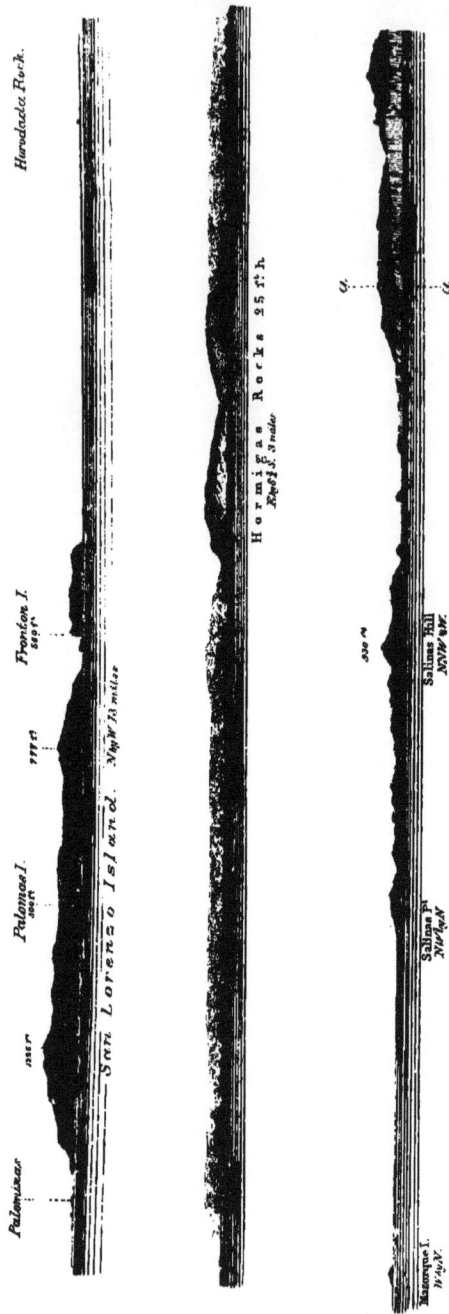

Harodada Rock.

Fronton I.

Palomoa I.

San Lorenzo Island.

Palomiaar

Hormigas Rocks 25 f.t

Salinas Hill
NNW¾W.

Salinas P.t
NW¾N.

sea breaks. It must be looked out for when going into Lurin.

About $\frac{1}{2}$ mile N. of Pachacamac are the two Viuda rocks; the sea breaks on them. The general direction of all these islets and rocks is NW. and SE. The channel between them and the coast is without danger, but subject to a heavy swell.

Viuda rocks.

From Lurin the coast continues to the northward, with a low, sandy beach, and high mountains in the interior. Landing is impossible for 10 miles, on account of the surf. This stretch of coast is called the beach of Conchan. Fish is abundant, and they are caught by the Chorillanos with hook thrown from the beach.

Beach of Conchan.

Conchan beach is terminated to the northward by point Solar, which is rocky and descends gradually to the sea. There is an islet a short distance from it, and in its vicinity are several rocks, on which the sea always breaks; it is covered with a light coat of guano.

Solar point.

Solar bay, limited by a hill of the same name, is to the northward of the point just described. The hill is a mass of blue rock, visible for a long distance; one of the spurs, running toward the sea, is 855 feet high, and has the appearance of an island when seen from the southward.

Solar bay and hill.

There is another bight, called Salto del Fraile, between the northern part of the hill and point Chorillos. The anchorage in both is 600 yards from the beach, in from 5 to 9 fathoms; but there is no shelter, and landing is often impossible. The point separating the two coves is called El Codo.

Chorrillos point is to the north of the cove of el Salto del Fraile; it projects from a mountain close to the north of Solar hill, called el Salto del Fraile. The surf continually beats on the rocks, which extend a short distance from it.

Chorrillos point

From Chorrillos point the coast forms the long and spacious cove of Chorrillos, completely surrounded by high, steep cliffs, which are very close to the edge of the sea. From the point these cliffs take a little turn to the east, and then trend north with a small inclination to the westward. Their profile is nearly horizontal, and they descend gradually farther north.

Chorrillos bay.

The best anchorage off Chorrillos is 600 yards from the

land, a little to leeward of the point as it commences to come in line with the extremity of the hill. The depth is 6 to 7 fathoms, bottom gravel and rock, in some places mixed with sand. The barracks situated on the heights will bear S. 74° E. Fitz Roy's advice is to keep point Codo open of point Chorrillos and anchor in 8 to 9 fathoms, as there is less swell than farther in. M. Le Clerc, from his soundings, gave the same anchorage as Captain Garcia. If several days should be passed in this place, it is best to moor head and stern, with the head to SSW., to stem the swell.

There is a wooden wharf in the cove to leeward of the point, which is the best landing-place. At the shore end of the wharf is a quay on which the bathing-houses are placed, outside of the wharf. An easy pair of stairs leads from this quay to the upper part of the ravine; half way up it divides into two branches, leading to different points of the town of Chorillos, and terminates in agreeable promenades on the cliffs. The town is on the plain at a little distance. It is frequented by the families of Lima from January to the end of April, but the baths are taken all the year round. It has fine buildings, and there are an abundance of supplies.

The town is connected with the capital by a telegraph and a railroad 9 miles long. It is lighted by gas.

Water was formerly obtained from the ravine of Agua Dulce, a little to the northward; at present it is conducted through the town by pipes.

Tides.   It is high water, full and change, at Chorrillos at 6ʰ; rise, 1¼ feet.

El Barranco.   The settlement and church of El Barranco is at the edge of the elevated plateau, 1½ miles north of Chorrillos, where a small gorge divides the coast; it is a watering-place with few resources; landing here is dangerous.

Miraflores.   The village of Miraflores, also a watering-place, is 2 miles farther to the northward. Its buildings can be plainly seen from Chorrillos. Beautiful gardens are cultivated around the dwellings in the town. As a country-place it is superior to Chorrillos, but inferior as a watering-place. There is no convenient landing, as the swell is continuous. There are 4 to 5 fathoms ½ mile from the land. It is one of the stations of the railroad between Chorrillos and Lima.

La Horada islet takes its name from being pierced from one side to the other; it lies off the center of Chorrillos bay N. 72° W. of Solar hill and S. 76° W. of Miraflores. It is 70 feet high and surrounded by rocks, the outermost of which are about 400 yards from it. At ⅓ mile from it there is no danger. Close to it there is a small islet. La Horadada.

The cliffs which surround the cove of Chorrillos continue to the northward of Miraflores, their direction being nearly NW. by W. for 7 miles, where they gradually descend to a low beach. A tongue of sand and stones extends to the SW. for 1½ miles, and forms point Callao. It is covered with bathing-houses, which are frequented during the summer; the communication with Callao being facilitated by horse-cars. Point Callao.

The name of La Mar Brava is given to the beach between Miraflores and point Callao, the heavy surf rendering it inaccessible the greater part of the time. The depth is 2½ to 3 fathoms ⅓ mile from the beach. La Mar Brava.

San Lorenzo island is separated from Callao point by a channel of 2⅓ miles. The island extends 4⅔ miles NW. and SE., and its greatest breadth is 1⅔ miles. Its greatest height is on the northern third of the island, about 1,284 feet. It is mountainous, barren, and terminates to the NW. in a steep, black-colored cliff called El Cabezo. When doubling it to the eastward, it is best to keep at a good distance to avoid calms. The upper sails must also be watched, as the sudden gusts may endanger the top-gallant masts. A very small point, with some rocks above water, extends 200 yards to the NW. from El Cabezo, (cape San Lorenzo.) San Lorenzo island.

From the light-house placed on the Cabezo, an octagonal wooden tower, 60 feet high, is exhibited a *fixed white* light, 480 feet above the level of the sea, visible 12 miles. It is not visible between the bearing N. 25° W. and N. 60° W., being hidden by the peak. Just open on the latter bearing it leads through the Boqueron channel in 4½ fathoms. It frequently happens that the light is obscured by the fog, or may appear through it as a star. Light-house: lat. 12° 04′ 00″ S.; long. 77° 16′ 30″ W.

Dock cove is on the north side of the island, one-third of the distance from the Cabezo to the eastern end. A floating dock sank in this place after having received the frigate Apurimac. The place of the accident is 371 yards east of Dock cove.

the factory buildings; it must be avoided in going through the Boqueron. This cove is a good anchorage; in 5 to 8 fathoms. There are on the shore some fishermen's huts, the ruins of a factory, and a two-story house designed for the men on the dock. There is no fresh water on the island.

There is also good anchorage in the different coves of that part of the island which faces the bay of Callao. Temporary wharves have been built in some, for loading the granite of the island, which is used in Callao.

Fronton island.   Fronton is an island 600 yards long and 503 feet high, lying off the SE. point of San Lorenzo island, with a channel between them 800 yards in width, but practicable only for boats, and these only by keeping close to the point of San Lorenzo. Six hundred yards to the NW. of Fronton is an islet, called, from its shape, Round island; between it and Fronton is a reef with several rocky islets.

Palominos islets.   The Palominos islets, 140 feet high, lie WSW. of the southern extremity of San Lorenzo; the outermost are $1\frac{2}{3}$ miles from the south extremity of the island, those nearest to the shore are $\frac{1}{2}$ mile, and all the rocks surrounding them are above water, the sea breaking over them; the guano they contain gives them a white and yellowish color. The channel which separates these islands is clear, and more than 1 mile wide.

Bay of Callao.   The island of San Lorenzo, point Callao, and the coast which extends from it to the northward, are the SW., S., and SE. limits of the commodious bay of Callao; it is first on the western coast of South America, not only from its security and importance, but also from being the nearest port to Lima, the capital of Perú.

The following remarks are by Lieutenant-Commander Edwin White, U. S. N.:

Anchorage.   At present men-of-war are anchored in line on a bearing nearly east and west, or on a line nearly parallel to the *Punta*, or Callao point, and nearest to it. They usually moor head and stern. In the rear of these, on the same bearing, is the line of mooring buoys for the Pacific Steam Navigation Company. To the northward of these all of the merchant sailing-vessels are anchored or moored head and stern. There are usually three or four lines of vessels moored close together, each line being about half a mile in length. Vessels returning from the islands to clear with

guano are permitted to select a temporary anchorage outside or to leeward of the lines.

The best and most convenient anchorage for a man-of-war
under the present arrangement is just outside of the line of
the Peruvian vessels, to the westward, in a position well
inside of the extremity of the Punta, and as close to the city
as possible. The following bearings will give an approximate idea of the location indicated :

San Lorenzo light-house W. by S.; east end of San Lorenzo island S. by W. ½ W.; tower of the old castle (rec-
tangular tower) E. ¾ N.. magnetic, in 5 to 6 fathoms of water; good holding-ground. This anchorage is well to windward, the wind being generally from the southward, and the
obnoxious odors from the shore are escaped. The anchorage to leeward, though a berth may be obtained nearer the
landing, is not considered healthy.

From the anchorage recommended, a vessel can leave the
port with a fair wind. From this point, also, a vessel may
exercise and maneuver her boats to advantage, the wind
during the afternoon being sufficiently fresh for sailing and
the sea smooth.

It is true that it never rains in Callao, but from the mid- *Climate.*
dle of April to the middle of November the atmosphere
contains a great deal of moisture, which at times amounts
to a heavy mist; dense fogs are prevalent; the moisture condensed on the rigging and spars drips in rain; this is particularly the case at night, though it often continues for days
in succession. Ships arriving then with wet sails may not
get them dry for weeks; at times the sun is not seen for
days.

During the months of January, February, and March the
weather is warm and pleasant, the sky is clear, and the air
is comparatively dry; frequently, however, a dense fog is
swept across the bay by the southerly winds. The mean
temperature of 1872, noted at noon of each day, was 66° 64′
Fahrenheit. There were two slight falls of rain, light showers,
between September 1, 1872, and April 1, 1873; these showers
produced considerable excitement, and, in accordance with
general prophecy, a slight shock of earthquake followed
each.

The harbor of Callao is not particularly unhealthy, if a *Health.*

proper anchorage is taken clear of the city. The city of
Callao is filthy beyond measure, due principally to a want
of proper drainage. The olfactory nerve of the native
Chalaco is not delicate, but a European who ventures off the
principal thoroughfares is soon satisfied.

The consequence is that when an epidemic occurs it is apt
to be fatal in its effects. There is not an efficient board of
health, and the mortality is never known. Certain regulations exist for preventing the spread of contagious diseases,
but they are by no means rigidly enforced.

During the months of March and April officers and men
are attacked with the tertiana, which generally yields to
treatment if the patient is careful and follows the advice of
the physicians. Though being of malarious origin, it sometimes assumes a malignant type.

Resources. The harbor of Callao contains a floating dock, the property
of a stock company under the title of the "Callao Dock
Company."

The dock, which is of iron, was built in Scotland, and
was put together and launched here in the year 1866. The
manager of the dock stated that the company will under-
take to dock vessels of five thousand (5,000) tons. The
following is an extract from the printed circular of the
company:

Rules for the use of the dock.

SAILING VESSELS.

	Sol.
First day, per register ton	*0.50
Each subsequent day, per register ton	.25

SHIPS OF WAR AND STEAMERS.

	Sol.
First day, per register ton	1.00
Four following days, each	.75
Each subsequent day	.50

Armor-plated vessels, and those entering with cargo or
an excess of ballast or other weights, will pay an additional
sum, according to the weight. The dock will not be sunk

* The Peruvian sol is about 8 per cent. less than the United States
gold dollar.

for less than one hundred and fifty sols, but two small vessels may occupy it at the same time.

Vessels stripped, calked, and metaled by special contract, or at fixed prices per sheet of metal nailed on, varying from 10 to 15 reales ($1 to $1.50) per sheet, including dock dues, labor, and all material excepting metal, nails, and felt.

The dock company will also furnish masts and spars, and they advertise to do all kind of ship carpentry as well as iron work. They furnish composition for the bottom of iron ships. The dock has the reputation of being well managed.

Besides the dock company there are several shipwrights who will contract for work, and will, if required, give bonds for a proper performance of contract.

The ship-carpenters are generally natives, and are not first-class mechanics. They seldom labor more than eight hours per day and receive large wages. This is true of all mechanics in Perú.

Calkers are plenty and do very fair work. They are slow but not lazy.

A steamer requiring repairs to machinery will find every facility. There are several private machine-shops, some of which are competent to turn a main shaft if necessary.

In addition to these are the works of the Pacific Steam Navigation Company, which are very extensive and complete; and though erected for the sole use of the company, the manager permits work to be done for men-of-war in case of necessity.

The Peruvian government possesses no navy-yard, but has an arsenal at Bella Vista, about 2 miles back of Callao, on the line of the Callao and Lima Railroad. The arsenal is under the control of the navy, and some work is usually going on. The works, however, are limited.

Materials for spars may always be obtained, but as everything of this nature comes from abroad, the price is often immoderate. Live-oak is extremely difficult to obtain. White oak may be procured, but not always of the best quality.

All kinds of ships' stores may be had, and generally of good quality. The manila rope is usually of English manufacture, and is good. The hemp is also English, but does

not compare with that furnished by our own Government
rope-walk. I think a great deal of "twice-laid stuff" is
sent to this market. On a recent occasion there was nothing else to be found.

Very fair canvas is always obtainable. Paints generally
come out from England, and are good when fresh.

Coal.   Good steaming coal may always be obtained. Cardiff
coal is always on hand, and at times American anthracite.
The usual price is from $15 to $18, United States gold, per
ton of 2,240 pounds, though at times, owing to a scarcity
or a speculation in coal, (not uncommon,) the price may
advance five or six dollars above those figures. A very
good steam-producing coal has been brought here from
Australia, which should be supplied at a much less rate,
but owing to the want of ships and of communication between this and Australia, the supply is by no means constant. The price above stated includes lighters, which are
numerous, each lighter carrying about twenty tons of coal.

Provisions.   All kinds of provisions may be obtained, and of fair quality, if care be exercised in their inspection. Imported provisions are very dear, and this is true likewise of fresh provisions, the demand being great and the country producing
but little. Bread of an excellent quality is obtainable at a fair
price, which seldom exceeds 7 cents per pound. The bread
is baked here from Chile flour. The cultivation of sugar-
cane has increased of late years to a very great extent,
causing the price of sugar to fall to a reasonable figure.
Coffee is high, as the duties are heavy. Launches are plenty
if wanted.

Wood and   Wood is scarce and dear; on shore it is sold at a certain
water.   price per bundle of sticks, and is used for kindling only.
Ships are generally supplied with refuse timber from lum-
ber-yards at about twenty dollars per cord. Advantage
may be taken of sailing-vessels discharging and intending
to reload with guano, by buying their wood used in stowage.
They find good market for it on shore, however, and generally dispose of it soon after their arrival.

Water may be obtained at all seasons from water-boats
supplied with good pumps and hose, but the quality of the
water is not good, and sometimes gives a ship's company,
upon their arrival, diarrhœa.

Men-of-war steamers frequently condense their own water in this harbor. Efforts are being made at the present time to establish new water-works, in which case the water will be brought from a different source and will be of better quality. The present cost of water is 2 sols and 40 centavos (about $2.25 gold) per tun.

The harbor of Callao is afflicted at times with what foreigners term the Callao Painter. It is always preceded by a whitish or milky appearance of the water, even when it is comparatively quiet and calm. The outside of the ship is covered with brown spots, and the recently-scrubbed copper turns black; whitewash becomes spotted, and silver turns black. Glossed paint escapes if it has not been ruined by scrubbing. Contrary to what would be naturally supposed, zinc-paint is also attacked by the Painter. In painting their boats the Peruvian men-of-war use white zinc, with a little oil and white varnish. By having the boats carefully wiped off after hoisting, their neat appearance is preserved for a long time, but when necessary to scrub them they become the prey of the Painter. Experience shows that the paint is much more easily cleaned on the second day after being attacked, and that less paint is scrubbed off.

The unit of money value established by law is the silver sol, a coin nearly equal in value to the silver dollar coined by the mint of the United States, but which, when compared with the American gold dollar, according to the valuation of American gold by the Peruvian law, is equal to ninety-two cents and six mills nearly, ($0.92592.)

The sol is divided into one hundred parts, each called a centavo, and represented by a nickel coin. In addition to the sol there are the following silver coins: the real, equal to ten centavos, the two-real piece, and the five-real piece, or one-half sol.

Gold has been coined, but is not in use. The gold coins comprise various denominations, from one to twenty sols. There is no scarcity of banks in Perú, and their notes have almost entirely superseded the use of gold.

By a decree of the government, United States gold is valued at 8 per cent. above Peruvian silver, which, as before remarked, is the standard. This decree is observed in all commercial transactions; American twenty-dollar gold

23 c

[margin note: "The Callao Painter.]

pieces, which are usually scarce, pass current at sols 21.60. To reduce Peruvian currency to the currency of the United States, divide by 1.08; to reduce United States currency to Peruvian, multiply by 1.08. English sovereigns (taken at the United States custom-house valuation) are worth about $8\frac{1}{2}$ per cent. above the Peruvian silver, or about sols 5.25 each. The French franc is worth about twenty centavos Peruvian, the twenty-franc piece being therefore valued at sols 4. For exchange on New York, banks and dealers demand four per cent. on American gold, or 12 per cent. over Peruvian sols. Exchange on London is usually 45d. per Peruvian sol; exchange on Paris 4.40 francs to 4.90 francs per sol.

Weights and measures. By decree of Congress, the French metrical system of weights and measures should have gone into effect on the 1st of January, 1873, but, in commercial transactions, the English system of weights prevails. In making purchases by the gallon, it is prudent to specify the gallon required, as various measures are used and misunderstandings sometimes arise. The Spanish vara is the most commonly accepted unit of linear measure. The vara is equivalent to 33 English inches.

Port charges. Merchant vessels arriving with or for a cargo are subject to the following charges: Mole dues, 12 cents per register ton every time a ship enters the port; 75 cents per ton weight or measurement on all cargo discharged; tonnage dues, 20 cents per ton register; light dues, $1\frac{1}{4}$ cents per register ton every time a ship enters the port; hospital dues, 4 cents per register ton every six months. Ballast, S. 1.50 per ton, which can be secured at any point on the coast at from S. 1 to S. 2.50 per ton. Discharging: Coal and heavy cargo is usually taken out at the rate of 45 tons per day; lumber at the rate of 25,000 feet daily.

The new mole. There is in process of construction quite an extensive system of docks or basins, wherein it is intended that vessels shall discharge. The works are being constructed at considerable expense, the Messrs. Brassey & Co., of England, being the contractors. As a matter of course, when these works are completed the facilities for discharging cargo will be much increased, as at present all cargo is landed by means of lighters.

The principal commercial feature of this port is the Pacific Steam Navigation Company. This company was organized in the year 1840, and commenced operations on this
coast with two steamers. At the present time, including
those building, their list contains sixty iron steamers, and
even this number is not sufficient to carry all their freights,
requiring them frequently to charter large ocean steamers to
carry extra freights from Liverpool to this coast. Since the
commencement of the year 1873 a weekly line has been in-
augurated between Liverpool and Callao, and so perfect is
the management that the steamers sail from here every
Thursday with the greatest regularity. The steamers com-
posing this line are fine specimens of naval architecture, of
from 3,200 to 3,500 tons, and from 500 to 600 horse-power.
The number of vessels of this class is nineteen, with two of
the same class building. Their engines are compound.
These steamers compare favorably with any vessels entering
the port of New York.

The trip from Liverpool to Callao is made in forty-four
days. The line is known as the "Straits Line of the P. S.
N. Co.," and was organized in the year 1869 as a monthly
line.

This company runs a line of steamers from Callao to Valparaiso semi-weekly, touching at twenty-three ports. These
steamers are of about 2,000 tons and 300 horse-power. Notwithstanding the fogs and the great number of ports, they
succeed in carrying out their time schedule with great regularity. Between Callao and Panama the number of trips
has been increased to four per month.

The steamers of this line touch at Paita and Guayaquil.
These steamers connect with the steamers of the Pacific
Mail Company from New York and San Francisco, as well
as with the various English, French, and German steamers
plying between the isthmus and Europe. Besides the regular lines of mail steamers above mentioned, the company
runs many intermediate steamers up and down the Peruvian
coast, both to the northward and southward of Callao, ab-
sorbing, to a very great extent, the trade previously carried
on by coasters.

The "Compagnie Générale Transatlantique" have a line
of steamers which make semi-monthly trips between Val-

paraiso and Panama, touching at Callao for freight and passengers. They connect at the isthmus with the steamers of the same line which run between Aspinwall and San Nazaire.

White Star Line. The White Star Line of Liverpool have established a monthly line between Liverpool and Callao.

German line. A monthly line of steamers has been established between one of the North German ports and Callao. The European steamers all stop at Valparaiso, Rio de Janeiro, and Lisbon, and, on stated trips, also touch at various other advertised ports.

Directions. Vessels bound for Callao from the southward can enter the harbor by two channels, by standing along the north coast of San Lorenzo, which is the best and safest route, or through the Boqueron channel, which separates San Lorenzo island from Callao point.

After making the island of San Lorenzo, which can be seen 20 to 25 miles in clear weather, approach it, keeping the Cabezo at least $\frac{1}{2}$ mile distant to avoid the calms, and steer for the anchorage, to reach which two or more tacks are generally necessary within the bay; if much to leeward a white sand-bank called the Ballena or Whale's back, which is 1 mile from point Callao, must not be approached; the bank on which it lies extends from point Callao toward San Lorenzo; a buoy was placed N. 70° W., $\frac{1}{2}$ mile from the Ballena, in 5 fathoms of water, but it disappeared years since and has never been replaced; the sea either breaks over it or else causes eddies. If entering with a free wind during the night or thick weather, after doubling the Cabezo, $\frac{1}{2}$ mile distant, an E. 6° S. course will lead to the anchorage.

The Boqueron. The passage through the Boqueron offers the following advantages: Avoiding the calms of the Cabezo, shortening the distance, and entering with a free wind.

With a vessel of heavy draught and when unacquainted with the port, it is not prudent to attempt this passage. In addition to the narrowness of the Boqueron channel, it is shoal in some places. As just stated, the Ballena bank is 1 mile from point Callao, near the middle of the Boqueron, and toward the center of the bay.

The part between this bank and the point is shoal; the bank extends 1 mile to the southward, and is known as the Carmotal bank, the depth in some places being but 6 feet at high tide.

Its SW. point is formed of coarse white sand and broken shells; it is steep-to, the water deepening from 3½ to 10 fathoms in a ship's length. The heavy swell during spring tides changes the limits of the bank and contracts the channel. The channel between the edge of the bank and the island of San Lorenzo is clean, the depth on the bearings being never less than 4½ to 5 fathoms, and its breadth is a little more than ½ mile. If it should be thought advisable to give the bank to the SE. of point San Lorenzo a wider berth, the bearing will have to be left, and the depth will not be less than 6½ fathoms, but the line of bearing must be taken again as soon as the bank is doubled.

The following should be observed in passing through the Boqueron: As soon as Fronton island is made, keep its southern extremity about a point open of the port cat-head. Keep this course until the castle of Callao is seen. The latter, with its two Martello towers, is on the inside of the peninsula which forms the point; when the castle is well open of Fronton, head for it, which course will lead between Fronton and la Horadada; keep the course until Horadada bears in line with Solar cove, that is, the center between Solar hill and the point of the same name. This line must be kept by heading N. 57° W., and it will lead through the middle of the channel. About 800 yards off the SE. point of San Lorenzo there are from 2½ to 3 fathoms. There are also off this point some rocks with shoal water between them; this reef has been called English bank; after doubling it there is no danger on the side of the island excepting off Dock creek. The bank will be passed when Round island, between Fronton and San Lorenzo, bears S. 22° 15′ W. When the western Martello tower is in line with the northern part of point Callao, the vessel can be headed N., and when this tower is at least 1 point open of the breakers of the bank, stand for the anchorage. When the light can be seen, it is a better leading-mark for the Boqueron, as the Horadada and Solar hill are often shut in by the fog. The light must at first be kept in line with the chimney of the old factory of the dock, but as this is not a permanent building, it cannot be counted upon.

As soon as the light is opened, the vessel must be headed N. 34° W., and run in the channel with the southern base of the light tangent to the high land. As soon as Round island closes, a N. by W. course must be taken until the cliff of the Cabezo of San Lorenzo is well open and bears N. 73° 30′ W. The anchorage must then be steered for. With these marks the castle is not needed; it is a bad mark, as it is often enveloped in fog.

Working through the Boqueron. When getting under way from the anchorage of Callao, steer for the northern peak of San Lorenzo, until Horadada opens from Callao point, then steer S. 61° 30′ W., heading for the sand bay containing the floating-dock, and, when the cliffs of Fronton and San Lorenzo touch, the vessel must be brought by the wind, tacking when the hole in Round island, between Fronton and San Lorenzo, is closed by the latter island. On the tack toward the Ballena a vessel must go about when the cliffs of Fronton and San Lorenzo open from each other.

Vessels drawing more than 23 feet must do so a little before, that is, when they open a remarkable dark spot on Fronton. When the red hill on San Lorenzo is in line with a saddle which is on the summit of the island, or bears N. 85° W., all the shoals will be cleared, and an easterly course can be taken.

Beating out requires great precision, but it is seldom necessary. There is but little current in the channel, but as that generally sets to the NW., it is important to watch the landmarks when standing to the eastward. The northerly current is at times considerable on the outside of Fronton, and with light winds the breeze is very variable near it; it should, therefore, be kept at a distance of ½ mile.

The first shoaling of the water found by the lead when working through this passage will give enough time to go about.

The steep appearance of San Lorenzo often leads strangers to mistake the distance from it, and, thinking themselves closer to it, they have grounded on Carmotal bank.

If it should become necessary to anchor in the Boqueron, it can be done on the line indicated, in 8 to 9 fathoms.

It is high water, full and change, in the Boqueron at 5ʰ
47ᵐ; rise, 4 feet.

Lima, the capital of Perú and of the department and
province of the same name, is on the river Rimac, 7 miles
from Callao, its port on the Pacific, in latitude 12° 03′ 16″
S., longitude 77° 06′ 35″ W.; its population, in 1871,
was about 160,000. The city, founded by Pizarro in 1535,
stands on a plain in a valley sloping gradually to the
sea; it is 500 feet above Callao, but the slope is so gradual
that the road appears absolutely level. To the W. and S.
there is no eminence to obstruct the view or break the
winds; 60 miles to the E. the Cordillera rises; its spurs,
trending toward the coast, pass close N. and E. of the city,
sheltering it completely. The city, 2 miles long and 1¼
wide, is divided by the river Rimac, which empties about 2
miles N. of Callao; the southern portion of the city is surrounded by strong walls, built in 1683; the streets are wide,
and cross at right angles; there are 33 public squares, the
most spacious of which is the Plaza Mayor. Fine public
edifices are numerous, the most remarkable being the
cathedral, the government and archbishop's palaces, and
the town-hall, all constructed by Francisco Pizarro, whose
ashes repose beneath the grand altar of the cathedral.
There are eight national colleges and about 70 public and
private schools; a public library, founded in 1822, now contains 40,000 volumes. There are numerous charitable institutions, many of them being sustained by foreigners. The
manufactures of Lima are very limited; the high price of
all kinds of labor rendering competition with foreign manufactures impossible.

Four lines of railroad lead from Lima to Callao, Chancay,
Chorrillos, and to Oroya, distances of 7, 60, 8, and 130
miles.

The climate of Lima is agreeable, and the range of the
thermometer remarkably small, varying from 73° Fah. in winter to 87° in summer. In the winter—from April to October—
a heavy mist hangs over the city in the mornings and even-
ings. Rain is of seldom occurrence; thunder and lightning
are unknown; earthquakes are frequent. The only disastrous epidemic recorded in the annals of Lima was the yellow fever in 1854.

Rimac river.   The river Rimac, running through the capital, empties about 2 miles north of Callao; it carries but little water to the ocean, most of it being consumed in irrigating the valley of Lima.

Carabayllo river.   The mouth of the Carabayllo river is in the same bay, 5 miles N. of that of the Rimac; it contains more water, although it is used as much for irrigating the valley. The water in its vicinity is shoal, and it is best not to approach the shore nearer than 1 mile until 9 miles north of Callao. The deposits of the river have formed banks of sand and mud, leaving only 8 to 12 feet of water at that distance from the shore.

Monton de Trigo.   The small, isolated hill, Monton de Trigo, is 9 miles from Callao, and may be seen above the low beach to the northward of the port. It is close to the sea, of moderate height, pointed, and remarkable for its position.

Point Bernal   Point Bernal is a small sand-tongue, which projects out a little abreast of Monton de Trigo.

Hormigas de Afuera.   The rocks called Hormigas de Afuera form a reef about 30 miles W. 7° N. from the north point or Cabezo of San Lorenzo. It consists of several low rocks, some of which are under water, of some isolated rocks, and an islet 31 feet high and ¼ mile in circumference, covered with a thin bed of guano; this islet, which is more to the southward, can be seen 6 or 8 miles during clear weather. Boats can land on its north part, but with great difficulty. The scattered rocks are in a circumference of two miles.

Vessels bound to Callao from the north and expecting to reach it in the morning must keep a good lookout for this reef during the night. Calms are frequent, and the current and heavy swell draw toward it. The noise of the breakers can be heard for 2 or 3 miles. As this reef is steep-to, its vicinity cannot be detected by the lead. Many vessels have been wrecked on it.

Point Pancha.   Point Pancha, 1,125 feet high, is 4½ miles NW. of Monton de Trigo; it projects but little; off it lies a white islet of moderate height, and some high and isolated rocks which stretch to the northward; they are all near the coast, above water, and covered with guano.

Point Mulatas.   Point Mulatas is 3½ miles to the northward of point Pancha; it is high, formed of several mountains and rocky hillocks, and it is clean in its vicinity. The passage between it and El Solitario, the nearest of the Pescadores islands, is deep and free from dangers.

Ancon is an excellent port, immediately to the N. of point Port Ancon. Mulatas. The anchorage in the SE. indentation is secure in 4 to 7 fathoms, sandy bottom, 500 to 600 yards from the shore.

It is one of the stations of the railroad from Lima to Chancay and Huacho; all the materials for it were landed in this port on a pier resting on wooden piles. Ancon will, without doubt, be not only one of the principal commercial places, but also a favorite watering-place for the inhabitants of the capital.

The coast surrounding this port has two different aspects— it is high, rocky, clear, and abrupt to the SW. and S., but low, with an extended sand-beach, to the SE. and E., and shoal in its vicinity. In the first, which is formed by the point to the eastward, is the small cove of Playa Hermosa, which has a fine and tranquil sand-beach.

In addition to the railroad, there is telegraphic communication. New buildings are being erected. Fresh provisions of all kinds can be obtained. The water comes from the interior, as that of the wells is a little brackish.

This port has served for several military operations during the war of independence and afterward. The last was the landing of the Chile-Peruvian expedition, which ended the Perú-Bolivian confederation, which was organized by General Santa Cruz.

The Pescadores islands are ten in number; they extend Pescadores islands. to the W. and SW. of point Mulatas; the nearest, which is ½ mile from the point, is called el Solitario. The others extend out to a distance of 4 miles, those to the westward being the highest and largest; they are of a whitish color, and in clear weather can be seen from Callao; they are clean, with plenty of water around them. To the eastward of the largest island is a sunken rock. If desirable, they can be passed to the westward or to leeward of the highest or northern one; they are covered with thin beds of guano.

The Hormigas de Tierra, two small islets, are 1 mile N. of Hormigas de Tierra. the western islets of Pescadores; they are without outlying danger.

Point Toma-
Calla.

From the northern extremity of Ancon bay high mountains approach the beach, and at some places end abruptly at the sea, without leaving even a path at their foot; they have a slight inclination to the E., and are overlooked by mount Stokes, 4,000 feet high. Through the middle of them is the line of the railroad from Lima to Huacho; at a distance of 5 miles is point Toma-Calla, which is high, rocky, and but little salient to the westward, and is formed of different-colored strata. Some rocks lie off it 400 yards, inside the bight, to the north, where there is a narrow beach. Boats can be beached here when it is not too rough.

Point Pasamayo.

The coast continues to the northward with the same aspect for 3½ miles, to point Pasamayo; the sea breaks on the rocks close to it. To the northward of it is a small cove, which is rendered inaccessible by the surf; the mountains run parallel to the coast.

Pasamayo river.

From point Pasamayo the coast follows a northerly direction, with a slight inclination to the W. The river Pasamayo empties 3 miles from it. It has an abundance of water during the summer, and the vegetation of its valley can be seen from a long distance.

Point Chancay.

The low beach which extends from the mouth of the Pasamayo river ends 2 miles from it in point Chancay, on whose extremity are three hillocks and a hill of moderate height, steep, and separate from the mountains of the interior.

Port Chancay.

After doubling point Chancay the port of Chancay is immediately to leeward. The depth is from 6 to 12 fathoms 600 yards from the land. It is formed of two interior coves on the E. side of the point; the westernmost is the most convenient for landing, and all trade is done through it; there are some store-houses on its heights. In the cove at the side of the point are some abandoned huts. These coves are separated from each other by a small hill of black rock, near which are some rocks above water. If a vessel is to remain a few days, it is best to get out stern lines, with her head to the SW., to avoid the disagreeable rolling. The swell often causes surf on the beach, and interrupts landing; during fine weather, boats can be beached on the eastern side of the cove.

The town is 1½ miles distant, on the plateau surmounting the cliffs to the northward. It is a station of the railroad

from Lima to Huacho, and is in telegraphic communication
with the capital. The place has the reputation of being one
of the best watering places, and has progressed rapidly since
the completion of the railroad.

Guano is brought here for the agriculture of the valley;
hogs, Indian corn, and other produce of the farms are taken
to Callao. Large quantities of sugar are exported.

From Chancay the coast trends in a curve nearly NW. for   Playa Grande.
22 miles, with an average height of from 390 to 490 feet,
after which it runs W. by N. for 5 miles. This stretch of
coast is called Playa Grande. It is barren, and rendered
inaccessible by the heavy surf. Nearly everything which
drifts away from Callao is thrown on this beach. In the
plain is Mount Millersh, 3,560 feet high.

Salinas point is the NW. extremity of the curve of the   Salinas point.
coast commencing at point Callao. It projects to the westward, beginning at a small hill of moderate height. There
is an islet close to it.

The name Herradura de Salinas is given to the small   La Herradura
bight to the northward of Salinas point. The depth in it is   de Salinas.
from 7 to 8 fathoms from 400 to 600 yards from the land; it
is nearly always subject to a heavy swell and surf.

From Herradura the coast trends nearly N., bordered by   Salinas mount-
mountains. The highest of these near the beach is Salinas,   ain.
which is in a remarkable position on the N. coast, somewhat
isolated, and ends in a point. Its height, 1,000 feet.

The captains of the English steamers Inca and Quito   La misteriosa.
report having seen the sea break on a rock which is not
marked on the charts. They reported it as equidistant
from points Salinas and Bajas, ¾ mile from the nearest shore,
Salinas mountain bearing N. 67° E., point Salinas S. 14° 30′
E., point Bajas N. 15° 30′ E. It is nearly on the course between Callao and Huacho, and should be given a good berth.

Point Bajas is the northern limit of the promontory of   Point Bajas.
Salinas, which extends 5 miles N. of Salinas point. It takes
its name from the low, flat islets which lie about ½ mile NW.
of it. The rocks in its vicinity are above water, and the sea
breaks on them.

Salinas bay opens to the NE. and extends 3 miles into the   Salinas bay.
land to leeward of point Bajas. It is bordered by a sand-

beach 5 miles in length, terminated by cliffs studded with rocks. The depth is everywhere good.

Creek of Playa Chica. The best anchorage in the bay is Playa Chica cove, in the eastern bight of the S. coast, off some huts, in from 4 to 8 fathoms, near the land. The celebrated salt of Huacho is shipped here by small boats, which land on the beach without difficulty. Loading is done rapidly, as the dépôt is always well provided.

The salt works are 5 miles distant, on the southern part of the plain which bounds the bay; the salt is brought to the coast by mules. A horse-railroad is being built. There is no water or resources, and no permanent residents.

Huara group. The chain of islands which forms the Huara group extends for 13 miles SW. of Salinas point. In the order of their distance from the shore, they are named as follows:

El Tambillo island. El Tambillo is the nearest to Salinas point; it is not high, and is clean, but very near it there is a shoal on which the sea always breaks. The channel between it and the point is 2 miles wide, deep and free from dangers.

Chiquitana islets. The Chiquitana are two islets 1 mile from Tambillo; they are larger, with no dangers in their vicinity; the passage between them and Tambillo is clear; the islets are covered with a thin covering of guano.

Bravo and Quitacalzones. The Bravo and Quitacalzones are 2 miles from Chiquitana; they form a group of three islets with some rocks on which the sea breaks; the passage between them and Chiquitana is clear; they are also covered with guano.

Mazorca island. Mazorca island is the largest of the Huara group; it is ¾ of a mile in diameter and 200 feet high, and lies 1 mile from Quitacalzones. It is of a yellowish color, from the guano deposit. Boats can land on its N. side.

Pelado islets. Pelado islet, 6 miles from Mazorca, is of a whitish color, and 100 feet high; it is nearly perpendicular, with deep water around it; it has no beach. The channel between the two latter islands is very deep, and is the one generally used by vessels from Callao bound to the northern ports.

Huacho point. Huacho point, 6 miles from point Bajas, is the northern limit of Salinas bay; it is of a dark color, high and rocky. Near it are some rocks which extend a little to the northward. A shoal on which the sea always breaks makes a short distance into port; there is another 216 yards S. 20°

Views off the Coast of Perú between the Lat.^8 11¼° and 10° S.

W. of the southern part of the point. They are above water, with plenty of water around them.

After doubling the rocks off point Huacho, port Huacho is to the SE.; the anchorage 400 to 600 yards from the SE. shore is in 4 to 7 fathoms. The best landing is off some small houses; some cultivated spots and the custom-house will be seen on the same side. The wooden wharf which was built some years ago is now nearly destroyed. Small crafts cannot run alongside of it at low water, as it is not long enough, and sand-banks have formed around its head. Port Huacho.

The town is on the cliffs to the northward, about 1 mile from the landing. The houses and churches can be seen distinctly from the anchorage, and even before reaching it. The road from the level of the sea to the town, though steep, is used by vehicles and horses.

Huacho is a small port of some importance; fruits, poultry, and fresh provisions are abundant. The farms of the neighborhood are rich and well cultivated, principally in sugar and cotton.

Landing is at times dangerous from the heavy surf. Water can be taken in near the landing by casks. The harbor is often frequented by whalers, who touch here for water and fresh provisions. The town has fine buildings, and the temperature is even and agreeable. It is connected with the capital by railroad and telegraph. Agriculture is the principal occupation in the vicinity of Huacho; its trade is growing steadily; the population is increasing, as is also the importation of merchandise for the interior and the transportation of salt to the mineral regions of Cerro de Pasco. The exports in fruits, fowl, chancaca, sugar, cotton, hogs, Indian corn, and other produce is large.

Gold was found in the vicinity of this port, at a place called Sanú, some years ago, but it is not worked, as it would not pay the price of labor.

When coming from the S., bound to Huacho, the land should be made near the peninsula and mount Salinas, to the northward of which, 8 miles inland, are the three double peaks of mount Beagle, 4,000 feet high. On approaching the island Don Martin will open from the land to the N., and after that the bay of Huacho will be seen under a brown cliff, whose summit is crowned by bushes and overlooked by the city. Directions.

Coming from the northward, make the island Don Martin and mounts Usborne and Beagle.

Point Carquin. The cliffs on which the town of Huacho is situated trend to the N., descending gradually for 2 miles, and form the low point Carquin, which is surmounted by a small hill. The sea breaks on some rocks above water, which are close to it.

Carquin bay. Carquin bay is to the northward of Carquin point. It is of no use, as it is always difficult of access, owing to the heavy surf. Along the coast is seen the luxuriant vegetation of the valleys of Huaura and Huacho, which at some points extend to the sea.

Carquin hill. Carquin hill, small and isolated, is at the northern extremity of the cove; it is of moderate height and close to the beach. Some rocks extend $\frac{2}{3}$ mile from it, but they are all above water, and the sea breaks over them.

Carquin islet. The shore to the northward of Carquin hill is low. Carquin islet is about $\frac{1}{2}$ mile from it, and 4 miles from point Huacho. It is rocky, but slightly elevated, covered by a thin bed of guano of very white color, and is a great resort of seal. There is plenty of water around it, and it can be passed as close to the westward as prudent. There are some breakers to the eastward of it.

Don Martin island. Don Martin island, 3 miles NW. of Carquin islet, is about $\frac{1}{2}$ mile from the coast. It is of moderate height, of a white color, and free from dangers, as the rocks off it are above water. It is a landmark for the ports of Begueta and Huacho.

Point Begueta. Point Begueta is $\frac{1}{2}$ mile N. of the island Don Martin; it is rocky, and surrounded by reefs which extend $\frac{1}{2}$ mile to the northward, and produce a continuous surf.

Port Begueta. Port Begueta is to leeward of Begueta point; it affords no shelter; the depth is from 5 to 8 fathoms, $\frac{1}{4}$ mile from the shore. About 3 miles farther north is a small point and cove, resembling the point and cove of Begueta.

From point Begueta the coast becomes of a dark color, with high mountains in the interior, the Beagle mountains, to the northward of which is a peak 4,200 feet high.

Point Atahuanqui. Point Atahuanqui is 8 miles N. of port Begueta, and a little to the west; it is clear, and formed of abrupt black rocks with white patches, particularly to the S. After doubling this point the vegetation of the valley of Supe will be seen.

The mountains which form the coast for 4 miles to the northward become gradually lower, trend to the westward, and then becoming higher again form the oval-shaped point Tomas. It is similar to point Atahuanqui, but has no white patches. It is clear, the few rocks off it being at a short distance and above water. Point Tomas or Supe.

Point Patillo is a small point ½ mile NE. of point Tomas, the intermediate coast being formed of high, dark rocks. Some rocks and a small white islet extend 100 yards from it, but they are either above water or the sea breaks over them. It is not necessary to approach this point; in standing for the anchorage it should be given a good berth. Point Patillo.

As soon as Patillo point is doubled, the low stretch of shore which forms the southern part of Supe bay is seen. It is clean, with a good anchorage in 4 to 6½ fathoms, 500 yards off the houses and huts. The town, which is 4 miles distant, is abundantly provided with fresh provisions. The best anchorage is about 200 yards from the rocks off point Patillo, with point Tomas just closed in by the former; this position is ¼ mile from the village. The anchorage farther out, in 6 or 7 fathoms, is exposed to the swell. Boats can easily be beached on the south shore, though the sea is generally rough. Canoes are used for loading and discharging, as there is no mole. The small point which limits the port to the northward has some rocks near it, and should not be approached. Bay of Supe.

Mount Usborne, 8,000 feet high, is the highest and most remarkable mountain of the interior chain, and is the best landmark for the bay of Supe; it bears N. 64° 30' E. from the anchorage. It has nearly the form of a bell, with three different elevations on its summit; the northern one of these is the highest; it can be seen distinctly, as there are no other peaks for some distance, the next in point of elevation, Mount Pativilca, being 20 miles NW. and 8 miles from the coast.

Barranca head is a small hill 2 miles N. of Supe; it projects a little to the eastward, is deep and clear; all the rocks off it are above water. To the southward and close to it is Barranca head.

a high rock, and to the N. a small islet, also close to. The southern rock is a good distinguishing mark for the anchorage of Supe and Barranca; it is white, and there is no other like it on this part of the coast.

Barranca bay. Barranca bay is to leeward of the head; it is rendered useless by the heavy surf. There are some cabins on the beach, which are inhabited during the bathing season. About ⅔ mile N. of the port are several rocks on which the sea breaks. They extend 400 yards to seaward.

River and valley of Pativilca or Barranca. The river Barranca empties 3½ miles N. of Barranca hill; it has an abundance of water during the summer months, and irrigates the beautiful valley of the same name. The cultivated ground reaches to the shore, and can be seen from seaward; there are but 5 to 6 fathoms 3 miles west of the valley.

Mountains of Horca and la Fortaleza. The valley of Barranca presents at its N. extremity, near the sea, two remarkable points, the Fortaleza, a mass of walls or the ruins of some Indian building, situated on a height, and resembling a fortress, and the mountain of Horca, which is close to the sea, isolated from the other heights, and of a dark color.

The 9 miles of coast between Barranca head and Horca is constantly beaten by the surf, the water being shoal in its vicinity. The height of the coast is about 100 feet. It is a cliff of clay, descending toward the N., and ending in brushwood.

Point Santander. Point Santander is 2½ miles from the mountain of Horca; the intermediate coast trends NW. It is low with dark rocks. Close to its western extremity is an island, which is clean, and can be passed at a short distance.

Pativilca or Darwin peak. Pativilca peak is in the interior, and about NE. by N. from Santander point. It is a pointed cone, 5,600 feet high.

Point Callejones. Point Callejones is high, projects but little, and is 9½ miles NNW. of point Santander; its surroundings are free from dangers; all the rocks are near it and above water.

Jaguey point. Jaguey point is low and dark, with some rocks close to it on which the sea breaks. It is 6 miles NNW. of Callejones.

Gramadal bay. Gramadal bay opens to leeward of Jaguey point; the anchorage is good, and is sheltered from the S., in 5 to 7 fathoms near the land. All the southern part is free from danger. It derives its name from the patches of grass which are seen on its shores. Landing is often difficult.

Fitz Roy advises anchoring in 6 to 7 fathoms, bottom sand, with the hill which forms the bay bearing S. 11° 30′ E.; this position is ½ mile from the land.

It is reported that a rock was seen from the ship Hercules Hercules rock. off this coast during the middle of the last century. The old Spanish charts place it in latitude 10° 23′ 30″ S., longitude 81° 35′ W., 75 miles from the land. The rock has often been searched for since, but without success.

Bufadero cliff is 8 miles N. of point Jaguey, it is steep Bufadero cliff. toward the sea, and formed of dark rock with white patches.

Mount de las Tetas is a short distance inland, and seen Mount de las Tetas. above Bufadero cliff. It is 1,620 feet high, and takes its name from the two elevations on its summit.

The coast runs N. for 2 miles from Bufadero, composed of Cabeza de Lagarto. dark cliffs from 200 to 300 feet high, terminated by the Cabeza de Lagarto, which is a projecting point of dark rock, steep, and resembling in shape an alligator's head. The land inside of it seems low when seen from the southward.

From Cabeza de Lagarto the coast is high, trends N. ½ Huarmey bay. mile, and takes a more easterly direction for more than ½ mile, where there is a stretch of low land forming the bay of Huarmey. Attention must be given to the rocks, which at some points extend 200 yards from the small point of the S. shore. On this shore, within the port and about 800 yards to the northward, is a white rocky island of moderate height. The passages on either side of it are clear. The best anchorage is inside of this islet, between it and the SE. beach. It is about 500 to 600 yards from the land, in 4 to 7 fathoms, bottom fine sand. This port is secure; boats can be run on the sand-beach. There are only a few huts and no resources. Shipping fire-wood and coal form the principal commerce. Large piles of wood ready for embarking lie on the shore. It is better and less expensive than at any other point of the coast. Large vessels carry it to Callao, where it is sold at a large profit.

The town of Huarmey is 2 miles NE. of the anchorage, Description. and is hidden by trees; it has but one street and about 500 or 600 inhabitants; there is an abundance of fresh provisions at low prices, but water cannot always be obtained,

24 c

although the river is full from March to the beginning of summer.

The tides are very irregular. It is high water, full and change, at port Huarmey about 6ʰ; rise, about 3 feet.

On coming from seaward the best course for making this port is to keep on the parallel of 10° 6′ S.; when some miles off the coast a mountain with a slender peak and marked with white patches will be seen. It is isolated, and situated to the northward of the port. The mountain gorge in which the river runs is high and rocky on the sides; the land to the northward of Cabeza de Lagarto is lower, and the large white island at the north extremity of Huarmey bay will be seen.

On entering, the little white island in the middle of the bay, called Harbor island, will be seen; by steering so as to keep the south coast 400 yards distant, the vessel will pass in mid-channel between it and Harbor island, and can then anchor in 4 fathoms, with the island bearing N. 37° W., and the ruins of the fort on one of the hills inland S. 84° 45′ E.; the water shoals regularly to the beach. This anchorage is ¼ mile from the landing. The usual landing place does not seem as good as another near a rock to leeward of the point.

The sea-breeze is often so strong that boats can hardly work against it, especially under the highlands, from which terrific gusts descend.

Point Culebras, 9 miles N. of Huarmey, is the most projecting point of the coast; it is free from dangers.

Culebras cove is to leeward of the point; it has from 6 to 8 fathoms from 500 to 600 yards from the S. shore, where there is a flag-staff near some houses; one of them is the property of the farmers of the valley, the only people who use this port. The beach is low; the landing is in the eastern bend, but landing is often dangerous. Some cotton is exported; there are no resources.

The valley which opens immediately to the northward shows some cultivation. The rivulet watering it has only water at its mouth during the summer.

Some islets and rocks lie 1 mile off the coast, which continues its northerly direction. The principal ones are the Erizos, which are 3½ miles N. of point Culebras, and the Conejos, of a very light color, 5 miles N. 10° E. of the same point.

The coast from Conejos island forms an inaccessible cove, which ends in point Mongoncillo, 6½ miles from Conejos; it can be easily recognized, as it is terminated by a hillock near the sea, which descends from the chain of high mountains to the northward; some small rocks above water are detached from it.

Point Mongon-
cillo.

Point Colina Redonda is 9½ miles to the northward of Mongoncillo point. The two hillocks which form it appear like an island when seen from the S.; its vicinity is clear.

Colina Redonda
point.

To leeward of the point is Colina Redonda cove, with an anchorage in 7 to 8 fathoms near the shore; there are no inhabitants or resources. There are some rocks off the northern extremity of the cove.

Colina Redonda
cove.

Mongon mountain is on the same parallel as the cove, and near the sea; it is a good landmark. Its elevation is 3,900 feet; it is separated from the mountains of the interior, and its summit terminates in several sharp peaks. Its slopes are generally covered by a belt of fog. When seen from the westward it appears round, but from the southward it looks like a long mountain with a peak at its extremity.

Mongon mount-
ain.

The mountains near mount Mongon, which follow the coast for 10 miles, are not high. Their profile is undulating, and they are terminated by Calvario bluff, which, formed of dark rock, projects but little into the sea; it is clean, and very precipitous.

Calvario bluff.

CHAPTER XIV

FROM THE BAY OF CASMA TO THE RIVER TUMBEZ.

Variation from 10° 00′ to 9° 00′ easterly in 1876, increasing annually about 1′.

Bay of Casma. After passing mount Calvario the coast curves to the NE., and keeps that direction with the same appearance for 2 miles, where there is an opening of 1⅔ miles, limited to the northward by mountains similar to those to the southward. These mark the entrance to Casma bay.

Fergusson rock. To the NW. of the southern shore of the entrance to the port, about 400 yards distant, is a rock, reported by Captain Fergusson, of H. B. M. S. Mersey, over which there is but 9 feet of water. It must be given a good berth; the sea breaks over it occasionally.

Standing in to the eastward, the low sandy beach, which gives the bay the appearance of a horseshoe, is seen. The best anchorage is in the eastern part of the bay, in the elbow which the inner point of rocks makes with the coast to the southward of it. There are from 4 to 6 fathoms between 200 and 300 yards from the shore. Vessels are obliged to anchor close to the shore to be protected from the heavy gusts. From 11 a. m. to sunset vessels should enter under short sail.

On the beach to the eastward of the southern high mountains are some huts; in front of them is a convenient wharf, which offers every facility for landing. It is not prudent to approach the SE. and S. shore within less than 700 yards, as the water is shoal to that distance.

On the E. and NE. shores are thick woods, but there is no water. Fire-wood and charcoal can be obtained; they are the principal articles of commerce in this locality. Indian corn and cotton are also exported; the cotton is of a superior quality.

The village of Casma is 6 miles from the port. Mules can be obtained at the huts; some stores and fresh provisions can be obtained at the port. Water is very expensive, being

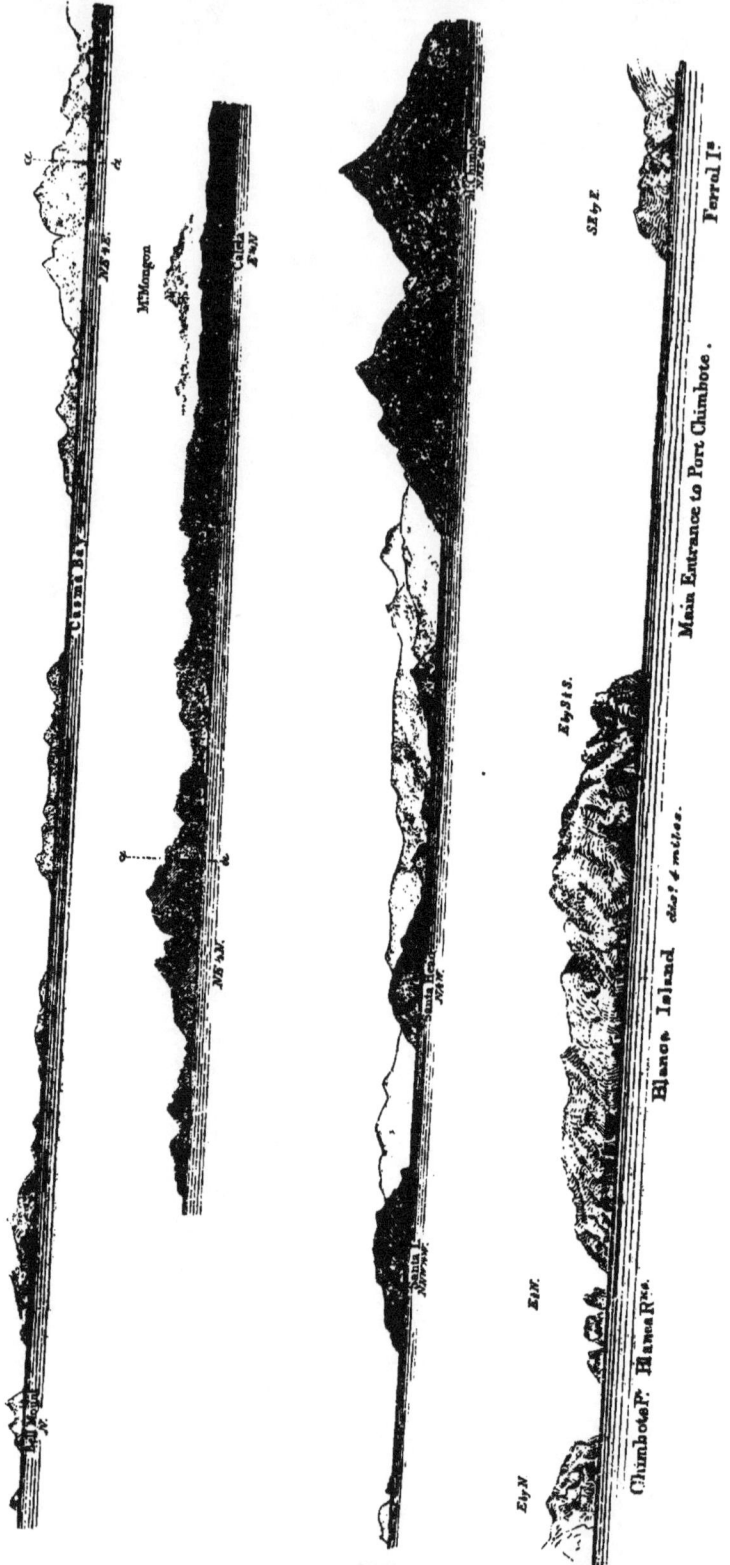

XVII

Views off the Coast of Peru between the Lat.º 10' and 8¾° S.

Main Entrance to Port Chimbote.

Chimbote P.   Blanca Rᵏᵃ   Blanca Island   dist! 6 miles.   Ferrol Iˢ brought from a long distance. Casma is a small port of the department of Ancachs; it is in telegraphic communication with the capital.

The rocky mountains which form the northern limit of the port are dark, having the same appearance as those to the southward. The rocks, which run out to a distance of 400 yards from it, are all visible, the sea breaking over them.

The best landmark for Casma is the sand-beach of the *Directions.* bay, with the sand-hills further in the interior; they contrast strongly with the dark rocks, called the Cheeks, forming the entrance. Black rock is also a remarkable point. To avoid the violent squalls which descend into the valley, a vessel should anchor so as to have the south cheek bearing S. 11° 40′ E., in 7 fathoms, bottom sand, ¼ mile off shore.

Black rock is a small islet 1 mile from the shore and S. *Black rock.* 87° W. from the north point of the entrance. It is steep-to, a few feet above water, of dark color, and the sea always breaks over it; close alongside of it there are 30 fathoms; the channel between it and the coast is deep and clean.

In 1854 the transport Mercedes, with 800 persons on board, left the harbor of Casma on a dark night in tow of the steamer Rimac. The tow-lines parted a little to windward of the rock, and before sail could be made the transport was thrown on the rock and sank in a few minutes. Over 700 persons were drowned.

From the bay of Casma the coast continues high, trending *Islets of La* northerly for 4 miles, when it forms a cove 4 miles in extent. *Viuda and La Tortuga.* Off it, and in the line of the coast, are the Viuda and Tortuga islets. The southernmost is the Viuda; both are of moderate height; the Viuda is clean; the Tortuga has some rocks to the SE., but the sea breaks on them.

The coast from the cove continues high and preserves its *Los Chinos islets.* northerly direction for 3 miles, where, 400 yards from the shore, are the two islets of Los Chinos, which are free from dangers.

To the northward of Los Chinos islets the coast forms a *Samanco point.* cove 1½ miles wide, with a sand-beach on which the sea always breaks. It is difficult and dangerous of access, excepting in the southern bend, in which boats can sometimes be beached. High and dark rocks run out to the westward of the northern extremity of the cove, which, from the chain in the interior, form point Samanco. There is a small islet to the southward of the point; both are clear.

Samanco head. Samanco head, a rocky promontory, dark and high, is the northern limit of the mountains, which form the point, and is about ½ mile from them. Some rocks, on which the sea breaks, run out from it about 200 yards.

Bell of Samanco. The Campana de Samanco is in the same group of mountains as the point and hill, but about 1½ miles NE. of them, of moderate height, with a large base; it is conical, and its oval surface resembles a bell. Alongside of it is another small mountain.

Bay of Samanco. The safe and spacious bay of Samanco opens immediately to the N. and E. of the point and hill of Samanco. It is about 6½ miles wide and 3½ deep; the entrance is 2¼ miles wide; it is deep and clean, and can be easily recognized by the high mountains on each side. The depth at the entrance is 20½ fathoms, which diminishes gradually to the northern part, where the land is low, and the depth but 3 fathoms ⅔ of a mile from the shore. The principal anchorage is on the SE. coast. Standing in to the eastward there is a fine cove, called Guambacho; at its entrance is a good anchorage in 4 to 7 fathoms, 500 to 600 yards off the houses on the low sand-beach. There are, generally, heavy gusts from 11 a. m. to sunset. There are but few resources. Water can be taken in at a brook which empties at this place; it is not very good; that in the well on the left bank is also brackish, but becomes better when kept on board.

The valley of Nepeña, in which the village of the same name and that of Guambacho are situated, commences here. The first is 15 miles east of Samanco, and the second 3 miles NE. of the same point. From this port are shipped firewood, charcoal, rice, corn, and cotton.

Inside of the bay, ⅔ mile from the NE. coast, is a small islet called de los Pajarros; it is of moderate height, and whitened by guano; there are some rocks 400 yards to the westward of it over which the sea breaks.

To the NW. of the cove is a very low, sandy spit, which gives the high hills forming the north shore of the bay and running to the southward the appearance of an island.

Lobo island. Lobo island forms the southern extremity of the high mountains which are the NW. limit of the entrance to the port of Samanco. It is about 400 yards from the coast, is high, formed of dark rocks with white patches, and has the same appearance as the immediate coast. The small islets and the rocks near it are all visible.

The chain of mountains to the northward of the entrance Mount Division. of port Samanco continues high; mount Division with its three peaks, 1,880 feet above the sea, to which it is very close, will be easily distinguished; this group of mountains forms a peninsula 5 miles in length, and is connected with the continent by the neck of sand already mentioned.

When bound for Samanco, coming from the north or the west, mount Division and the Bell should be made; the latter is easily seen when to the northward of the bay. To the eastward, a little in the interior, is mount Tortuga, higher, but resembling the Bell.

When coming from the south, point Samanco must be kept some hundred yards distant; as soon as the bay is open, Leading bluff, a rocky mass in the midst of the sand, will be seen; the south coast must then be followed without approaching it too close, until the anchorage is reached.

The three Ferroll islets, almost in a straight line, lie off Ferrol islets. the hills which descend from mount Division toward the north; they are tolerably high and their surroundings clear; the passage between the northern and middle Ferrol islet is about $\frac{1}{2}$ mile wide, with a plenty of water.

Blanca island, $1\frac{1}{8}$ miles NNE. of the northernmost of the Blanca island. Ferrol islets, is about 1 mile long and close to the N. coast; its color is owing to a thin covering of guano. About $\frac{2}{3}$ mile west of this island is a reef, some of the rocks of which are above water. Over it there is always a heavy swell and eddies.

The following remarks are by Commander J. N. Miller, U. S. N., commanding U. S. S. Ossipee:

The harbor of Chimbote is the northern portion of Ferrol bay, between Blanca island and the main-land to the eastward.

Ferrol bay is a large indentation of the coast, of an oval Ferrol Bay. shape, about 7 miles long and 4 miles wide, with several islands between the main-land at mount Division and that at mount Chimbote. The group of three islets nearest the point at mount Division is called the Ferrol islets; the large

island north of these, Blanca island; and for the sake of
distinction we have named the two rocks between Blanca
island and the main-land to the northward, Blanca rocks.

There are three passages to Ferrol bay available for vessels. The main passage, between Blanca island and the
northern Ferrol islet, is 1¼ miles wide, and free from rocks
or shoals, with plenty of water close to both islands. The
passage between the northern and middle Ferrol islets is
about ½ mile wide, with plenty of water, and is used by the
steamers of the Pacific Steam Navigation Company in coming into the port, but it is not deemed advisable for sailing-
vessels to use it on account of the swell, and as but little
distance is saved by it. The northern passage, between
Blanca rocks and Chimbote point, is about ¼ mile wide,
and is also used by the steamers of the Pacific Steam Navigation Company, but it would not be prudent to use it in
a sailing-vessel on account of its narrowness and the liabil-
ity to be set ashore by the eddy, currents, and swell. The
passage between Blanca rocks and Blanca island, between
middle and southern Ferrol islets, and between southern
Ferrol islet and the main-land, should not be used by any
class of vessels.

The appearance of the land on entering any of these passages is that of a low sandy beach, with high hills and
mountains rising in its rear. That portion of the bay surveyed, north of middle Ferrol islet, which would be used by
vessels bound to Chimbote, is perfectly free from rocks and
shoals; it is sandy bottom on the eastern side of the bay,
but on the western, toward Blanca and at the anchorage, it
is partly muddy, with good holding-ground.

During the full and change of the moon a swell sets in
through the main passage, making it rough at the mole for
landing, but never dangerous for vessels at anchor.

The rise and fall of the tide is slight, about 2½ feet, and
no currents of importance were noticed in any portion
of the bay. Vessels bound to Chimbote, after passing
through the main or Ferrol passages, should steer direct for
the mole and anchor a short distance from it, in 4¾ or 5
fathoms of water. In coming in through the northern passage the soundings on the chart will indicate the course to
be pursued.

As the prevailing winds and currents are from the south-ward, vessels bound to Chimbote from distant ports should make the land well to the southward of the port, and after recognizing some marked point, follow the land along until opposite the passages, taking care under no circumstance, if it can be avoided, to fall to leeward, for, with a head wind and current, they will be much delayed in beating up.

The prominent landmarks when near Ferrol bay are Bell mount and mount Tortuga, on Samanco bay; mount Division, between Samanco and Ferrol bays; mount Chimbote, Santa island, and Santa head, to the northward.

The general direction of Blanca island and Blanca reef to the westward of the island are not correctly laid down in the sketch by Captain Folger. Blanca reef consists of a few projecting rocks which are awash at times, about $\frac{1}{8}$ mile westward of the middle of Blanca island, and extend-ing out not farther than $\frac{1}{2}$ mile from the island. The position is laid down on the chart with tolerable accuracy, but until more thoroughly surveyed vessels should keep at a safe distance. Soundings are laid down on the chart around the reef, and along the western side of Blanca island.

The town of Chimbote, the terminus of the Chimbote, Huaras and Recuay railroad, is situated in the province of Ancach, in latitude 9° 04' 40" S., longitude 78° 32' 15" W. from Greenwich, on the NE. shore of Ferrol bay. The old Indian village, consisting of a few huts, still exists on the western edge of the town. *Chimbote*

It is situated on a sandy plain at the foot and to the east-ward of Chimbote mountain. It is regularly laid out with streets at right angles to each other, and with the usual plazas of Spanish towns. From the anchorage it presents the appearance of a few board houses, built on a desert of sand, and without the sign of a tree or vegetation. At present the town consists of a ship-chandler's store, the dwelling of the superintendent of the railroad, with offices for the employés attached, two or three pulperias, a butcher's shop, a bakery, a store-house for materials and supplies for the road, two or three buildings for coolies, and a very small, rough, board hotel. The custom-house which is being built is a frame building, much larger than the commerce of the place will require for some years to come.

A mole is being constructed 500 meters long, and it will have the necessary derricks for discharging heavy weights. About 300 feet have already been finished. The prevailing winds come through the main passage in line with the mole, and during the full and change of the moon, when the swell is greatest, it will sometimes be difficult to land cargo, even when the mole is completed. I think a better position for the mole would have been at Red knoll.

Supplies.

Water is obtained from a stream emptying into Coisca bay, between Chimbote point and Santa, by means of a water-lighter of twenty-eight tons, which is kept running as occasion requires. It is sold to vessels at about two cents per gallon, and is probably as good as Callao water. Water is also obtained from a well about 1 mile back of the town, which is said to be good and the supply ample. Fresh beef and mutton can be obtained at about the same prices as at Callao; vegetables and fruits are scarce and dear. Ships' supplies and stores can be obtained at moderate prices. At present there is no wood in the harbor, but it could be readily procured should there be a demand for it. There are no spars or materials for repairing damages to be obtained at this place.

Ballast for vessels can be obtained at three sols per ton, delivered alongside, or it could be obtained for nothing by anchoring near the north end of Blanca island, and loading with the crew.

Mails and tele-
graph.

At present the mail is received from Callao by steamer on Saturday of each week, and mails are sent every Thursday from Santa, which is 10 miles from Chimbote. The steamers do not touch at Chimbote on their return to Callao.

There is a line of telegraph between Santa and Lima, and in a short time connection will be made with Chimbote, but the line is often down or not in working order.

Railroad.

The railroad is to extend from Chimbote to Recuay, a distance of 164 miles. passing through Huaylas and Huaras. The contract was taken by Mr. Meiggs for constructing the road for 24,000,000 sols. It is to be a narrow-gauge road, 3 feet 6 inches wide, and the specifications require a well-constructed road.

Throughout its entire extent, I believe, it follows the valley of the Santa river, with the exception of a few miles from the mouth of the river, where it branches off to the southward, and strikes Ferrol bay at Chimbote.

The object in constructing the road is to open up the mineral and agricultural resources of this portion of the country. Mines of gold, silver, copper, and coal are said to abound in the mountains near the line of the road, and the valley of the Santa is said to be capable of producing large quantities of cotton, rice, coffee, cocoa, and other agricultural products.

It is high water, full and change, at Ferrol bay at $5^h 50^m$; rise, 2 feet. *Tides.*

The Chimbote mountains are the northern limits of the bay of Ferrol; they are two dark peaks, which can be easily recognized; they commence at the edge of the sea and end in sharp peaks, the northern being the lowest. They spring from one base, and their surface has dark bands rising from north to south. *Chimbote mountains.*

The coast north of the bay of Ferrol, which is formed by the high mountains mentioned, is, for $3\frac{1}{2}$ miles, very steep and clean, when a small bend of low beach forms the bay of Coisca. There is an anchorage in the SE. indentation, in 5 to 6 fathoms, 400 yards from the land, off a fishing village. The beach is always exposed to a heavy swell, which renders landing difficult. On the north side of the bay is an islet without any dangers around it. The valley of Santa, whose fields can be seen from the sea, commences here. *Bay of Coisca.*

Santa island is opposite the bay of Coisca, separated from the nearest land by a channel $1\frac{2}{3}$ miles wide; it extends north and south $1\frac{1}{2}$ miles. Off its northern end is a small islet, which appears as if connected with the main island, and two large rocks, 20 feet high, surrounded by smaller ones of moderate height and of a grayish color, are $\frac{2}{3}$ of a mile SW. of the island; the channel between it and the coast is clear and deep. *Santa island.*

The low, sandy coast which forms the bay of Coisca trends nearly W. for $2\frac{1}{2}$ miles, and is limited by a promontory of isolated rocks called Santa head, which resembles an island when seen from the southward. About 200 yards from the low, rocky point is a rock on which the sea does not always break. *Santa head.*

Bay of Santa.   The bay of Santa lies to leeward of the point. Though small, it is tolerably protected. It has a good anchorage in 4½ to 5 fathoms, about 800 yards from the shore, on the side of the point, and another in 4 fathoms ½ mile NNW. of a hamlet on the beach.

The best place is when the NW. extremity of the point bears S. 55° W. Landing is effected by beaching the boats in front of the huts.

The town is 2 miles distant, and abounds in fresh provisions, in which a brisk trade is carried on with the island of Guañape. It is in telegraphic communication with the capital. The principal articles of export are wood, charcoal, rice, cotton, and other produce. Water can be obtained, but at a high price.

Santa river.   Santa river empties on the beach, which extends north 4 miles from the port. It is one of the largest on the coast of Perú. The surf almost renders this coast inaccessible.

Corcobado Islet.   Corcobado islet is remarkably white; it lies 3 miles from the coast, and 3½ miles NW. of Santa point. Its west side is steep, but it slopes gradually to the east. About ½ mile NNE. is a small reef, on which the sea always breaks.

La Viuda Islet.   La Viuda islet is N. by W. 3 miles from Corcobado and 1 mile from the coast; it is smaller than the preceding, and of a dark color. A low rock, on which the sea always breaks, lies N. 43° W. of La Viuda, and about 2 miles from the coast.

Valley of Santa.   The valley of Santa is on both banks of the river of that name, and is about 10 miles long; its fields can be seen from the ocean. It is bordered by hills with sharp peaks, and serves as a landmark for this part of the coast.

Chao hill.   The low beach, which extends for 13½ miles to the N. and NW. of the river Santa, is rendered inaccessible by the heavy surf. It has some elevated projections in its northern parts, the highest and most projecting being Chao; the water around it is free from dangers.

Point Chao.   The small point of Chao is 1½ miles north of Chao hill; it is clean, and offers nothing remarkable.

Chao island.   Chao island is 1½ miles west of the coast, between Chao point and hill. It is 120 feet high, and is rendered white by a covering of guano. The channel between the island and the coast is clear and deep. There are some small islets

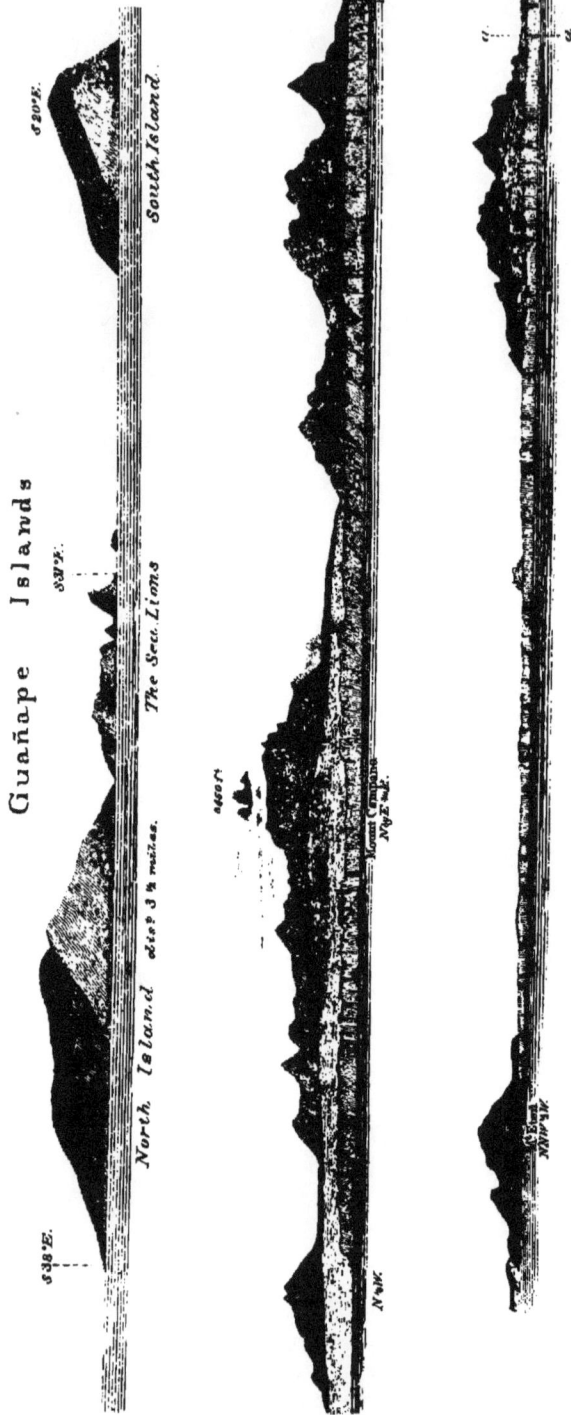

Views off the Coast of Peru between the Lat.s 8¾° and 7° S.

XVIII

Guañape Islands

North Island dist? 3½ miles.   The Sea. Lions   South Island.

S 38° E.   S 20 E.

S 87 E.

resort   N 9 E.

Church   E ½ N.

N ½ E.

N ½ W.

close to it to the NW., and on the same side is a reef 2 miles long, on which the sea always breaks.

Coscomba cove.

Just to leeward of point Chao the coast line is slightly convex for a short distance and then trends north for 4 miles, where there is a high point, on the north side of which is Coscomba cove. It has a poor anchorage in from 5 to 7 fathoms, ½ mile from the land; there are no inhabitants or resources.

Anchorage of Chao.

The anchorage of Chao is off the mouth of the river Chao, 2½ miles north of Coscomba cove. The best anchorage is 1½ miles from the land, off the huts which are near the water, in 6½ to 7 fathoms. The surf and swell are always heavy.

Ship's boats cannot be used. After noon the wind always freshens, rendering work difficult. This anchorage is only visited by vessels which load fire-wood, charcoal, and produce from the neighboring farms. The provisions come from the interior; some are sent to the Guañape islands. The Chao river empties here.

Virú river.

From Chao hill the coast trends NW. for 11½ miles, where the Virú river, a small stream, empties on the beach.

Guañape head.

Three and one-half miles farther to the NW. is Guañape head, a stony promontory 700 feet high; it is easily distinguished, as it is very pointed, and the shore to the north and south is so low that it resembles an island. Its surroundings are clean.

Guañape islands.

The Guañape group consists of two large islands named North and South, in accordance with their position; of two islets between them, and of some large rocks NW. of North island. All of them are within a circle of 3 miles diameter. The channel between them and the land is clear.

North island is low and the nearest to the land, being 5½ miles S. 26° W. of Guañape head.

South island is 1,640 feet high, and lies 1½ miles S. by W. from North island. The western part of it is almost perpendicular; the opposite side, which slopes gradually, is covered with guano. The channel between the two has a depth of from 16 to 22 fathoms. The center islets are high, and almost in the middle of this channel.

Both of these islands have good anchorages, South island on its NE. coast, and North island on the side toward the main land, close to the island, in from 15 to 20 fathoms.

Godard shoal.   The large rocks to the NW. of the North island are called the Sea Lions, and are free from outlying dangers, but vessels must not pass between them and the islands, as a dangerous shoal, called the Godard, lies 492 yards N. 55° W. from the NW. point of the North island, and 197 yards N. 30° W. from the small islet in the center of the passage; it has only 3 feet of water over it at low tide, and it should be avoided even by boats, as the sea breaks over it when least expected. These islands are the best landmark for vessels bound to the ports of Salaverry or Huanchaco. Guano has only lately been taken from these islands, and large and costly works had to be undertaken before it could be shipped.

During the months from January to April fogs are very frequent, and often so thick that the communication between the two islands is not safe.

Winds from E. to S. prevail among these islands; sometimes they are fresh. From April to June they increase in force; the swell is heavy and loading difficult. The South island is more exposed than North island.

The governor's house and custom-house are on the South island, but there are also many houses and sheds on the North island. There is a captain of the port, and some troops to maintain order. Fresh provisions and water can be obtained, the former at moderate prices, the latter at about $8 a tun. Vessels must get their license for loading at Callao, and clear either at Paita or Callao.

Guañape cove.   Guañape cove is on the east side of the small point which runs out from the north part of Guañape head. The best anchorage is ½ mile from the land, off the huts on the low beach. The swell is generally heavy, and it is always imprudent to land in ships' boats. Large quantities of wood, cotton, and other produce from the farms are shipped here. There is a water-pipe which reaches to the sea to fill the small vessels, which carry it to the islands. Fresh provisions are plenty.

Carretas hill.   From Guañape head the coast is low, and trends N. by W. for 13 miles to the small hill of Carretas. In the interior the mountains are from 4,000 to 5,000 feet high. Carretas hill is overlooked by Mount Garita, 3,716 feet high. The surf on all this stretch of coast is heavy, the depth 8¼ to 10 fathoms, 2 miles from the land.

Port Salaverry is immediately north of Carretas hill. The depth in it is 5½ to 6 fathoms, bottom sand, ½ mile from the shore, to leeward of the small point which runs out from the hill. There is a *tasca*, and the breakers commence some distance from the shore; it is, therefore, best not to use ships' boats for landing. With the native boats landing is more convenient than at Huanchaco. Salaverry has been substituted as a port of entry in place of Huanchaco; a wharf to extend outside of the breakers is being constructed, and the port can soon be frequented with security. A custom-house has been built, and some fine houses in the new town, which is being supplied with water. A railroad is contemplated to Trujillo, 7½ miles distant.

The coast trends NW. by W. 6½ miles, to the mouth of the Moche, which contains but little water.

Trujillo is in the interior, NNE. 1½ miles from the mouth of the rivulet Moche. Its church steeples and houses can be distinctly seen from the sea. Trujillo was founded by Francisco Pizarro, who named it after his birthplace in Spain. It is built in the valley of Chimú, where the ancient Incas had one of their most important towns, the ruins of which can be seen, covering a large extent of ground; it is in telegraphic communication with the capital. The population is about 8,000. It is the capital of the department of Libertad.

There are some huts on the beach opposite Trujillo, 1 mile NW. of the mouth of the Moche, which compose the hamlet of Huaman. It was at one time intended to make this the port of Trujillo, as the construction of a mole was considered easier here and less expensive than at Huanchaco; but the project was abandoned, as its advantages were not equaled by those of Huanchaco and the inconveniences were greater. There is a bad anchorage, without shelter, in 5 or 6 fathoms ⅔ mile from the shore. The coast is unbroken, and constantly beaten by the surf; it is also considered unhealthy, being surrounded by swamps.

The coast from Carretas hill runs nearly NW. 12½ miles, to the port of Huanchaco, which has such remarkable points that it can be easily recognized by persons visiting it for the first time. It is between the low beach to the southward and the steep cliffs to the north. The anchorage is bad, with a depth of from 5 to 7 fathoms 1 mile from the land; bottom muddy sand. There is no shelter from the continual swell, and the surf often extends a long distance out.

The town is situated on a small plateau in front of the cliffs, on which a church, painted white, with a tower, has been built. The isolated and high position of this building makes it conspicuous at a great distance. It is the best landmark for the port, as the remarkable mountains of which mention will be made are often enveloped in fog.

In order to reach the best anchorage vessels must pass 1 mile from the south point of the port, and then steer for a spot of red earth which is on the cliffs to leeward. This direction and distance must be kept until the light is seen through the openings in the front and back of the tower; when they commence to shut in, the anchor should be dropped and 70 fathoms of chain veered. This anchorage, called El Pozo, is the best in the port. The surf reaches out farther on the windward side, and the bottom there is full of scattered rocks, which often cause the loss of anchors.

The heavy swell often forces vessels to sea; it is, therefore, well to make sure that the anchor is clear, so that there will be no trouble in getting under way.

It is not prudent to attempt landing in ships' boats, even in calm weather; the shore boats, manned with nine men, all familiar with the *tasca*, come off as soon as the vessel is anchored. If none are seen approaching, it can be concluded that the sea is too rough. The communication with the shore is often interrupted for 2 or 3 days, especially from June to September. The inhabitants have some very narrow balsas, whose front part is high, called *caballitos;* they are used for fishing or going to vessels in very rough weather. They are moved by a double paddle, used by one man, who sits on his feet. These men cross the roughest *tasca*, and though they are often washed out, they always quickly regain their caballito, as they are excellent swimmers.

Boats are beached near the custom-house.

Until lately this port was much used for importing foreign merchandise and for exporting rice, cotton, grain, chancaca, starch, and silver in bars, but this commerce has been transferred to Salaverry, which, as before stated, was made a port of entry in place of Huanchaco.

The town is 6 miles from Trujillo, and is used as a watering-place by the people of that city. Provisions can be obtained at Huanchaco. Water is scarce and expensive, and it is hard to ship; this also renders ballast expensive.

Running for port Huanchaco a vessel should approach the land on the parallel of 8° 7' S., which is 7 miles to windward, and make mount Campana, which is 7 miles N. 10° E. from the roadstead. To the south of this mountain, and on the north shore of the valley of Chimú, the sharp peak of Huanchaco will be seen, and soon after the church and the vessels at anchor. The village cannot be seen until north of the point. *Directions.*

The peak of Huanchaco is an isolated mountain east of this port, and 6 miles in the interior; it is conical with a very pointed summit. It can be seen from a long distance, and can easily be distinguished by its shape and position. *Peak of Huanchaco.*

The bell of Huanchaco is the most remarkable mountain of this coast; it is 6 miles north of Huanchaco and 5 miles inshore. It resembles an old bell resting on the ground, its peaks being the lugs. It is 3,450 feet high, and owing to its position can hardly be confounded with any other. There is often a belt of fog lying on it which leaves both its base and summit uncovered. *La Campana de Huanchaco.*

The coast to the north of Huanchaco trends nearly N. 50° W., with steep cliffs near the sea, leaving only a very narrow beach. It is exposed to the heavy swell, and there are from 8 to 10 fathoms 3 miles off shore. At 14 miles from Huanchaco the cliffs are cut by the fine valley of Chicama, whose vegetation reaches to the sea. A river of the same name runs through the valley, which is dry at its mouth from May to November. *Valley and river of Chicama.*

Following the coast to the north for 3 miles from Chicama valley, and nearly west of the town called Magdalena de Cao, whose houses, situated a little inside of the heights, can be seen from the sea, is Brujo cove, also called San Bartolomé, which was used as a shipping point. The anchorage in 5 fathoms, 1 mile from the shore, is bad and full of rocks, with no shelter, and the swell and surf are constant; these disadvantages have closed it to commerce. *Brujo cove.*

25 c

Mount and point
Malabrigo.
The coast to the northward of the valley of Chicama
trends NW., and has the same aspect as that to the south.
At the distance of 13 miles it projects to the westward,
where the isolated mountain Malabrigo rises to the height
of 800 feet, resembling an island from a distance. It descends gradually, and some spurs, running to the sea, form
the point of the same name, which is clean to the SW. and
W., but not so to the NW. and N., where the water is shoal
and there are some rocks on which the sea breaks.

Macavi islands.
The Macavi are two islands nearly S. 15º W., 6 miles from
point Malabrigo. They can be called North and South
islands from their respective positions. The North island is
the highest and smallest, and is separated from the other
by a channel of an average breadth of 120 feet, which can
only be used by boats in fine weather. The North island
contains the least guano, and is 100 feet high and of small
area.

The best anchorage is to the NE. of it, in 9 to 18 fathoms,
bottom sand, 100 yards from the north point of the island.
All the approaches of this island are perfectly clear; there
are from 11 to 20 fathoms 300 yards from either side. It is
not prudent to anchor farther out than indicated, as there is
no shelter.

The South island is lower, and is entirely covered with
guano, which gives it a dark-yellow or bronze color. There
are some large rocks near it, but they are all very close to
the island, which is otherwise clear. There are from 12 to
20 fathoms 300 yards off. This island will be first distinguished when coming from the south. It does not seem
prudent to anchor near this island, as a vessel would be entirely exposed. Vessels bound to these islands for guano
approach them on the east side, and when within 1 mile
steer so as to preserve that distance until the channel opens,
when they will be under the NW. island, and should anchor
as soon as the two islands are in line. When there are
vessels at the anchorage, it is better to anchor to leeward of
them than without the shelter of the islands. Fresh provisions are brought from the neighboring port of Malabrigo.

During the night the winds blow from the SE. to E., and
during the day from the S. to SSE. It is seldom calm. Fogs
are frequent and very dense from December to April; under

such circumstances vessels should keep off until noon, when the fog generally lifts. These islands could be passed easily without being seen, and, as the current at springs often runs 2 knots, it is hard to regain them.

Malabrigo cove is to the eastward of and to leeward of Malabrigo point. In running for it the coast must be kept ¼ mile distant, as within that it is shoal. The best anchorage is at the head of the bight, between the houses on the SE. shore and the small point of black rocks called Observatory point. The depth is from 4 to 4½ fathoms ⅔ of a mile from the land. The center of the village will bear about S. 57° E. Ships' boats should not be used, as the water is very shoal a long distance from the beach and the breakers extend far out. The boats of the inhabitants are beached in front of the huts.

In 1869 Commander E. Nares, R. N., commanding H. B. M. S. Reindeer, examined the anchorage of Malabrigo, and gave the following information relative to Garcia rock, a shoal rocky patch near the usual anchorage:

Garcia rock is the shoalest part of a rocky patch 400 yards in extent, consisting of pinnacles with sand between. The Garcia rock lies on the following bearings:

SW. extremity of point Malabrigo S. 51° 30′ W.; Observatory or Black point, S. 18° 30′ W.; White cross west of village, S. 37° 30′ E.

There are from 9 to 12 feet of water on a space of from 4 to 6 feet of this rock; the depth then increases rapidly to 4 and 4½ fathoms. A vessel is in the line of the rock when the northern street of the village is open.

From Reindeer rock, with 12 feet of water over it, the SW. extremity of point Malabrigo bears S. 55° 30′ W.; Observatory or Black point, S. 20° 30′ W. From Sixteen-feet rock, covered by 16 feet of water, the center of Malabrigo mountain, which will be seen above the point, bears S. 53° W.; Observatory or Black point, S. 21° 30′ W. Vessels may anchor in Malabrigo road in 4 fathoms with the center of the village bearing ESE., distant ¾ mile; not bringing Observatory point to bear farther to the W. than S. 5° W., to avoid the rocky ground. All the cove of Malabrigo is shoal; there are but 5 to 5½ fathoms 1 mile from either shore.

At the anchorage the breezes are fresh with strong gusts

Malabrigo cove.

Garcia rock.

from noon until sunset. Vessels should, therefore, enter
under short sail. Heavy swells are frequent, and they com-
pel vessels to drop a second anchor or force them to sea.

The village, situated in the SE. part, consists entirely of
store-houses and huts. There are neither resources nor water.
The principal articles of export are rice, grain, sugar, cotton chancaca, starch, and other produce of the valley of
Chicama; lighters can only load near high water.

There is telegraphic communication with the capital. A
concession was granted for a railroad of 20 to 25 miles between this port and the towns of Paijan and Ascope. A
pier is also proposed.

Point Arcana or Puemac.
Following the coast to the northward, point Arcana is 12½
miles from Malabrigo; it projects but little, and can be easily
recognized from the number of high wooden crosses upon
it surrounded by frames. As there are several rocks on
which the sea breaks, off the point, it should not be approached closer than 400 yards.

Puemac cove.
Puemac cove is to leeward of the point; its depth is from
4 to 4½ fathoms, 600 yards from the land, in front of some
huts on the south shore. These constitute the village, and
are inhabited by fishermen. There are no resources. The
shelter is insufficient, and boats can only land during fine
weather.

Point Pacas-mayo.
Point Pacasmayo is 7 miles to the NW. of point Arcana.
The intermediate coast is low, sandy, and beaten by a heavy
surf. The fields of the valley of San Pedro or Pacasmayo
are seen toward the interior. The land of the point is the
highest, and it descends gradually toward the sea, where it
terminates in a low and extensive curve. For ⅓ mile outside of the point the water is very shoal and the sea breaks
with a heavy swell.

The steamer Arica of the P. S. N. Company was totally
wrecked on this point when bound south.

Port Pacas-mayo.
The large cove of Pacasmayo, penetrating 2 miles to the
NE., is immediately N. of the point. When in the bay, after
doubling the point, the SE. coast can be approached within
500 yards, and the houses of the village will be seen when
they bear N. 50° E.; before that they are shut in by point
Pacasmayo. The best anchorage is about 900 yards from
the west extremity of the village, in 4 to 5 fathoms of water

bottom sand and mud. The point will then bear S. by E., and the village E. The boats of the inhabitants are an-chored at this place, and are a good leading-mark for ves-sels visiting the port for the first time. The anchor can be dropped alongside or outside of them. This distance must be kept, as with a heavy swell the breakers extend out a long distance.

In making the port of Pacasmayo from seaward, it is best Directions. to keep on the parallel of 7° 25'; the hill of Malabrigo can be distinguished when within 18 miles; it has the appear-ance of an island with gradual slopes.

Mount Arcana, which is rugged, will be seen north of Malabrigo; it is 7 miles south of point Pacasmayo; finally, in clear weather, mount Sullivan, 18 miles inland and 5,000 feet high, can be seen. On approaching, the low, yellow cliffs will be distinguished, those to the north of the road being the highest. On their summit, on the north side of the point, is a square detached building which shows distinctly. It is seldom that ships' boats can be used for landing, and it is always best to use the shore boats. The road is sub-ject to fresh winds, accompanied by heavy gusts, after 11 o'clock in the forenoon, during which time vessels must enter under short sail.

The principal commerce of Pacasmayo consists at present Description. in large shipments of rice, grain, chancaca, anise-seed, coffee, and cotton, which is largely cultivated. Wood can be ob-tained.

The village has increased much in size and importance. At its north end there is a fine new building, painted white, containing machinery for pressing cotton, &c.; it can be seen a long distance when coming from the northward. A small brook containing fresh water empties near the houses. Fresh provisions can be obtained in abundance at the vil-lage of San Pedro, which is 5 miles distant. Steamers stop here weekly.

There is telegraphic communication with the capital, and a railroad contracted for from Pacasmayo has been finished to La Viña, as has also a fine pier, extending outside the tasca, 2,538 feet in length. Guadalupe, which is distant 15 miles, is celebrated for the fair held there the first nine days in December of each year; it is visited by more than 20,000 people from all parts of the republic.

Valley and river of Jequetepeque. The valley of Jequetepeque is 3 miles north of Pacas‧ mayo. The fields and vegetation can be seen from the sea. Through its center runs the river of the same name, which is well supplied with water in the summer, but dries between May and December. On the shore N. and S. of the valley are some huts, which are occupied by fishermen or people visiting the sea-shore.

Zaña point. Zaña point projects considerably, and can be seen from Pacasmayo, from which it is 15½ miles distant; the intermediate coast trends about NNW. There are some rocks close to the N. shore on which the sea breaks; another lies 600 yards off shore in the same direction, and can only be seen with a heavy swell.

Chérrepe cove. Immediately to the north of Zaña point is a slight indentation forming Chérrepe cove. The anchorage is ⅛ mile from the land, in 5 to 6 fathoms; the heavy surf prevents landing with boats. On the shore are some store-houses and huts. Those who guard them are the only inhabitants. This cove is used for shipping the produce of the plantations of Cayaltí, Ucupe, and others in the vicinity, which is the same as those of the plantations of Pacasmayo. There are no resources.

Progreso cove. Progreso cove is 7 miles north of Chérrepe; from it are shipped wood, charcoal, and the produce from the plantations; it is only frequented by vessels for this purpose. There are no resources, and the anchorage, which is 1 mile from the land, in 6 to 7 fathoms, offers no shelter. A heavy swell sets in, and the beach is only accessible to the boats of the natives in very fine weather.

Eten head. * Eten head is 17 miles N. 38° W. of point Zaña, this bearing being the mean direction of the coast; it is formed by two connected hills, very near each other. The southern one, the highest, is 640 feet high. They are both separated from the mountains in the interior, are very close to the sea, and form a small point. They are surrounded by low land, which gives them the appearance of an island when seen from a distance.

The north side is white, and can be easily distinguished.

---

* Lieut. A. Miller, U. S. N., gives the longitude of Eten head as 1 mile to the eastward of its position on the British Admiralty charts.

In the interior are two peaks, 7 and 14 miles distant, 1,900 and 2,440 feet high.

Port Eten is immediately north and to leeward of the hill just described. It is everywhere clean, with depths from 4 to 5 fathoms 800 yards from the shore, and 6 and 7 fathoms at a distance of 1½ miles. The anchorage is open to the SW. swell, which often causes a heavy surf, extending 400 and 500 yards from the shore. The coast which surrounds the port is formed of steep cliffs, which in many places do not leave any beach. Port Eten.

A railroad is in course of construction, which from the port of Eten will extend to Monsefú, Chiclayo, Lambayeque, and Ferreñafe, a distance of 28 miles. A branch of 7 miles will run from Chiclayo in the direction of Tuman. For this reason the port has been opened to commerce. Vessels must anchor 1½ miles from the point in 6 to 7 fathoms, bottom gravel. The best anchorage is with point Eten bearing S. 41° E., church N. 32° E., flag-staff N. 66° E.; or farther out, with point Eten bearing S. 49° E., church N. 47° E., flag-staff N. 70° E.

There is a fine pier building, which will extend outside of the breakers; when finished there will be no danger in landing. Vessels can approach to within ¾ or ½ mile of the shore in 5 to 5½ fathoms. The anchorage is safe with one anchor and 50 fathoms of chain; there is no danger, though the heavy swell is disagreeable.

This new port has, without doubt, great advantages over those of Pimentel and San José, which are now used by the provinces of Chiclayo and Lambayeque. The principal of these advantages are, its greater depth, allowing vessels to anchor nearer the shore; the less breadth of the *tasca*, and the shelter which the hill will give to the pier. This work was to have been finished in 1871, at which time the other ports were to be closed.

A village is being built, and water will be brought to the mole, and, as soon as the communication with the interior is established, every kind of supply will be abundant. A flag-staff on a white pyramid serves as a leading mark during the day; at night a light is hoisted on it.

The islands of Lobos are distinguished by the names Lobos de Afuera and Lobos de Tierra, or as High island and Low island. Islands of Lobos de Afuera.

The first group is composed of two principal islands, separated by a channel 120 feet wide with a depth of 4 fathoms. The greatest length of the islands from N. to S. is 2½ miles, and the greatest breadth is 1½ miles. Near them are several islets and large rocks. The mean elevation of the islands is 100 feet; they are of a white color, being covered with guano. Immense flocks of birds, and also seals, visit them from October to April. At their western extremity is a bank which runs nearly ¼ mile to seaward, and close to the NE. side is another bank. These islands have several landing-places; one of them known to the fishermen under the name of Puerto Grande, is on the east side. The anchorage, close to the land, is good for any vessel, the depth being 14 fathoms, bottom sand. These two islands form a cove to the southward and another to the northward. The former is entirely open and cannot be used as an anchorage; the latter, on the contrary, has a convenient and secure anchorage, where there is never other than a light wind. In it vessels can anchor as close to the rocks as prudent; there are several good creeks for landing, the principal one being to the SE., where a stone landing was built by the fleet in 1852. The guard occupy some wooden houses on the beach.

There is a dangerous bank in this port, about 100 yards long and 16 yards wide, in 8 feet of water. It is called Bajo Gamarra, its position having been determined by the officers of the Peruvian brig of war of that name. It is situated at the entrance. The following remarks will give an idea of its position: bring the most projecting extremity of the NW. islet in line with the opening which is formed by the NE. islet, named Quita-Calcal, and the principal island; on this line mark a point in the center of the port; a little to the eastward of this is the bank. It can be said that it is in the middle of the entrance with a slight inclination to the eastern side and a little inside of the point. In order to enter, therefore, the tacks in the west channel must be short, until to windward of the position of the bank; the channel being very narrow. The surest plan is to anchor a boat on the bank; the current would not allow it to be kept in position by oars. Steamers can enter freely without other precaution than to keep close along the rocks on either shore. There are no other hidden dangers.

XIX

Views off the Coast of Perú between the Lat.s 7° and 5° S.

It must not be forgotten that this group is situated nearly west of Eten head, 49 miles distant and 45 miles S. 77° W. from port San José, so that precaution is necessary when leaving the latter port in the evening, bound south, to guard against the probability of the wind falling light.

The south and west coasts of the island are inaccessible. Excellent fish are abundant; the fishermen from San José are met with here on their balsas. The exportation of guano from these islands is forbidden. A governor and a few guards reside on the islands; there is a total absence of resources.

The soundings in the channel between this group and the coast are regular; the bank which separates them has on it not less than 50 fathoms, and 3 miles from the continent there are from 6 to 7 fathoms, the depths being greater near the islands.

Lobos de Tierra is 5½ miles long N. and S., and is sepa- *Lobos de Tierra.* rated from the main land by a channel 10 miles wide; it is 28½ miles N. 13° 30' W. of the islands de Afuera; close to it are some islets and large rocks. The SW. and S. coasts have no landings, but there is good anchorage the length of the NE. coast, near the land, in 8 to 11 fathoms, bottom sand and broken shells. The passage between the island and the coast is clear.

This island, like the preceding, has large deposits of guano. It is the residence of the guard, who, like those of High island, are dependent on the custom-house of San José. All provisions and water are brought to them. The island is visited by the fishermen of Lambayeque.

When navigating in the vicinity of the Lobos islands, especially north of point Aguja, great attention should be given to the longitude, as the currents appear to be at times strong and irregular. Vessels have been drifted 36 miles to the westward in 24 hours, and others the same distance to the eastward. It can be taken as a general rule that after a strong south wind the current is west.

The coast N. of Eten is a sand-beach; it is very low, and *Coast.* the sea breaks on it heavily. The noise of the breakers can be heard for 8 or 10 miles.

The hamlet Santa Rosa is on the beach, 4 miles N. of *Santa Rosa.* Eten hill; it is inhabited by fishermen, and during the bathing season is visited by the families from Chiclayo. The beach is inaccessible to boats.

Port Pimentel.   Port Pimentel is 7½ miles to the northward of Eten. It has all the inconveniences of San José, which is close to it. The store-houses of sugar-cane which form the village can be seen on the heights. The low beach to the southward of the hamlet extends into the sea and forms a shoal, on which it always breaks, producing the rough water always experienced at the anchorage. The anchor can be dropped in 5 fathoms 1½ miles from the land, the flag-staff on the office of the captain of the port, which is in the center of the village, bearing N. 71° 25′ E.; a flag is hoisted when a vessel is seen approaching. The anchorage, devoid of all shelter, is open to the SW. swell. The dangers and inconveniences of this port have caused it to be almost abandoned. There are no resources excepting water, which is found in a brook close by. The nature of the bottom, as in that of San José, is shifting sand, which renders the construction of a mole almost impossible, and it would have to extend 1 mile to reach beyond the tasca.

Port of San José de Lambayeque.   San José de Lambayeque is 13 miles to the NW. of Eten head, at the head of a deep bay formed by the coast from the south and that which trends to the WNW. This port offers no shelter; it is an open roadstead, where vessels may anchor in the open ocean off a low and dangerous sand-beach.

Vessels bound for this port should make Eten head and follow the coast, keeping 3 or 4 miles from it, until the town is seen. This is at the edge of the sea. As its store-houses and huts are built of straw and cane, it would be readily confounded with the beach were it not for the church with its steeple and the burial-place, two remarkable points, both painted white; the latter being situated on a hill at the north extremity of the village admits of its being seen a long distance. They are leading marks for the anchorage.

When the church steeple bears N. 83° 40′ E., when 2½ to 3 miles from the coast, the anchor must be dropped in 5½ to 6 fathoms. The best anchor should be used, with 70 fathoms of chain. The bottom is sand and poor holding-ground. Fitz Roy advises the use of two anchors. It is a bad anchorage, entirely exposed to the SW. swell. The sea sometimes

breaks for 1½ miles from the beach, and often forces vessels
to make sail in order to avoid the loss of the anchors. The
chains should always be buoyed.

Ship boats should never be used for landing. The balsas,
peculiar to this place, are the same as those found at Tumbez by Pizarro. They are used for loading and discharging,
and are manned by ten or twelve Indians. Communication
with the shore is often interrupted for two or three days by
the bad state of the sea.

In spite of its inconveniences, San José is one of the larg- *Description.*
est ports of export of Perú. All the produce of the rich
provinces of Chiclayo and Lambayeque is shipped from here,
consisting principally of rice, chancaca, sugar, maní, tobacco, and cotton, as also, on a small scale, hides, tallow, grain,
turkeys, and other articles of produce and manufacture.

The traffic of this port will undoubtedly pass to Eten as
soon as the railroad and mole are finished.

The population is small, and composed principally of
Indians; the town is surrounded by sand-dunes, which are
always encroaching on it. At a distance of 6 and 7 miles are
the towns of Lambayeque and Chiclayo, which are connected with the capital by telegraph. Fresh provisions and
fruits are abundant. Water can be obtained by landing
small casks on the balsas, but there is no running water at
the port.

The balsas do not come out immediately, as they wait for
low tide and a favorable wind. If they are not alongside in
24 hours it can be inferred that it is too rough. There are
almost always some coasters here at anchor.

Lambayeque river empties 3 miles to the NW. of the town *Lambayeque river.*
of San José. It has but little water the year round, and is
subject to freshets during the months from January to May.
There is some vegetation on its banks.

The shore from San José runs WNW., low and sandy,
and keeps this general direction, with more or less undula-
tion, for 33 miles, to False point Aguja.

This sandy shore is the limit of the desert of Sechura,
which extends 120 miles N. and S. without a sign of vegetation.

At the end of this low coast there is a peninsula and *Mount Illescas.*

some tolerably high hills, the most remarkable of which is mount Illescas, which ends in a sharp point.

False point Aguja. False point Aguja is, as before mentioned, the western limit of the low coast which commences at San José. The point is low, projecting but little, and offers nothing remarkable.

Point Aguja. Point Aguja, a low sand-spit, runs into the sea 5½ miles N. of False point Aguja; some rocks and small islets extend 800 yards from it.

Point Nonura. North of point Aguja the coast trends to the NNE. 3½ miles to point Nonura, which is high and free from dangers. There is more water around it than to the north of it.

Point Pizura. From Nonura point the coast inclines more to the NE., to the high point of Pizura, which extends from several mountains and hills which are close to the coast. The rocks in the vicinity of the point are all above water.

Sechura bay. The spacious bay of Sechura opens to the E. and NE. of Pizura point; throughout the bay the depth is good, but there are two anchorages which are generally taken, that off the village and the Salina.

Vessels bound to the bay for salt should approach from to windward, and, after doubling point Pizura, keep along the shore to the eastern bight, where there are some huts on the beach called the Salina. Vessels can anchor 3 miles from it in 5 to 6 fathoms; the beach is always quiet and boats can land easily. Large vessels must keep at this distance, as there is a shoal in the vicinity of the land, with little water, leaving hardly a passage on the S. side for crafts of 40 to 50 tons. The salt is loaded by means of sailing-balsas, many of which are always fishing in the bay. There are no resources at this port. Vessels wishing to reach the village must steer for the NE. part of the bay; the church-towers will be readily seen, and are a good guide; one of these towers has a strong inclination to the north, and at a distance resembles a tree more than a stone building. The anchorage is 1¾ miles from the beach, in 5 to 6 fathoms, with the church-steeples bearing nearly E. It is a good anchorage, but exposed to fresh winds and a heavy swell, which produces a surf on the beach. Ships' boats, however, can be used, with a little care.

The river Piura empties into this part of the bay, and in the summer months can be ascended to the village Sechura; the bar is sometimes dangerous, and it is always best to have an Indian pilot. All traffic is carried on by balsas. Fishing is the principal occupation of the inhabitants; they go in their balsas to the ports of Ecuador and Colombia to sell their salt and salt-fish; on shore they are muleteers.

The village of Sechura, which can be seen from the sea, is 4 miles inland, on the banks of the river. It has no resources. Its church is celebrated on account of the great value of its ornaments in gold and silver.

The coast, the northern limit of the bay, trends nearly N. 40° W., terminating at the distance of 28 miles in the high and abrupt point Foca. Close to the NW. of it are some rocks. *Point Foca.*

Foca island, of moderate height, is about 1 mile NW. of Foca point. The channel between it and the coast is full of sunken rocks, and impracticable. The sea always breaks on the rocks, which make out for a short distance to the N. and NW. of the island; they are all above water. *Foca island.*

La Silla de Paita, 1,300 feet high and about 2 miles from the sea, is nearly W. of Foca island; it is the commencement of a chain of mountains which terminates a short distance to the SE. This mountain, black or yellow according to the position of the observer and that of the sun, is an excellent landmark for vessels bound to Paita; it is easily distinguished from its isolated position and its shape, the upper part resembling a pack-saddle. When coming from the southward it will not be seen until bearing N. 55° E. *The saddle of Paita.*

From point Foca the coast makes a slight curve, back of which is the saddle; 5 miles from this is Rocky point, which projects to the westward, in which direction, as also to the northward, it is surrounded by rocks and shoals extending from it ½ mile. *Rocky point.*

Should Paita be left during the night, bound south, care must be taken not to confound this point with Foca island, as they are very similar. The black rocks of the point stand out in relief against the bays of sand on either side. This in an uncertain light makes it look like an island. If the mistake should be made, and the vessel kept on her course to point Aguja, she would soon be among the rocks and shoals of the island.

Point Paita.   Point Paita is 9 miles N. by E. from point Foca, and 4 miles from Rocky point; there are some rocks above water near it. At 2 miles from this coast there are about 38 fathoms of water, bottom mud.

Tierra Colorada or False bay.   This small cove is immediately to the E. of point Paita. It takes its name from some red spots on its shores. The anchorage is safe, and boats can be beached easily.

Telegraph point.   Telegraph point is $1\frac{2}{3}$ miles from point Paita, and is the eastern limit of Tierra Colorada cove. There are some rocks 200 yards from it, which must be avoided, as the sea is smooth over them in fine weather, and shoal water seems to extend farther from the point than is marked on the charts. There is a signal-mast on the point.

Port of Paita.   The port of Paita is to the SE. and E. of Telegraph point; the anchorage is good everywhere, the depth varying between 10 and 5 fathoms near the mole. Fitz Roy advises not to anchor in less than 6 fathoms, as the water shoals suddenly toward the land. The winds which come from the town commence every day about 10 a. m., and last until evening; the land is so near that it does not raise any sea.

Vessels standing in should take in their upper sails before rounding Paita point, as the gusts there and opposite Colorada cove are heavy. The anchorage can seldom be reached in one tack, but as there is plenty of room the anchor can be dropped whenever convenient.

There are two small wharves; about 100 yards off the larger one is the wreck of a vessel which was burned.

Description.   Paita is the principal port of the department of Pinra ; it is 30 miles from the city of San Miguel de Piura, and in telegraphic communication with the capital ; its population is about 6,000. A railroad from Paita to Piura, 62 miles, is in course of construction.

An abundance of provisions and naval stores can be obtained; good workmen can be had for repairs. Wood is expensive. Water is also expensive, as it is brought from the village of Colan, 6 miles distant. The government has contracted for building an aqueduct to bring the water from the river Chira, through the ravines N. and E. of the port. This work will fill one of the greatest necessities of Paita, and be of great importance to the vessels which visit it.

Whalers visit this port to take in provisions and water,

to ship their oil, and to receive their mails. The custom-house gives them some privileges, by allowing them to land certain articles for exchange.

The principal commerce of the port consists in the importation of foreign merchandise and in the export of its products of industry, which consist in straw hats, called *catacaos;* cotton of very good quality, hides, and some articles of less importance. Large numbers of cattle are shipped to Callao.

The town is small, the streets narrow and irregular, owing to the small extent of the plateau on which it is built. The climate is hot and dry, but healthy; rain is almost unknown, and the houses are lightly built.

The port is inclosed to the E. and N. by a cliff 200 feet high, steep, and so close to the sea as to leave no beach in many places. On the upper part is a plain. The English steamers from Panama to Callao touch at Paita to coal.

It is high water, full and change, at Paita at $3^h 20^m$; rise, 3 feet. *Tides.*

After making the Saddle of Paita, steer for Paita point; *Directions.* rounding it at a moderate distance, the bay of Tierra Colorada will open; this is rocky, and should be passed without approaching Telegraph point nearer than 450 yards. Here the wind is frequently baffling; the sandy shore of the bay will be seen, and the vessel must be worked up to the anchorage.

The town cannot be readily distinguished, as the houses are of the same color as the cliffs.

Colan is a village at the foot of the cliffs, 6 miles N. of *Colan.* Paita; it is without importance, and its beach can only be reached by balsas; water is carried to Paita on them.

From Colan the ravines extend into the interior, and the fine valley of Chira, which descends to the sea, opens to the north. Off it is a low beach, in the center of which the river Chira empties; it has plenty of water in the summer; at its mouth is a bar which closes the entrance; there are days, however, when balsas and canoes can enter. The valley is very fertile, and can be seen a long distance from the land; about 1 mile from the beach there are from 7 to 8 fathoms. There is generally a heavy surf on the beach.

Pariñas point. The sand-beach which commences at the valley of Chira takes a NW. direction from that point; it is bordered by a chain of hillocks and cliffs for 18½ miles, and terminates in point Pariñas, a projecting point 78 feet high, and formed of dark rocks. This point is the western extremity of South America.

Some islets and rocks lie off it, some of them extending out ½ mile; they are above water, and the sea breaks on them.

Point Talara. From Pariñas point the coast takes a nearly north direction for 24 miles, inclining, if anything, a little to the eastward. Twelve miles from point Pariñas is point Talara, a small point, surrounded by low rocks above water. This part of the coast is generally formed of cliffs, those to the northward having a white color.

Point Talara is composed of two parts: the southern is a cliff 78 feet high, off which there is a small black rock; the northern part is much lower, and surrounded by breakers. To the N. of the point is a shoal bay, after which the coast is again formed of high cliffs.

Cape Blanco. It has been mentioned that the coast from point Pariñas trends nearly N. with a slight inclination to the E. for 24 miles; there it is terminated by a moderately high, round, and white hillock, called cape Blanco, which is remarkable for the strong breeze always blowing in its vicinity; it is said that the wind is fresher here than at any other point on the coast of Perú; it never commences in squalls, however, but rises gradually with very clear weather from the middle of the day until sunset; more than two reefs in the topsails are seldom required. Some rocks, above water, extend from the point ¼ mile; the most remarkable one is a small islet of moderate height.

To the N. of cape Blanco the coast takes a NE. direction with some long curves.

Mountains of Amotape or La Brea. The Amotape mountains are high, lying back of the coast between points Pareñas and Blanco. They form a chain running NE. and SW., with a mean height of 3,000 to 4,000 feet. Large quantities of rosin are drawn from the trees on these mountains.

Los Organos. Los Organos is the name given to a high cliff 6 miles N. of cape Blanco, its western side resembling the pipes of an

organ, from which its name. It is evidently a basalt formation.

Following the sand-beach, which is backed by high cliffs,
4 miles to the N. of Los Organos, there is a small point, to
leeward of which is Máncora cove. Vessels bound for this
point must necessarily recognize Los Organos, which will
facilitate finding the cove. It has an anchorage in 4 to 6
fathoms, 200 to 300 yards from the shore. On the beach
are some cane huts, and sometimes large masses of fire-
wood, which is largely exported to Callao. This is shipped
by balsas, which are hauled forward and backward between
the shore and the vessel. There are no resources. The
store-houses belong to the plantation of the same name, and
the order for wood is obtained from the agent of the proprietor at Paita.

The gorge of Máncora, which is small, opens 8 miles to
the NE. of Máncora cove. At certain seasons of the year
a rivulet runs through it.

Point Sal, 24 miles from cape Blanco, is the northern limit
of Máncora cove. It is 120 feet high, and projects but little. Its approaches, as that of the preceding reach of coast,
are safe.

The coast to the northward is formed of cliffs for 7 miles,
and then again becomes low, similar to that at Máncora.

In the interior is mount Castro, 1,200 feet high.

Point Picos is 19 miles NE. of point Sal. It is low, but
little salient, and, as the intermediate coast, clear. It is an
inclined hill bordered by a sand-beach. Nearly E. of the
point is a chain of heights trending N., all the summits of
which are sharp peaks, rising 700 feet above the sea.

Boca de Pan is a small cove 4½ miles NE. of point Picos.
It has a good anchorage in 4 to 7 fathoms very near the
land. The beach is of sand, and convenient for landing.
On it are some huts and store-houses. Large quantities of
fire-wood are shipped from here.

NE. 5½ miles from Boca de Pan, 9¾ miles from point Picos,
is Malpaso cove. The anchorage is excellent in 4 to 8 fathoms 300 yards from the land. The cove is surrounded by
an abrupt cliff, with a small plateau near its center, on which
are some huts and store-houses, which can be distinguished
from a long distance. It was the principal establishment of

26 c

the plantation of Máncora for gathering and shipping the orchilla. Below the houses is a mine which is said to contain coal of good quality, but it has not been fully explored. There are many petroleum springs in this vicinity.

Point Malpelo. After making a slight curve, the thickly-wooded coast runs NNE. for 12½ miles from Zorritos to point Malpelo. On its southern part, besides the numerous trees, it is entirely covered by vegetation. It is very salient, but its approaches are shoal, there being only from 2 to 3½ fathoms 1½ miles from the land.

This point is the southern limit of the Guyaquil river, the ebb and flood of which are felt on all this part of the coast. The island of Muerto, or Santa Clara, 20 miles distant, and in the middle of the gulf, can be seen from the point in clear weather.

Bay and river of Tumbez. To the E. of Malpelo point is the spacious bay of Tumbez, in which the river of the same name empties. The best anchorage is 3 miles from the E. point, off the mouth of the river, and nearly N. and S. of it. The land should not be closely approached, as the water is shoal ½ mile from the shore. This place is called El Pozo; the holding-ground is good in 5½ to 6 fathoms 1 mile from the land.

During the freshets, the river has several mouths; the western ones are sometimes dangerous. The broadest one, that off the anchorage, is very quiet, forms no bar, and can be entered easily by boats. The river can be ascended 6 miles to the village of Tumbez; the banks are entirely covered with varied vegetation, and offer a picturesque view. Water can be taken in easily by going 1½ or 2 miles up the river. From this, and the abundance of sweet-potatoes and other vegetables, this port is more frequented by whalers than any other on the coast of South America; they enjoy the same privileges as at Paita. There are sharks at the mouth of the river, and inside alligators abound, and the mosquitoes are intolerable; all this country is full of impenetrable forest.

At the town the orchilla is received and a large trade in wood is carried on with Callao, and in mangroves with the whalers; up the river there is plenty of wood for building.

The Spanish forces under Pizarro, which conquered Perú, first landed at Tumbez.

The remainder of the coast to the north of the bay is equally fertile, and covered with mangrove and other trees. There are small estuaries running into the interior, which, during the freshets, are in communication with the arms of the river Tumbez.

The north branch of this river is the coast boundary between Perú and Ecuador.

It is high water, full and change, at the island of Santa Tides. Clara at 4^h; rise, about 2 feet.

ADDENDA.

Shoal near cape Carranza, p. 191.

The commander of the French vessel of war L'Infernet reports that in passing the Carranza rocks, which lie off cape Carranza, he observed a shoal spot, distant about 650 yards N. 12° E. from the outermost of the rocks, which he judged to be covered by not more than 2 fathoms of water.

Dangers off point Toro, p. 196.

Mr. Clement Mossop, master of the British brig Corouilla, reports that he passed within 250 yards of a rock lying N. 1° E. of point Toro, 3 miles distant. The sea seldom breaks upon it; and he estimates the depth over it to be about 4 feet at low water. Position by Mr. Mossop, latitude 33° 43' 10" S., longitude 71° 48' 43" W.

The Chilian government has given notice of a reef about 800 yards in length in an east and west direction, and about 550 yards in width, 6½ miles N. 21° E. from point Toro.

Port Tongoy, p. 221.

...e best anchorage for large vessels at port Tongoy is in 7 fathoms of water, with Range peak and the church-spire in line bearing N. 35° E., and chimney in line with Morgan rock N. 10° E.

Rock under water off point Lengua de Vaca, p. 220.

The steamer Bolivia, with a draught of 18 feet, on a passage from Lebu to Tongoy, struck on a sunken rock about 1½ miles from point Lengua de Vaca; point Lengua de Vaca bearing S. 3° E., and the peninsula of Tongoy S. 64° E.

COAST OF PERÚ.

Point Grueso, p. 301.

New deposits of guano have been discovered, amounting, as reported, to about 3,000,000 tons, about 10½ miles south of Iquique, on a bluff promontory making out from the base of mount Tarapacá, called point Grueso or Grande, in latitude 20° 23′ S., longitude 70° 16′ W. There is no harbor nearer than Iquique, where vessels will have to await their turn for loading. The loading-place is an open roadstead, sheltered from the prevailing southerly winds, and is said to be remarkably smooth with these winds, but it would be rough with northerly weather. A vessel can lie near the beach in 6 fathoms of water, and could be towed there from Iquique in two hours.

As yet the commissioners appointed by the government of Perú to examine these deposits have not visited the locality. (Lieutenant Commander Keyser, U. S. N.)

ALPHABETICAL INDEX.

www.ingramcontent.com/pod-product-compliance
Lightning Source LLC
Chambersburg PA
CBHW020903210326
41598CB00018B/1757